Flavor Analysis

ACS SYMPOSIUM SERIES **705**

Flavor Analysis

Developments in Isolation and Characterization

Cynthia J. Mussinan, EDITOR
IFF

Michael J. Morello, EDITOR
The Quaker Oats Company

Developed from a symposium sponsored by the Division of Agricultural and Food Chemistry at the 214th National Meeting of the American Chemical Society, Las Vegas, Nevada, September 7–11, 1997

American Chemical Society, Washington, DC

Library of Congress Cataloging-in-Publication Data

Flavor analysis : developments in isolation and characterization / Cynthia J. Mussinan, editor, Michael J. Morello, editor.

p. cm.—(ACS symposium series, ISSN 0097-6156; 705)

"Developed from a symposium sponsored by the Division of agricultural and Food Chemistry at the 214th National Meeting of the American Chemical Society, Las Vegas, Nevada, September 7–11, 1997."

Includes bibliographical references and indexes.

ISBN 0-8412-3578-3

1. Flavor—Congresses.

I. Mussinan, Cynthia J., 1946– . II. Morello, Michael J., 1953– . III. American Chemical Society. Division of Agricultural and Food Chemistry. IV. American Chemical Society. Meeting (214th : 1997 : Las Vegas, Nev.) V. Series.

TP372.5.F52 1998
664'.07—dc21

98-23312
CIP

The paper used in this publication meets the minimum requirements of American National Standard for Information Sciences—Permanence of Paper for Printed Library Materials, ANSI Z39.48–1984.

Distributed by Oxford University Press

PRINTED IN THE UNITED STATES OF AMERICA

Advisory Board

ACS Symposium Series

Foreword

THE ACS SYMPOSIUM SERIES was first published in 1974 to provide a mechanism for publishing symposia quickly in book form. The purpose of the series is to publish timely, comprehensive books developed from ACS sponsored symposia based on current scientific research. Occasionally, books are developed from symposia sponsored by other organizations when the topic is of keen interest to the chemistry audience.

Before agreeing to publish a book, the proposed table of contents is reviewed for appropriate and comprehensive coverage and for interest to the audience. Some papers may be excluded in order to better focus the book; others may be added to provide comprehensiveness. When appropriate, overview or introductory chapters are added. Drafts of chapters are peer-reviewed prior to final acceptance or rejection, and manuscripts are prepared in camera-ready format.

As a rule, only original research papers and original review papers are included in the volumes. Verbatim reproductions of previously published papers are not accepted.

ACS BOOKS DEPARTMENT

Contents

ANALYTICAL CHARACTERIZATION

SENSORY CHARACTERIZATION

Indexes

Preface

Flavor is a primary consideration for food selection. Consequently, we strive to enhance desirable flavor, minimize or delay flavor loss, and avoid off-flavors. If we are to meet these goals, we need to characterize the compounds that stimulate our perception of flavor. However, flavor characterization is often very challenging.

This book focuses on techniques utilized in flavor analysis. It is based on the symposium *"Challenges in Isolation and Characterization of Flavor Compounds"* that was held as part of the 214th ACS National Meeting in Las Vegas, NV; in September 1997. Researchers from Australia, Germany, Great Britain, Israel, New Zealand, Canada, Switzerland, and the United States showed how challenges in flavor isolation, separation, and characterization can be overcome using both classical and state-of-the-art techniques.

An overview of the state of analytical flavor chemistry is given in the first chapter. The remainder of the book is arranged into three sections: Isolation; Analytical Characterization; and Sensory Characterization. The section on Isolation covers techniques ranging from classical extractions on a micro scale to state-of-the-art extractions using supercritical fluids and solid-phase microextraction. The section on Analytical Characterization covers topics ranging from chiral separations and countercurrent chromatography to isotope labeling and LC-MS. Finally, the Sensory Characterization section includes discussion of the gas chromatography—olfactometry as well as the new Selective Odorant Measurement by Multisensor Array (SOMMSA) technique.

Acknowledgments

We acknowledge with sincere appreciation the financial support of the following sponsors: The Agricultural and Food Chemistry Division, American Chemical Society, Alex Fries, Dragoco, Inc.; Elan; Hewlett Packard Company; IFF; J&W

Scientific; McCormick & Co.; Nestle; The Quaker Oats Co.; and Quest International. Without the generosity and support of our sponsors, the symposium would not have been possible.

CYNTHIA J. MUSSINAN
IFF R&D
1515 Highway 36
Union Beach, NJ 07735

MICHAEL J. MORELLO
The Quaker Oats Company
617 West Main Street
Barrington, IL 60010-4199

Chapter 1

Challenges in Flavor Chemistry: An Overview

Roy Teranishi

89 Kingston Road, Kensington, CA 94707

This chapter gives an overview of some of the challenges facing flavor chemists today and how some of these challenges are met by internationally known scientists who participated in this symposium. Although the fundamentals of modern organic chemical methods used in flavor chemistry have been established, details of modifications and extensions of existing methods must be worked out in order to solve specific problems. Isolation methods such as headspace trapping, cold trapping, extraction, etc. are described, as well as special applications of gas and liquid chromatographic separations. Some applications of mass spectrometry and sensory characterizations are discussed.

Flavor chemistry is simply a specialized application of organic chemistry. Organic chemistry started in the 19th century when it was shown that chemical transformations took place without supernatural influences. The first era consisting of the theory of vitalism died, and the second period of organic chemistry began with the birth of structural theory, to explain how atoms were organized to form molecules, the network of carbon to carbon bonds. In the third era of organic chemistry, the present modern era, there are three major developments (1):

1. Electronic theories of valences and bonds.
2. Understanding of mechanisms of organic reactions.
3. Development of instruments for separating, analyzing, and identifying organic compounds.

Chemical industries utilizing these developments have developed enormously to produce dyestuffs, explosives, medicines, plastics, elastomers, fibers, films, perfumes, flavors, etc. Studies of natural products, volatiles from food materials, have been the backbone of flavor chemistry.

In the 1950's, only about 500 compounds had been characterized for their flavor properties (2). With the advent of liquid and gas chromatography, infrared, nuclear magnetic resonance, and mass spectrometry, after a latent period of a few years in which application techniques were worked out, the number of compounds which have been characterized has rapidly reached about 15,000 (3). It must not be forgotten that micromanipulation methods which are routinely used today were developed by many pioneers, beginning with Michael Faraday, the father of micro-organic chemistry methods (4). Concurrently with the advent of modern instruments, there has been a development of sensory analyses utilizing odor thresholds (5). Although there are still many isolation and identification problems to be solved, the shift of emphasis is towards the correlation of chemical structures to sensory properties.

The fundamental concepts of isolation methods, such as distillation, liquid and gas chromatography, have been established, but there are still many variations, extensions, and developments necessary to meet the challenges of isolation and characterization of compounds contributing to various flavors. There are still many problems to be solved. In the early stages of flavor research, most of the emphasis was on development of methods to establish chemical identity of constituents found only in trace quantities. Because flavor chemistry is a special application of organic chemistry, the methods worked out in flavor chemistry can be applied in other fields in which small amounts of organic compounds can have profound biological effects, such as in nutrition, air or water pollution, plant and animal hormones, insect attractants, etc. Now that many of the constituents have been identified, the task remains to determine biological activities; that is, which constituent, or constituents, is/are contributing to the characteristic sensory properties of the food product being investigated. This symposium covers some of these applications and challenges which the flavor chemist, and other natural product chemists working with trace amounts, faces today.

Although the fundamental chemistry used in flavor chemistry is based on information gained by fundamental organic chemical research, some of the analytical methods now used by scientists in other disciplines were developed first in flavor research. Two examples are: 1. the use of programmed temperature capillary gas chromatography (6), and 2. the use of gas chromatography combined with mass spectrometry (7). Both methods are widely used in many disciplines and fields. Similarly, various methods developed in flavor chemistry research can be utilized in studies of water and air pollution, insect attraction, toxicology, etc.

ISOLATION METHODS

The first area covered by the symposium and in this book is the fundamental and broad topic of isolation. No analytical method is valid unless the isolates represent the material being studied. No logical sensory conclusions can be made if the isolates do not have the characteristic sensory properties of the food product being studied. The importance of the isolation step cannot be overemphasized. Sample size requirements of analytical instruments have become smaller and smaller with new developements, especially with the application of computer methods. Nevertheless, great care must be taken in concentration and isolation so that isolates have sensory properties of the food being studied. Care must be taken so that heat labile compounds are not destroyed by harsh conditions, highly volatile compounds not lost in concentration by distillation, or low solubility compounds lost in extractions.

Most isolations use some form of distillation or extraction, or a combination of both. These methods utilize differences in vapor pressures (distillation) and solubilities in different solvents (extraction and chromatography). Specific applications have special names. Headspace and vacuum headspace trapping are forms of distillation. A combination of distillation and extraction can be used in a simultaneous distillation of aqueous material and extraction with an organic solvent. Polymeric particles or a probe coated with adsorbant material can be used to trap volatiles. Gas chomatography utilizes both vapor pressure and solubility. Countercurrent chromatography utilizes solubilities in different immiscible organic solvents. So flavor chemists have many techniques and various combinations of such methods at their disposal.

Headspace analysis in the early stages was done simply by sampling vapors from various products with a syringe and injecting the vapors in a gas chromatograph (8). Now there are many sophisticated applications and variations of this method: purge-and-trap, cold trap, trap with polymeric solid-phase, etc. This method permits analyses with minimum delay from sampling to analysis, with the minimum of introduction of artifacts developed or introduced during sampling. Unless one resorts to vacuum distillation, distillation usually means that the sample has been exposed to temperatures above ambient. At high temperatures, some of the thermally unstable compounds are destroyed and some artifacts are produced by combination of fragments formed. Thus, headspace analyses permit analyses of volatiles emitted by plants or food products to provide data representing fresh flowers, fresh fruits and vegetables, of products as they are, not data containing artifacts formed during distillation and/or extraction. Some of the analyses discussed in this symposium utilizing headspace methods are: stored and fresh citrus, strawberry, pear, peach, cherry, pineapple, an Australian truffle-like fungus, dry pet foods, milk, meat products, volatile Maillard reaction products, etc.

Solubilities and reactivities on microscale samples can be used to reduce some of the complexity of volatiles isolated from natural products. Odor assessment of each stage of extraction is made to help indicate compounds of critical importance.

CHROMATOGRAPHY

Gas chromatography, first proposed by James and Martin (9), is a very powerful separation method utilizing very small samples. Since the early days of flavor research, more stable columns with greater separation resolution are now commercially available. Previously, each research group had to rely on some member of the team to fabricate its own columns. Gas chromatography became even more powerful and useful with high resolution capillary columns used in combination with mass spectrometers (7).

Separation of enantiomers by gas chromatography and isotope studies with nuclear magnetic resonance and mass spectrometry are very effective and powerful methods for detecting adulteration. Reducing adulteration of our foods has far-reaching implications in keeping our foods pure and safe. Novel chiral sulfur compounds have been isolated by capillary chromatography utilizing chiral stationary phases. Such columns also permit the study of the efficiency of enantioselective syntheses of important optically active lactones.

Liquid partition chromatography, especially countercurrent liquid chromatography (CCC), is needed in studies of flavor precursors, which are usually water-soluble low volatility compounds containing sugar moieties. The principles of this method of chromatography were reported as early as 1941 (10). Because this separation method is done at ambient temperatures, it is a very powerful tool for isolating labile compounds. Current capabilities and applications of countercurrent chromatography to flavor analyses are discussed. Progress and data accumulated in studies of flavor precursors (11) have been possible by the use of liquid chromatography, especially countercurrent chromatography. This is very important because such data are necessary in biotechnological development of more flavorful foods. If genetic engineering is to be used to produce foods with more flavor, it is necessary to know which precursors must be increased in concentration in plants. Also, knowledge of precursors will permit processing parameters to be adjusted for maximum yield of flavor.

MASS SPECTROMETRY

Mass spectrometry was used by petroleum chemists long before flavor chemists used it. Possible usefulness and applications of mass spectrometry in flavor research (12) were recognized before the advent of gas chromatography (9). Problems of handling microsamples and of purification of samples held back progress until the combination of programmed temperature capillary gas chromatography and mass spectrometry (GC-MS) was introduced (7). Direct introduction of samples separated by temperature programmed gas chromatography capillary columns into fast-scan mass spectrometers solved the problems of purification and manipulation of microsamples. The early work showed some general possibilities but are crude by today's standards. The challenges today are those encountered in fine-tuning parameters to obtain precise and maximum information in specific application flavor problems. Precise and reproducible control of instruments is now possible with application of computer technology. Storage and retrieval of data are immensely facilitated by use of computers. Some of the challenges which face the flavor chemists today are in the improvement in computer handling of data for more rapid and concise use of the enormous amounts of data accumulated.

In this book, compounds purified by gas and liquid chromatography which have been analyzed by mass spectrometry range from acetals, furans and furanones and other compounds from the Maillard browning reaction, to nonvolatile bio-active and flavor-active compounds such as phenols, flavonoids, glycosides, saponins, etc. Capillary gas chromatographs coupled to on-line Isotope Ratio Mass Spectrometers (GC/IRMS) are used to determine the origin of various flavors, especially citrus, peppermint, vanilla, etc., to determine whether the flavors were synthesized by plants or by man. Samples separated by high performance liquid chromatography (HPLC) analyzed with either atmospheric pressure chemical ionization-tandem mass spectrometry (APCI-MS-MS) or inductively coupled plasma-mass spectrometry (ICP-MS) can be rapidly analyzed for un-derivatized organosulfur or organoselium compounds found in *Allium* plants.

SENSORY ANALYSES

Ever since gas chromatographs equipped with non-destructive detectors were built, scientists have been sniffing the effluents from gas chromatographs. Simple and crude sniffing methods have been developed into sophisticated sensory methods. Sniffing effluents from capillary gas chromatographs and water solution odor threshold determinations were used to indicate the most important compounds contributing to tortilla flavor. Gas chromatographic and mass spectral data were used for chemical identification, and olfactometric techniques were used to characterize which compounds are the important contributors to a specific variety of apples.

This method of using capillary columns to separate materials and sniffing effluents to establish olfactory properties (13-15) is important because sensory data, qualitative and quantitative, can now be obtained on very small samples. There should be many applications of these methods in the future to provide information heretofore very difficult to obtain.

Chemical and sensory analyses were utilized to determine what compounds contribute to citrus off-flavors. Selective odorant measurement by a multisensor array (SOMMSA) technique has been used to sort out chemosensors for the selective detection of key odorants. The ratio of concentration to odor threshold was used to characterize the key odorants formed when cysteine and carbohydrates were heated. Variations of odor thresholds of various unsaturated branched esters are reported.

Umami, a "delicious" and savory taste, is a topic of much interest and importance. This taste component is very important in the production of soups, stews, soup stock, gravies, and many commercial food products. A small linear peptide in an extract of papain hydrolyzed beef has been reported as a flavor enhancer. The terms flavor *potentiator*, flavor *modulator*, and flavor *enhancer* are discussed.

This symposium covered how some of the challenges, problems facing flavor chemists in basic and applied research, have been met. Information gained in these investigations will be useful in many applications for industrial flavor chemists. Experimental methods described herein will be useful in other fields in which minor organic chemicals have important biological functions.

Literature Cited

1. Cram, D.J.; Hammond, G.S. *Organic Chemistry*, McGraw-Hill, New York, **1964**, pp. 4-5.
2. Weurman, C. *Lists of Volatile Compounds in Foods*, First Edition, Division of Nutrition and Food Research TNO, Zeist, The Netherlands, **1963**.
3. Basset, F. *Informations Chime n° 300*, Givaudan-Roure, Dübendorf, Switzerland, Décembre **1988**, 207-209.
4. Michael Faraday, Tube chemistry, In *Chemical Manipulation,* John Murray, Albemarle Street, London, **1830**, pp 386-419.
5. Teranishi, R. In *Proceedings of the 5th Wartburg Aroma Symposium*, Rothe, M.; Kruse, H.-P. (Editors), Eigenverlag University of Potsdam, Potsdam, **1997**, in print.

6. Teranishi, R.; Nimmo, C.C.; Corse, J. *Anal. Chem.,* **1960**, *32* (11), 1384-1386.
7. McFadden, W.H.; Teranishi, R.; Black, D.R.; Day, J.C. *J. Food Sci.,* **1963**, *28* (3), 316-319.
8. Teranishi, R.; Buttery, R.G. In *Proceedings of the International Federation of Fruit Juice Producers Symposium*, Juris-Verlag, Zurich, **1962**, pp. 257-266.
9. James, A.T.; Martin, A.J.P. *Biochem. J.*, **1952**, *50*, 679-690.
10. Martin, A.J.P.; Synge, R.L.M.; *Biochem. J.*, **1941,** *35*, 1358-1368.
11. *Flavor Precursors: Thermal and Enzymatic Conversions*, ACS Symposium Series 490, Teranishi, R.; Takeoka, G.R.; Guentert, M. (Editors), American Chemical Society, Washington, DC, **1992**, 269 pp.
12. Turk, A.; Smock, R.M.; Taylor, T.I. *Food Technol.*, **1951**, *5* (2), 58-63.
13. Acree, T.E.; Barnard, J.; Cunningham, D.G. *Food Chem.* **1984**, *14*, 273-286.
14. Fischer, K.-H.; Grosch, W. *Lebensmittel-Wissenschaft und Technologie*, **1987**, *20*, 233-236.
15. Ullrich, F.; Grosch, W. *Zeitschrift für Lebensmittel-Untersuchung und Forschung*, **1987**, *184*, 266-287.

Isolation Techniques

Chapter 2

Applications of a Microextraction Class Separation Technique to the Analysis of Complex Flavor Mixtures

Thomas H. Parliment

Kraft Foods Technical Center, 555 South Broadway, Tarrytown, NY 10591

The purpose of this work is to describe applications of a new technique, which was developed in our laboratory for the isolation and simplification of samples prior to gas chromatographic analysis. This technique is easy to perform, requires minimal of equipment and can generate reproducible results that are particularly appropriate for organoleptic analysis. It is of particular value when the organic sample is highly complex or where there are a number of components present at high concentration. In the latter case such high concentration peaks will overwhelm smaller peaks during GC/MS or organoleptic analysis. Examples are provided demonstrating the utility to a coffee, a cheese, and a Maillard system.

Sample preparation techniques in the flavor field have been discussed in books since 1971 (*1*) to as recently as 1996 (*2*).

Difficulties may arise in flavor research for a number of reasons:

- Concentration Level
  The level of the aromatics is generally low: Typical levels are ppm, ppb or ppt.
- Matrix
  The sample frequently contains non-volatile components such as lipids, proteins, carbohydrates, which complicates the isolation process.
- Complexities of Aromas
  The aromatic composition of foods are frequently very complex. For example, coffee currently has 800 identified components and wine has more than 600.
- Classes of Compounds
  The aromatic composition of food is composed of a variety of compound classes that cover the range of polarities, solubilities and pH's.
- Variation of Volatility
  The volatility of the aromatic components of food range from those with boiling points well below room temperature to those that are soild at room temperature.
- Instability
  Many components in an aroma are unstable and may be oxidized by air and degraded by heat or extremes of pH.

One of the more common sample preparation techniques employed today involves steam distillation followed by solvent extraction. The distillation may be carried out at atmospheric pressure or under vacuum. The primary advantage of this technique is that the distillation step separates the volatiles from the non-volatiles; distillation is also simple, reproducibile, rapid, and can handle a large variety of samples.

The process described in this paper expands upon a technique previously suggested by this author (*3*). The technique is particularly useful if the organic phase is limited in quantity and involves a set of sequential experiments. In general terms, the sample is placed in an extractor, the pH made basic, and the sample extracted with solvent. Sufficient sample is removed for gas chromatographic analysis, e.g. 1 μl. The aqueous phase is made acidic and the sample re-extracted with the same solvent and gas chromatographic analysis repeated. Finally the sample may be treated with other reagents, re-extracted, and the ethereal phase re-analyzed. In this fashion a series of different analyses can be made from the same sample in a short period of time and subjected to GC/MS and organoleptic analysis.

Introduction

Sample Preparation by Distilation Techniques

In direct distillation the sample is placed in a flask and dispersed in water. The aqueous slurry is then heated directly to carry over the steam distillable components. Problems can be encountered due to scorching of the sample, and bumping may occur when the sample contains particulates. Stirring may prevent these problems.

Indirect steam distillation has many advantages over the direct technique. It is more rapid and less decomposition of the sample occurs since the sample is not heated directly. The steam and volatiles are usually condensed in a series of traps cooled with a succession of coolants ranging from ice water to dry ice/ acetone or methanol.

If sample decomposition remains a concern, then the steam distillation may be performed under vacuum. In this case inert gas should be bled into the system to aid in agitation. A number of cooled traps should be in line to protect the pump from water vapor and the sample from pump oil vapors. Another simple method to generate a condensate under vacuum is to use a rotary evaporator. Bumping is normally not a problem in this case.

Extraction and Concentration

Once the vapors have been condensed, the dilute sample is extracted. When relatively large amounts of aqueous samples are available, then separatory funnels or commercial liquid-liquid extractors may be employed. A large number of solvents have been summarized by Weurman (*4*) and reviewed by Sugisawa (*5*). The solvents most commonly used today are: diethyl ether, diethyl ether/pentane mixtures, hydrocarbons, Freons, and methylene chloride. The latter two have the advantage of being non-flammable. Solvent selection is an important factor to consider and the current status has been summarized by Reineccius (*6*). Slow and careful distillation of the solvent will produce the aroma concentrate.

As an alternative, a simultaneous distillation extraction (SDE) or Likens/Nickerson (L/N) process may be employed to produce the organic concentrate.

Analysis

Gas chromatographic analysis is normally used to separate the sample. After separation the individual components may be detected by various GC detectors (FID; mass selective detector) or by sensory analysis. When, as is often the case, the gas chromatographic column does not resolve all components then further refinements must be made. This particular problem frequently occurs when the sample is highly complex.

Sample Simplification

One solution to the problem of a highly complex sample is to pre-fractionate the sample based upon solubility class prior to gas chromatographic analysis. The general technique was first proposed in a recent book covering techniques of sample preparation (*2*) and the technique was refined in a subsequent work (Parliment, T.H. In *Proceedings of the Ninth International Flavor Conference;* Limnos, Greece; July 1997, Elsevier, The Netherlands, in press).

Applications of this technique are described here. It will be demonstrated that sequential removal of classes of compounds progressively simplifies the analysis and reveals the presence of additional compounds. This procedure is carried out in one extractor, thus minimizing sample loss throughout the process.

Manipulation of the Aqueous Phase

In the current process which will be described in detail, the sample is placed in a Mixxor™, the pH is adjusted with base, and the sample is extracted with diethyl ether. Sufficient sample is removed for gas chromatographic analysis, e.g. 1 μl. The aqueous phase is made acidic and the sample re-extracted with the same diethyl ether and gas chromatographic analysis repeated. Finally the sample is made neutral and saturated with a reagent to remove carbonyls and re-extracted. The ethereal phase is re-analyzed. In this case three different analyses can be made from the same sample in a short period of time and subjected to GC/MS, and organoleptic analysis.

Adjustment of the pH of the aqueous phase before extraction may accomplish two goals. First, emulsions may be broken permitting phase separation to take place rapidly. Second, class separation will take place that may simplify the gas chromatographic pattern. Frequently small peaks are concealed under larger ones and the smaller ones may be revealed for organoleptic evaluation or identification.

This chemical manipulation of the aqueous phase can be extended even further. Many food aromatics contain carbonyl compounds. By adding sodium bisulfite to the aqueous phase it is possible to selectively remove the aldehydes and the methyl ketones by forming their water soluble bisulfite addition complexes. Alternatively, 2,4-dinitrophenylhydrazine may be used to remove all carbonyl compounds from the sample.

Experimental

Extraction

Use of the Mixxor™ (New Biology Systems Ltd., Israel) in sample preparation has been described by Parliment (*7*) and its value described in a number of publications (*8,9*). Such a device is shown in Figure 1. These extractors are available with sample volumes ranging from 2 mL to 100 mL. The smaller capacity extractor is a particularly convenient capacity for flavor research. Briefly, the aqueous phase is placed in Chamber B and the assembly is cooled and a quantity of a low density immisible solvent containing the analyte is added. The solvent may contain an internal standard. The system is extracted by moving chamber A up and down a number of times. After phase separation occurs, the solvent D, is forced into axial chamber C where it can be removed with a syringe for analysis.

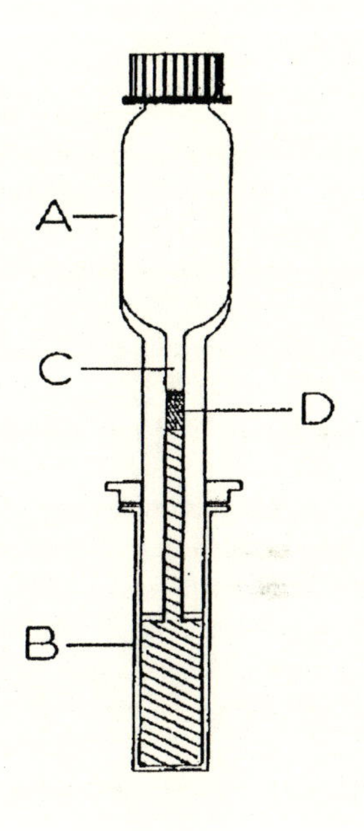

Figure 1

A less sophisticated alternative exits. The organic sample may be placed in a screw-capped centrifuge tube and a small amount of aqueous reagent added. After exhaustive shaking, the tube can be centrifuged to break the emulsion and separate the layers. The organic phase can be sampled with a syringe.

Generalized Fractionation Procedure

Total Sample. 1 mL of the sample in diethyl ether was extracted with 1 mL water in the Mixxor™ and the organic phase analyzed by GC/MS.

Removal of acids. Sufficient sodium hydroxide was added to make the aqueous phase alkaline to phenolphthalein, and the sample remaining from the initial extraction was re-extracted with the same ether in the same system.

Removal of bases. The same system, from the removal of acid step, was taken and the aqueous phase made acid (to UI paper) with an inorganic acid. The extraction was repeated.

Removal of Carbonyls. The ethereal phase from the third step was taken and combined with 0.5 mL of 2,4-DNPH solution. (The reagent was prepared by dissolving 3 gm of 2,4-dinitrophenylhydrazine in 15 mL conc. sulfuric acid. The 2,4-DNPH solution was added, with stirring, to a mixture of 20 mL water and 70 mL of 95% ethanol.) After a 10 min reaction period, water was added and two phases formed. The organic phase was analyzed.

GC/MS Analysis of Fractionated Samples

The gas chromatographic analysis of the samples was performed on a Hewlett-Packard 5890 gas chromatograph interfaced to a HP 5972 MSD detector. The column was a HP-5 (5% phenyl methyl silicon liquid phase): measuring 30 m x 0.25 mm with a 0.25 µm film thickness and a helium flow rate of 1 mL/min. The oven temperature was held 2 min @40 °C, then increased in a linear rate of 7 °C/min to 250°C. Final hold was 10 min. Normally 1 µl of sample was injected using a split injector.

Sensory Analysis of Fractionated Samples

Three microliter samples of the various samples were injected into a Hewlett-Packard 5890 gas chromatograph modified by DATU (Geneva, NY) for sensory evaluation of odors. The column was a DB-5 (5% phenyl methyl silicon liquid phase): measuring 30 m x 0.53 mm with a 3 µm film thickness and a helium flow rate of 3 mL/min. The oven temperature was held 2 min @40 °C, then increased in a linear rate of 10 °C/min to 250 °C. Final hold was 4 min.

Fractionation of Model System

A model system was prepared to represent a variety of classes of organic flavor compounds. Twenty mg each of 2-hexanone, 2-octanone, 3-octanone, 1-heptanol, heptanoic acid, limonene, octanal, 2-hexenal, methyl anthranilate, eugenol, anethol, 2-acetyl pyridine, and ethyl nonanoate were placed in 0.5 mL methanol as a stock solution. Ten microliters of the above were placed in 1.0 mL ether to represent the flavor system. The sample was separated as described in the generalized procedure above.

Fractionation of R&G Coffee Sample

Roasted and ground coffee (100 gm) was placed in 1 L of distilled water. This was atmospherically steam distilled using a Likens/Nickerson extractor for 1 hr with 1:1 pentane:diethyl ether as the solvent. The solvent was carefully distilled to 1.0 mL. The procedure followed was similar to that described in the previous section. The primary difference was that after each extraction in the Mixxor™ and analysis of the organic phase, the aqueous phase was removed. New reagent (dil base, dil acid, 2,4-DNPH) was added for each step. This was necessitated since the coffee sample is a highly complex one.

Fractionation Procedure for Swiss Cheese

Swiss cheese (235 gm) was homoginized in 1.2 L of distilled water. This was atmospherically steam distilled using a Likens/Nickerson extractor for 1.75 hr with diethyl ether as the solvent. The solvent was carefully distilled to about 1.0 mL. Since Swiss cheese was not expected to contain appreciable bases, the solvent extract was not extracted with acid. The aqueous phase from each extraction was removed, and the original organic phase was further worked up by removing the acids and then the carbonyls. Each extraction produced a progressively simpler sample and was performed in a 2 mL Mixxor™

Fractionation Procedure for Maillard Reaction

A solution consisting of 2 gm ribose, 2 gm glucose, 0.3 gm leucine, 0.4 gm cysteine, 0.25 gm thiamin in aqueous ammonium hydroxide was heated at 80 °C overnight. The solution was neutralized and extracted with diethyl ether. The ether was dried and concentrated to 1.2 mL. This product was sequentially separated into a number of fractions. The ethereal phase was extracted with 1 mL of a saturated aqueous solution of sodium chloride and analyzed. Concentrated hydrochloric acid was added and the sample re-extracted. The organic phase was analyzed as the neutral/acidic sample. The phases were separated and (1) the aqueous phase made alkaline and re-extracted with ether to give the basic compounds while the organic phase (2) was treated further. This organic phase (2) was re-extracted with aqueous base. The organic phase contained the neutrals, while the aqueous phase was acidified and re-extracted with ether to give the acidic fraction.

Results and Discussion

Model System

The four chromatograms representing the four stages are presented in Figure 2. Table 1 gives the identifications and retention times of the components.

Table 1. Compounds and Retention Times of Model System

Compound	Retention Time
2-hexanone	4.5
2-hexenal	5.9
3-octanone	7.9
1-heptanol	8.6
2-octanone	9.1
octanal	9.4
limonene	10.0
2-acetyl pyridine	10.1
heptanoic acid	11.3
anethol	15.5
ethyl nonanoate	15.6
methyl anthranilate	16.7
eugenol	16.9

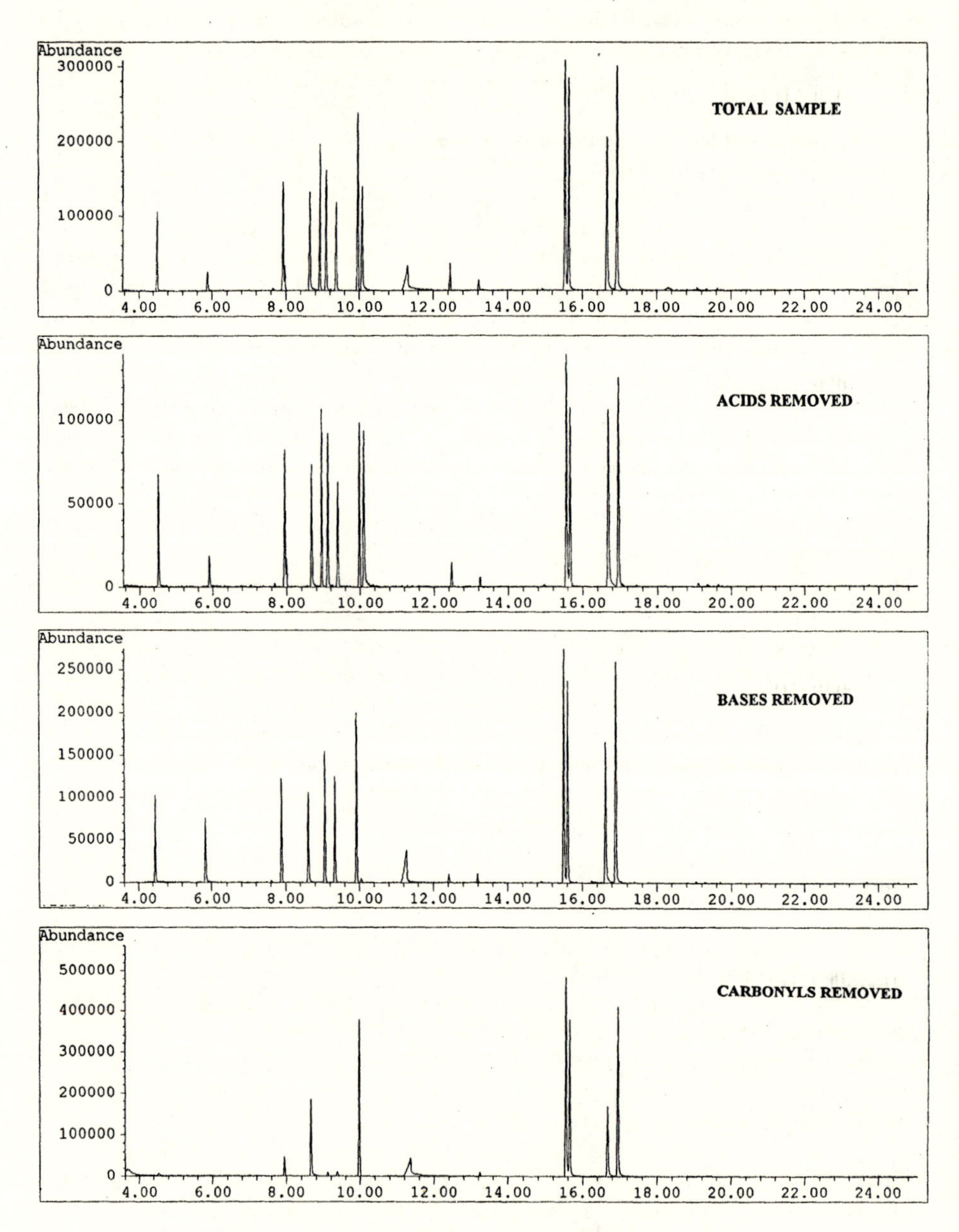

Figure 2. Four stages of fractionation of model system.

Figure 2a is the pattern for the original mixture. The next Figure (2b) shows the elimination of heptanoic acid at Rt of 11.3. when the sample is treated with base. Figure 2c represents the removal of the bases. In this case, the acetyl pyridine is eliminated.

The final figure (2d) shows the result of removing the carbonyls and the bases. All carbonyls were removed, except for 3-octanone, which was markedly reduced. The 1-heptanol, heptanoic acid, limonene, methyl anthranilate, eugenol, anethol, and ethyl nonanoate remained. The fact that the sample became progressively simplified led to an experiment on a real-world sample.

Coffee Sample

The L/N extract of R&G coffee was concentrated to about 1 mL. Since coffee is a much more complex sample, a somewhat different procedure was followed. In this case, the aqueous phase from each extraction was removed, and the original organic phase was extracted further. Thus each extraction produced a progressively simpler sample.

The first analysis was performed on the intact sample to give the complex chromatogram in Figure 3a.

The organic phase was taken and extracted with ca 1 mL of dilute NaOH in a 5 mL Mixxor™. The organic phase, now freed of the acids, was analyzed (Figure 3b). In the Rt region of 6.5 min 2- and 3-methyl butanoic acids elute. They are cleanly eliminated by the alkaline extraction.

The aqueous alkaline phase was removed, and the organic phase was extracted with dilute sulfuric acid to remove bases. The resulting chromatogram is shown in Figure 3c. The group of two carbon-substituted (i.e. ethyl and dimethyl) pyrazines, which were located at Rt 7.0 to 7.5 min have been eliminated. Now revealed at Rt 7.2 min is acetyl furan; in addition, the very important coffee flavor compound furfuryl mercaptan is now readily evident at Rt 7.17 min. Also lost is methyl pyrazine at Rt 5.1 min.

The aqueous phase was removed and the organic was treated with 2,4-DNPH and then partitioned against water. The chromatogram of the residual organic solvent is presented in Figure 3d. At Rt 5.4 the compound furfural has been removed, revealing under it methoxy methyl furan and benzyl alcohol. In addition, acetyl furan (Rt 7.2) has been removed and furfuryl mercaptan is much more clearly evident.

It is clear that this procedure sequentially removes classes of compounds. If the removed compounds are of interest, then the aqueous phase may be retained and re-analyzed. For example, if the alkaline phase from the earlier step is acidified and re-extracted with ether, the acid fraction may be analyzed. If the dilute sulfuric acid is made alkaline and re-extracted then the bases such as the pyridines and pyrazines are available for qualitative analysis. Such a procedure is employed for the Maillard sample.

Sensory Evaluation of Coffee Sample

The samples were odor evaluated by a trained flavorist. An appended 2 min version of the evaluation is presented in Table 2. A number of observations can be made. At 9.0 min a fatty acid elutes; this character is removed by the base extraction revealing a fruity character.

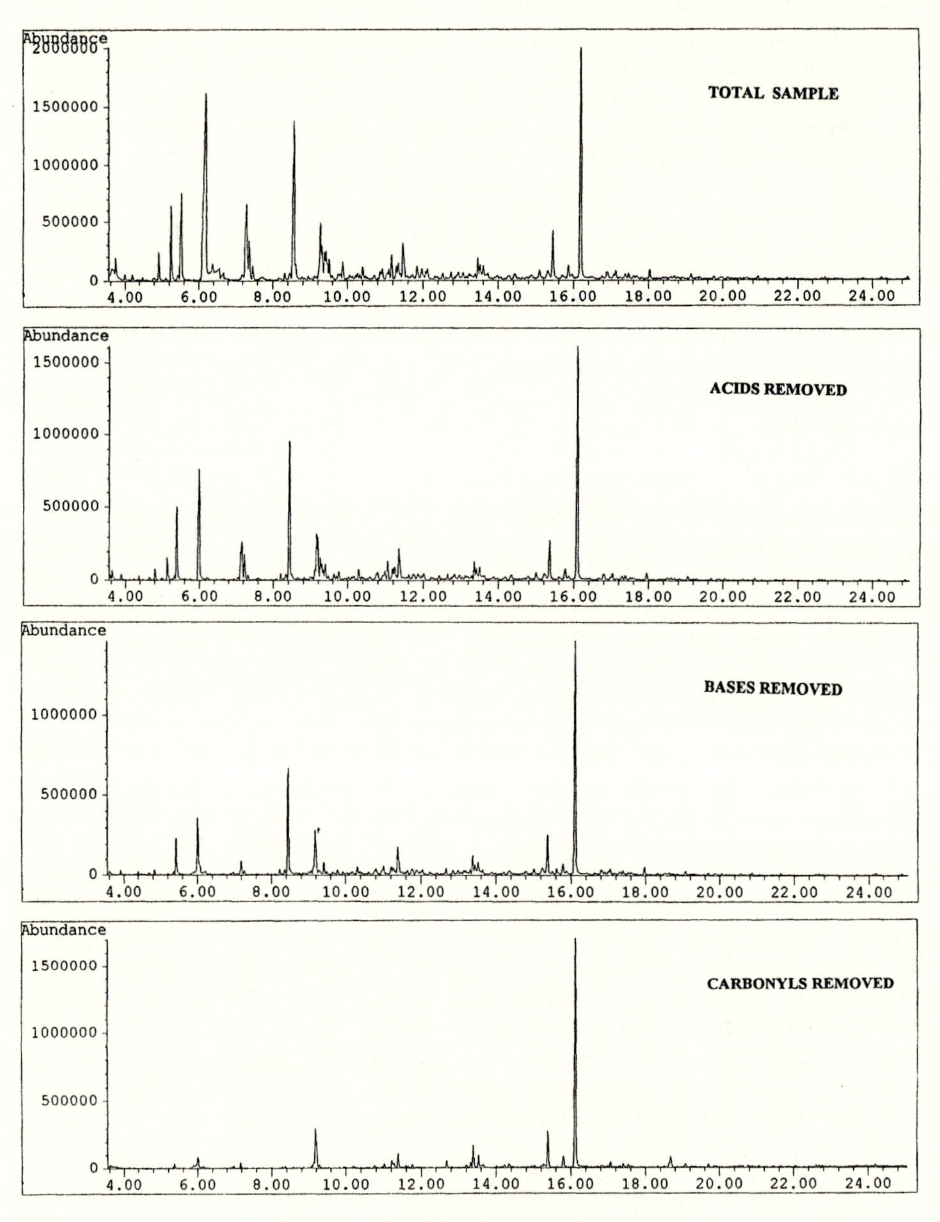

Figure 3. Simplification of coffee flavor

Table 2. Odor Assessment of Coffee Fractions

Retention Time min	Intact Sample	Base Extracted	Acid Extracted	2,4-DNPH Treated
9	Valeric acid	Fruity		Sl fruity
9.2	Cheese acid			
9.4	Cheese acid			
9.6	Sharp green	Green pungent	Sharp green	Sharp green
9.8	Cheesy	Cheesy	Cheesy	
10	Skunk	Sulfury	Sulfury	Skunk, rubber
10.2	Sweaty		Sharp green	Solvent
10.4	Potato	Potato	Potato	Potato
10.6	Sulfury	Sulfury, phenolic	Coffee	Coffee
10.8	Roasted nut	Peanut		
11	Sl sharp	Sour	Sour	

At 9.8 min an interesting cheese aroma elutes. This aroma remains through acid and base extraction; it is eliminated by the 2,4-DNPH. Thus this component may be a carbonyl compound.

The initial impression of the peak at Rt 10.6 was unimpressive. Only after the acids and bases were removed was a distinct coffee aroma evident. Thus this sample would be the choice for identification studies.

A pleasant roasted peanut aroma was observed at Rt 10.8 min. This character disappeared when the sample was acid extracted. Thus it is probably a nitrogen heterocyclic compound. It would be possible to take the aqueous acid solution and make it basic and re-extract. This would produce a basic fraction, less complex in character, more amenable to odor and mass spectral analysis.

Swiss Cheese

The first analysis was performed on the intact sample to give the complex chromatogram in Figure 4a (upper). Numerous peaks at high concentrations are present.

The organic phase was taken and extracted with dilute base in a 2 mL Mixxor™. The organic phase, now freed of the acids, was analyzed as Figure 4b (middle). At the Rt of 9.2, 13.4, 17.2, 20.6, 23.7, and 26.6 the homologous series of even carbon acids eluted. They are cleanly eliminated by the alkaline extraction as is the propanoic acid at 4.0 min.

The aqueous phase was removed and the organic phase was treated with 2,4-DNPH and then partitioned against water. The chromatogram of the residual organic solvent is presented in Figure 4c (lower). The major components, which were the C7, 9, 11, and 13 methyl ketones are removed by this process.

It is clear that this procedure sequentially removes classes of compounds. If the removed compounds are of interest, then the aqueous phase may be retained and re-analyzed. For example, if the dilute aqueous basic phase from the earlier step is acidified and re-extracted with ether, the acid fraction may be analyzed.

Odor assessment of the intact sample yielded numerous acid-notes, progressing from characteristic short chain length propionic to the soapy dodecanoic and tetradecanoic acids. Interspersed were the unmistakable aromas of the methyl ketones, with their

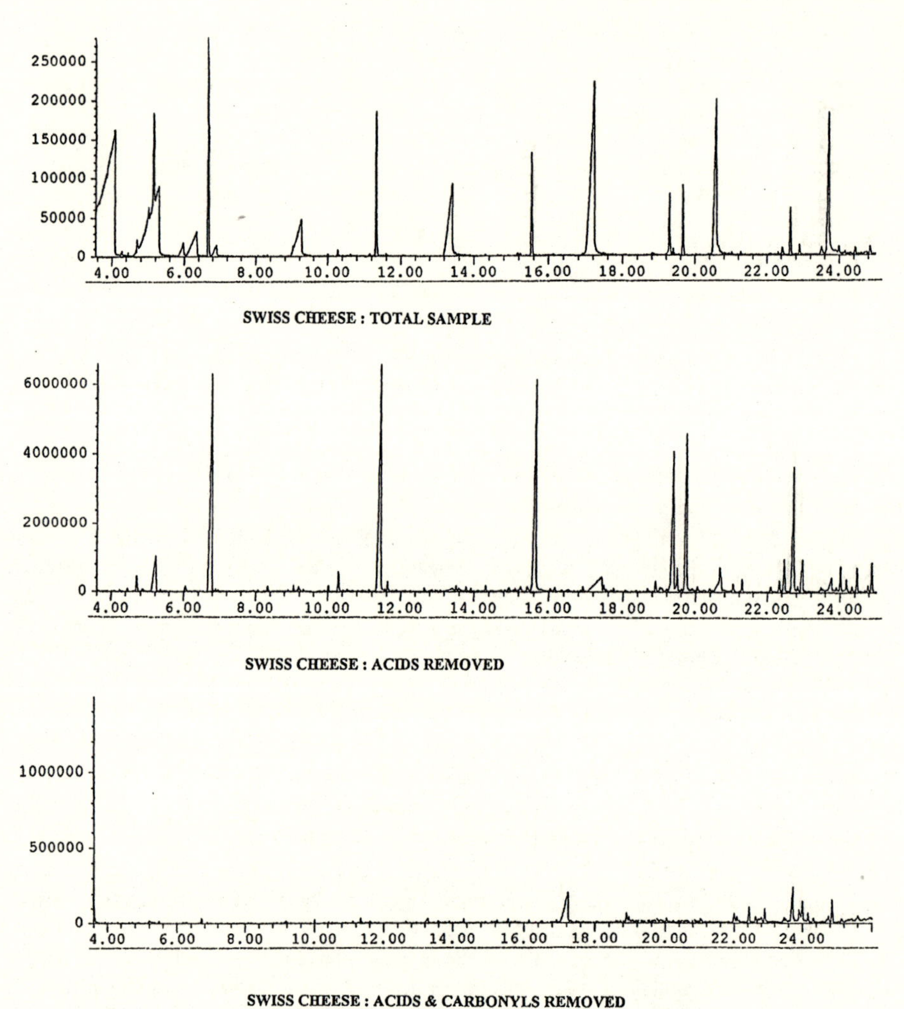

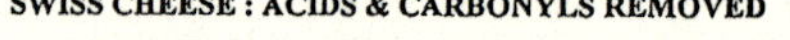

Figure 4. Fractionation of Swiss cheese sample

cheesy, blue cheese character. Few other notes were observed, due to the overpowering character of the previous.

Removal of the free acids left a chromatogram dominated by the methyl ketone aromas. When the carbonyl compounds are removed, few discernible aromas were evident. One class of compound which remained were the lactones. In the original mixture an odor described as spicy, nut-like eluted at 19.4 min. After 2,4-DNPH treatment, a pleasant dairy, creamy, cocoanut aroma was observed that was identified by GC/MS as delta decalactone.

Maillard Reaction

The sequential separation of the Maillard reaction is presented in Figure 5. In that Figure the upper curve represents the total sample before fractionation. The sample contains the pyrazines, sugar breakdown products, acids and thiazole derivatives normally expected.

Removal of the bases eliminates a number of peaks including those at 5.1, 7.1, 15.3 and 17.5 min. Observation of the composition of the basic fraction shows that methyl pyrazine elutes at 5.1 min. a group of C2 pyrazines elute at 7.1 min., and 4-methyl-5-thiazole ethanol elutes at 15.3 min.

The acid fraction contains furancarboxylic acid at 11.79 min, and thiophenecarboxylic acid at 14.0 min. The neutral fraction contains primarily furfural at 5.4 min.

The odor of the separated sample was evaluated by a professional flavorist. An appended version of the results are given in Table 3. The intact sample had a meaty, savory, sweet aroma. In the 8.8 to 10 minute region there were aromas described as burnt sugar, meaty, and popcorn which eluted in rapid succession. The most interesting sample (in the 8 to 10 min GC Rt region) was the basic fraction. In succession were nut, savory, and popcorn. This region is much easier to analyze now that the large component eluting at *ca* 9.4 has been removed. The sweet notes are found in the acidic and or neutral fraction. These compounds likely possess the -enolone structure, thus extracting into the acidic fraction.

From the above samples it is apparant that separation of flavor samples on a micro scale can be used to generate samples for subsequent GC/MS or organoleptic evaluation.

Table 3. Odor Assessment of Maillard Reaction

Retention Time, min	Total Sample	Neutral& Acids	Bases	Acids	Neutrals
8.8	Burnt sugar		Nut	Sweet	Sweet
9	Meaty, savory		Meaty , savory		
9.2	Green	Sulfury		Sulfur	Green
9.4					
9.6	Popcorn	Chemical	Popcorn		
9.8					
10	Sugar cookie	Sweet cookie		Sweet	Fatty

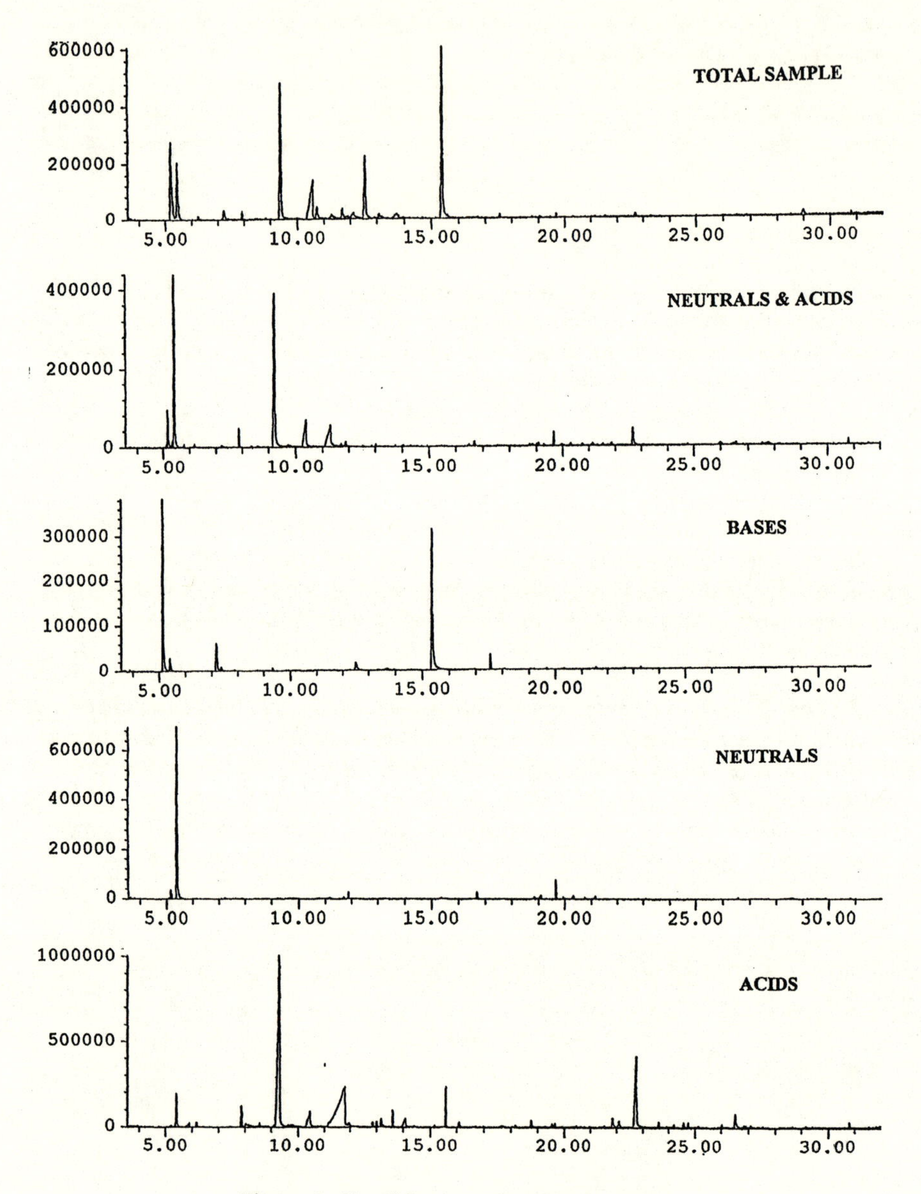

Figure 5. Simplification of Maillard system

Literature Cited

1. Teranishi, R.; Hornstein, I.; Issenberg, P.; Wick, E. *Flavor Research: Principles and Techniques,* Marcel Dekker, New York, **1971**

2. Marsili, R. *Techniques for Analyzing Food Aroma;* Marsili, R., Ed.; Marcel Dekker, New York, **1996**

3. Parliment, T. H. In *Techniques for Analyzing Food Aroma,* Marsili, R., Ed.; Marcel Dekker, New York, **1996**; pp1-26

4. Weurman, C., J Agric Food Chem, **1969**, *17*, 370

5. Teranishi, R.; Flath, R.; Sugisawa, H. In *Flavor Research, Recent Advances*; Marcel Dekker, New York, **1981**, pp 27-31

6. Leahy, M.; Reineccius,G., In *Analysis of Volatiles. Methods. Applications.*; Schreier, P. Ed.; de Gruyter,NY, **1984**, pp 19-48

7. Parliment, T.H., Perf Flav, **1986**,*1*,1

8. Parliment, T.H.; Stahl, H.D., In *Sulfur Compounds in Foods*, Mussinan, C.; Keelan, M., Eds; ACS Symposium series #564; American Chemical Society; Washington, DC, **1994**; pp 160-170

9. Parliment, T.; Stahl, H., In *Developments in Food Science V37A Food Flavors: Generation, Analysis and Process Influence*; Charalambous, G. Ed; Elsevier; New York, **1995,** pp 805-813

Chapter 3

Evaluation of Purge-and-Trap Parameters: Optimization Using a Statistical Design

Mathias K. Sucan[1], Cathryn Fritz-Jung[1], and Joan Ballam[2]

[1]Ralston Purina Company, Checkerboard Square, 2RN, St. Louis, MO 63164
[2]Ralston Analytical Laboratories, 2RS, St. Louis, MO 63164

Literature studies emphasize the importance of experimentally optimizing all purge-and-trap parameters to obtain adequate amounts of representative compounds for gas chromatographic analyses. Conventional "one variable at a time" evaluation of method parameters calls for exhaustive experimentation, systematically altering one variable at a time until an optimum point is attained. Response surface methodology was utilized in this study to determine the best combination of sampling parameters that maximized the amounts (with respect to proportion) of compounds isolated from dry pet foods. Purge time, sample size, and their interaction had the greatest effects on response. Quantities of volatile flavors isolated could be optimized and the sampling time could be reduced using response surface methodology.

One objective for the quantitative and qualitative analyses of flavors is to correlate sensory data with instrumental results to determine flavor components of organoleptic importance. Volatile flavor compounds occur in foods in defined proportions, and they exhibit their desirable sensory attributes in given concentration ranges. Therefore, the analytical methods used to assay flavors must yield data that are reflective of the original composition of a sample. Unfortunately, the isolation and characterization of flavor compounds is very challenging because flavor compounds have trace level properties, complex nature, and diverse chemical classes.

Purge-and-trap, also referred to as dynamic headspace sampling (DHS), is a means of enriching headspace gas prior to GC analyses. DHS with thermal desorption has been shown to be highly sensitive and relatively inert in the analyses of fragrances and food flavors (*1-3*). Cessna and Kerr (*4*) commented that the use of purge-and-trap systems for the analysis of aroma compounds has grown in popularity because of greater sensitivity in detecting both volatile and semi-volatile organic compounds.

A purge-and-trap system involves a purging gas (He, N_2), a sample holder, and a trap at the exhaust of the system. Flavor compounds are stripped off the sample matrix by the purge gas and are entrained on a trap packed with adsorbing materials. Tenax-TA (2,6-diphenyl-*p*-phenylene oxide) is the most widely used porous polymer adsorbent in flavor studies, because it provides the best overall trapping efficiency for organic compounds (*5*). In developing purge-and-trap methodologies, it is critical to evaluate method parameters. Hartman and coworkers (*6*), and more recently Sucan and Russell (*2,3*), emphasized the importance of experimentally optimizing all variable parameters to obtain data that are most reflective of the original composition of the food sample.

Typically, sample matrices range from a few grams to several hundred grams, sampling time ranges from 15 to 90 min, sampling temperature ranges from ambient to 200 °C, and purge gas flow rate ranges from 10 to 80 mL/min. Adsorbent traps are usually thermally desorbed for periods of 3 to 10 min at temperatures ranging from 50 to 350 °C depending on the analysis (*2,3,6*). Previously, desorption parameters were optimized using traps of similar sizes under similar conditions (*2,3,6,7*). All of these parameters affect the quantity and quality of flavors isolated from foods. Therefore, they need to be optimized to prevent altering flavor proportions, which are critical for the specific traits of a particular flavor note.

In the conventional "one variable at a time" approach, one parameter is altered, while the others are held constant, until an optimum point is attained. This calls for exhaustive experimentation, and it does not provide any information about the interactions between parameters. Response surface methodology (RSM) can be utilized to evaluate all the parameters at once, which allows determination of interactions; it can also be used to find the best combination of sampling parameters, which maximize the amounts (with respect to proportion) of compounds isolated from foods. This statistical technique utilizes quantitative data from designed experiments to obtain and solve multivariate equations (mathematical models) as functions of input conditions. The method has found wide application in the food industry; it was used in the optimization of beef aroma volatile flavors (*8*), in sensory evaluation (*9*), in food formulation (*10,11*), and in model Maillard reaction systems (*12,13*). To our knowledge, RSM has not yet been reported for optimization of sampling parameters in aroma analyses.

The objectives of this study were: (1) to optimize purge-and-trap parameters using both the conventional "one variable at a time" and RSM approaches, and (2) to compare the usefulness of the two methods for the isolation of volatile flavors.

EXPERIMENTAL PROCEDURES

Isolation of Flavor Compounds. Approximately 90 g of dry pet food were ground in a coffee grinder (Braun, Denver, Co) for 1 min. Varying amounts of the ground sample were packed in the center of a glass tube and secured by silanized glass wool at each end of the tube. Subsequently, the tube was installed in a solid sample purge-and-trap apparatus (Scientific Instrument Services, Inc. (SIS), Ringoes, NJ). Nitrogen gas was used to purge the volatile flavors into an SIS silanized glass-lined stainless steel desorption tube (3.0 mm i.d. x 10 cm length) packed with 100 mg, 60/80 mesh of Tenax-TA adsorbent.

GC Analysis of Volatiles. The desorption tube (containing the volatiles), which was prepared from the isolation procedure, was connected to an SIS model TD-3 Short Path Thermal Desorber unit. Before thermal desorption, the tube was purged at ambient temperature with N_2 gas (50 mL/min) for 5 min to eliminate moisture. The tube was then linked to the GC injection port via a 26-gauge stainless steel needle, and the volatiles were thermally desorbed at 220 °C for 5 min with a carrier gas flow rate of 1.5 mL/min. During the desorption process, the inlet section of the analytical column was held at -70 °C by means of an SIS cryotrap, which utilized liquid CO_2 to cryogenically trap the volatiles as a narrow band at the head of the column. At the end of the desorption process, the needle was retracted from the injection port, and the volatiles were analyzed by GC. The GC injector was operated in splitless mode at 220 °C; the carrier gas was helium with a linear velocity of 25 cm/sec. The GC column was a nonpolar fused silica open-tubular capillary column (30 m x 0.32 mm (i.d.), 0.25 µm thickness, DB-5MS; J&W Scientific, Folsom, CA). The GC oven temperature was linearly programmed from 40 to 200 °C at the rate of 5 °C/min. The FID (300 °C) was interfaced to a computer for capturing and storing chromatographic data and reports. Varian Star GC workstation software for system automation controlled the entire assay and showed a real-time display of the detector response. Chromatographic data were converted to ASCII format, which facilitated subsequent processing by appropriate analysis software programs.

Sampling Temperature. Higher temperatures generate greater vapor pressure; however, degradative reactions do proceed more rapidly at elevated temperatures. Therefore, it is important to determine a suitable temperature for aroma isolation. For example, Izzo et al. (*14*) utilized a temperature of 80 °C to assay wheat flour aromas. In this study, temperatures ranging from 50 to 90 °C were evaluated to determine the effect on recovery of volatiles from dry pet foods (see Figure 1). Greater amounts of flavors could be collected at higher temperatures. However, artifact formation, through non-enzymatic browning and other heat-catalyzed mechanisms, was evident at high temperatures. Artifacts were a significant portion of the total aroma isolated at 90 °C. A temperature of 70 °C was found adequate and used for the remainder of the study.

Optimization of Sampling Parameters.

"One Variable at a Time" Approach. The effects of sample size, trapping time, and flow rate on total area counts were initially studied using the conventional "one variable at a time" approach. Each factor was varied across a range of interest while the other 2 factors were held constant at a mid-point setting. The mid-point settings for each variable were 7.5 g for sample size, 30 min for time, and 30 mL/min for flow rate. The experimental ranges for each variable were 2.5 to 12.5 g for sample size, 10

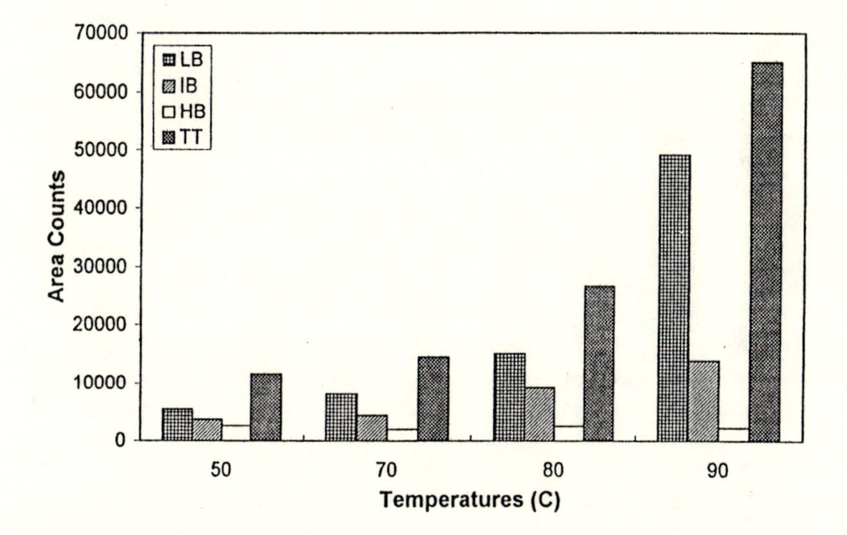

Figure 1. Effect of sampling temperature on area counts (LB = low boilers; IB = intermediate boilers; HB = high boilers)..

to 50 min for trapping time, and 10 to 50 mL/min for flow rate. The 15 experimental trials required to study the 3 sampling variables are given in Table I.

Response Surface Approach. This approach studied the three sampling variables using response surface methodology. The total area counts and proportions of volatile flavors at varying levels of time, sample size, and flow rate were modeled statistically using data from a spherical central composite design. The ranges used were the same as described for the "one variable at a time" approach. Fifteen experimental trials were required to study the three sampling variables plus one trial was replicated on each of the three days of experimentation (see Table I).

Stepwise regression was used to build simple, predictive regression equations for each response from all possible terms for a full cubic model. The linear effects of sample size, time, and flow rate were forced into each model, and a significance level of $P < 0.10$ was required for an additional term to enter and remain in the regression. After modeling each of the 4 responses, a desirability function (*15*) was optimized that simultaneously maximized total area counts while holding the proportion of peaks to within 0.5% of predicted optimum. Design-Expert software (STAT-EASE, Inc. *Design-Expert*, Minneapolis, MN, 1996) was used to perform the statistical analysis.

RESULTS

The chromatograms from GC analyses showed approximately 100 resolved peaks. Each chromatogram was divided into three sections because food volatiles exhibit a wide range of boiling points. The compounds eluting within 10 min were referred to as low-boilers (LB). Those eluting between 10 and 20 min were termed intermediate-boilers (IB). The high-boilers (HB) were those compounds with retention time greater than 20 min. The sum of the integrated areas of all peaks in peak-types LB, IB, and HB; the total area counts in each chromatogram (TT); and the proportions of LB, IB, and HB with respect to TT were used to evaluate method parameters.

"One Variable at a Time" Approach. The total area counts and proportion of each of the three eluting ranges were plotted separately against each of the three sampling variables that were studied. Varying amounts of replicate ground samples were used to evaluate the impact of sample size on volatile flavors isolated from dry pet foods (Figure 2). Larger sample sizes generated greater area counts; optimum area counts were obtained at a sample size of 10 g. Changes to the proportions of LB, IB, and HB were just noticeable at a sample size of 5 g. Complete alteration of proportions occurred at larger sample sizes. Therefore, the sample size of choice lay below 5 g.

Investigation of the effect of trapping time on volatiles isolated from the pet food (Figure 3) showed that the overall amount of volatiles recovered did not reach an optimum level within 50 min. When longer sampling time was evaluated, volatile recovery was optimized within 60 min. Although longer sampling times resulted in greater area counts, the amount of LB decreased after 60 min. Decrease in LB was probably due to breakthrough from the Tenax cartridge, or the LB may have been displaced by the IB and HB which may bind more tightly to the polymer adsorbent.

Table I: Experimental Trials for Studying Trapping Time, Sample Size, and Flow Rate Using the One Variable at a Time Approach and Response Surface Methods.

Day of Trial	*Trapping Time (Min)*	*Sample Size (g)*	*Flow Rate (ml/min)*	*One variable at a time*	*Response Surface Design*
1	30	5.0	30	x	
1	18	10.5	42		x
1	30	7.5	10	x	x
1	40	7.5	30	x	
1	42	10.5	42		x
1	30	7.5	30	x	Replicate
1	30	7.5	50	x	x
2	30	12.5	30	x	x
2	18	10.5	18		x
2	30	7.5	30	x	Replicate
2	10	7.5	30	x	x
2	42	4.5	42		x
2	30	10.0	30	x	
2	18	4.5	42		x
2	30	7.5	40	x	
3	30	2.5	30	x	x
3	42	4.5	18		x
3	20	7.5	30	x	
3	42	10.5	18		x
3	30	7.5	20	x	
3	18	4.5	18		x
3	30	7.5	30	x	Replicate
3	50	7.5	30	x	x

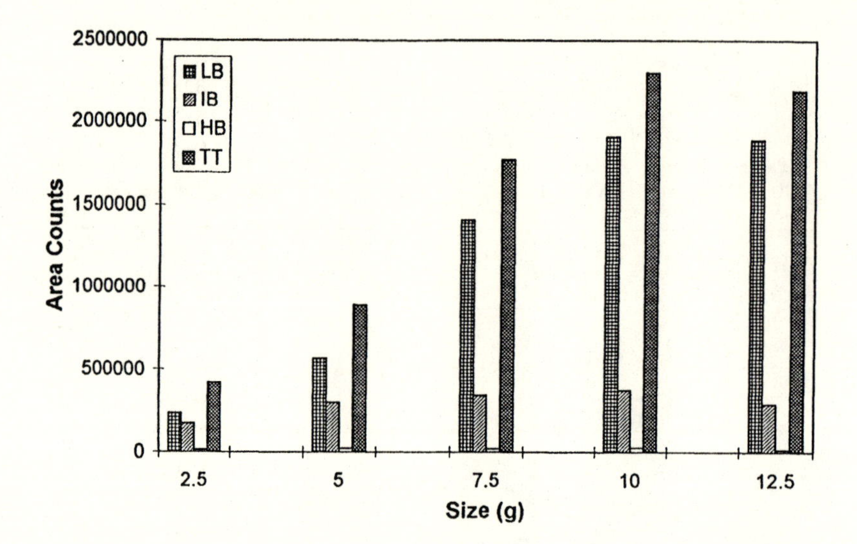

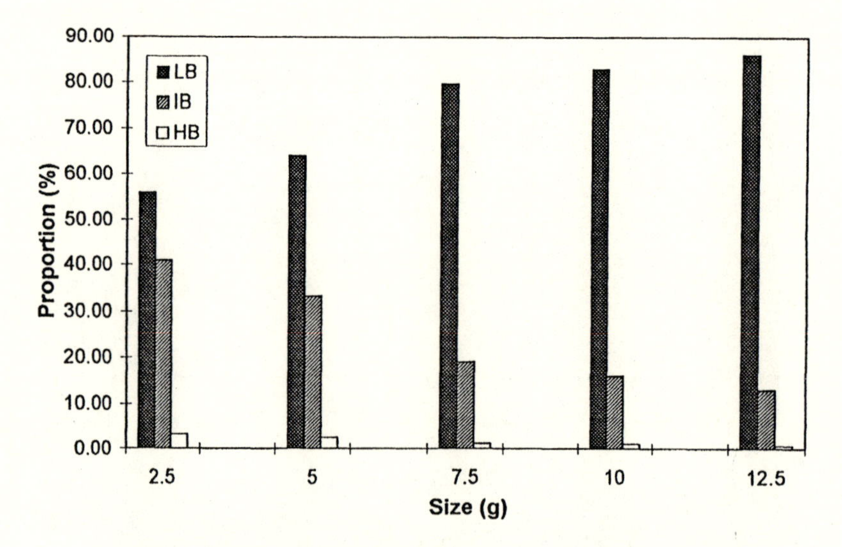

Figure 2. Effect of sample size on area counts and proportions.

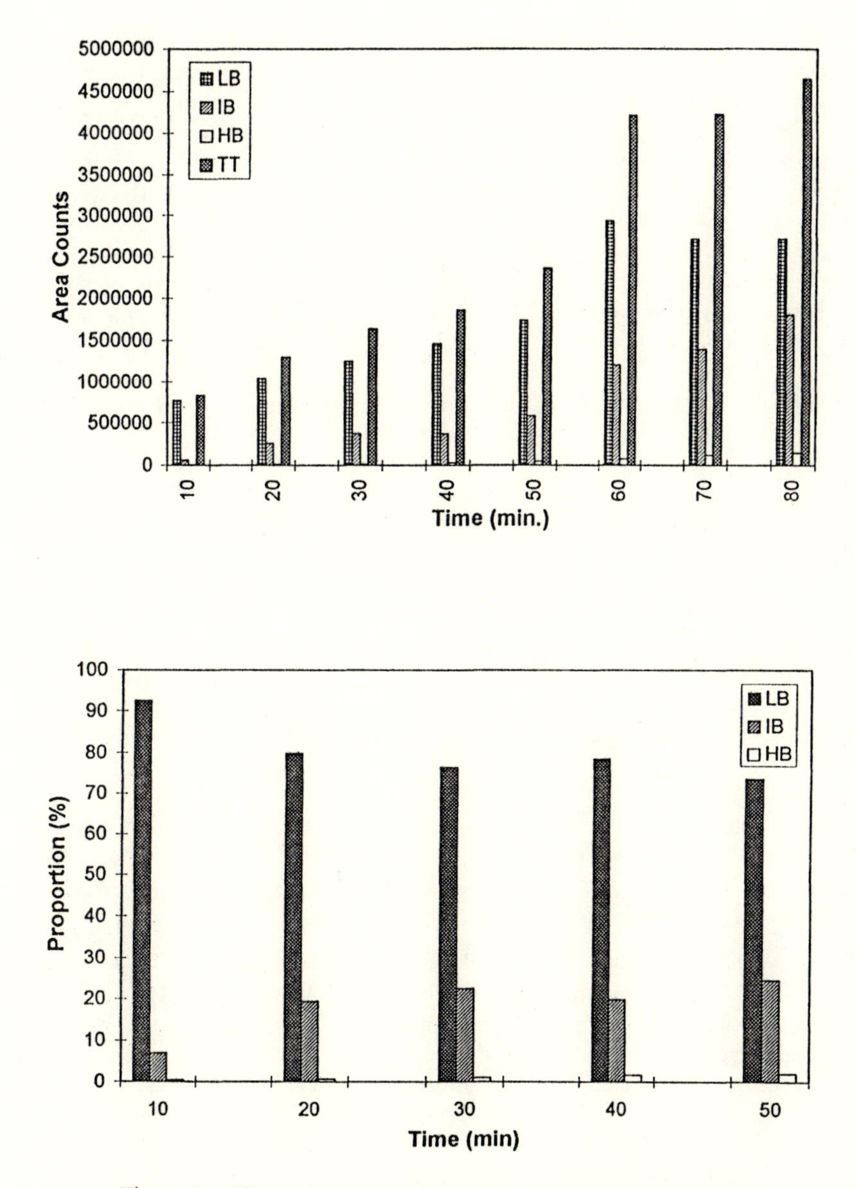

Figure 3. Effect of sampling time on area counts and proportions.

This phenomenon has been observed elsewhere (*2,3*). The changes in proportions of LB, IB, and HB increased with trapping time. The changes were less noticeable up to 30 min. Evidently, proportion did not agree with area counts for sampling time evaluation.

The plot of area counts vs. flow rates (Figure 4) has a bell shape with a maximum at 30 mL/min for LB and total volatiles. IB and HB continued to rise in amounts with increasing flow rate. The decrease in LB may have been due to a breakthrough phenomenon that took place during sampling. The distortion in proportion with respect to purge gas flow rate (Figure 4) was just visible at 30 mL/min. Higher flow rates resulted in more pronounced distortion. The optimum flow rate here was 30 mL/min and was in agreement with the optimum area counts.

Results from this evaluation indicated that optimum conditions for analyzing dry pet food were 30 mL/min for purge gas flow rate, but sample size and sampling time could only be estimated with great difficulty at 5 g and 30 min, respectively. Because of these difficulties, a response surface methodological approach was considered.

Response Surface Approach. Total area counts and proportion of each peak-type were modeled separately. Regression equations for each model and levels of significance are given in Table II. Sampling time and sample size were the most influential factors on total area count. Sample size and flow rate were the most significant predictors of the proportion of LB, IB and HB. Contour plots for responses are given in Figures 5a and 5b.

The proportion of each peak-type at the lowest level of sampling intensity was considered to be most representative of the original sample, because of the proportional distortion induced by sampling parameters. The lowest levels tested in the design were a sampling time of 18 min, sample size of 4.5 g, and a flow rate of 18 mL/min. Under these conditions, the regression equations predicted proportions of 74.6, 23.6, and 1.7% for LB, IB, and HB, respectively. These predicted proportions were used as constraints when maximizing total area counts.

Optimization of total area counts, while simultaneously holding the proportion of each peak-type to within 0.5% of its predicted value, was performed using a desirability function (*15*). The solution was at a sampling time of 48 min, sampling size of 7.9 g, and flow rate of 32 mL/min with a predicted total area count of 2.4×10^6 (Table III). Alternately, a solution that would reduce the time to run the method was sought. With the time restricted to 30 min, a total area count of 2.0×10^6 was predicted with a sample size of 10.7 g and flow rate of 46.5 mL/min (Table III). The solution at 30 min was outside the original spherical design space, but it was within the individual ranges set for sample size and flow rate.

To confirm the predicted data, two different bags (A & B) of the same type of dry pet food were analyzed for their proportion of volatile flavor compounds, using the resolved sampling parameters above. Area proportions from the confirmation runs differed somewhat from the predicted proportion (see Table IV). The largest difference was noticed among the IB. The amounts of IB increased about 29%, from samples used to build the model to confirmation samples. This may have been caused by increased lipid oxidation products (IB-type compounds) in the confirmation

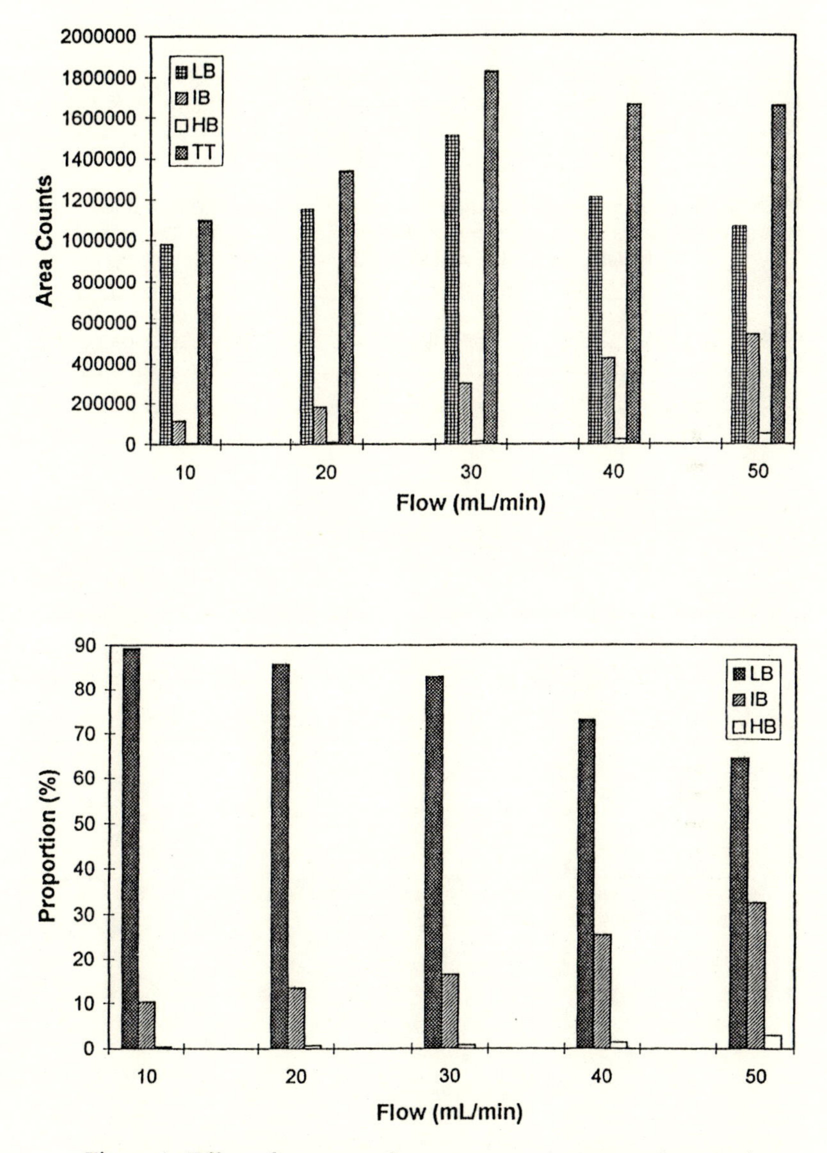

Figure 4. Effect of purge gas flow rate on area counts and proportions.

Table II: Regression equations relating total area count and proportion of low, intermediate and high boilers to purge and trap conditions. Probability values indicate the level of significance for each term.

Regression Term	*Proportion LB*		*Proportion IB*		*Proportion HB*		*Total area*	
	Coef.	*Prob.*	*Coef.*	*Prob.*	*Coef.*	*Prob.*	*Coef.*	*Prob.*
Intercept	96.71		2.31		2.45		$-1.26x10^6$	
Sampling time	-4.44	0.34	4.11	0.35	0.31	0.14	$-1.30x10^4$	0.01
Sample size	8.24	0.01	-7.38	0.01	-0.85	0.01	$1.94x10^5$	0.01
Flow rate	-0.57	0.01	0.52	0.01	-0.055	0.01	$7.52x10^4$	0.04
Time*size							$6.18x10^3$	0.02
$Time^2$	0.14	0.26	-0.13	0.20	-0.011	0.23		
$Size^2$	-0.34	0.02	0.31	0.02	0.040	0.01	$-1.52x10^4$	0.06
$Flow^2$					0.0020	0.01	$-1.02x10^3$	0.08
$Time^3$	-0.0015	0.04	0.001 4	0.04	0.0001 3	0.01		
$^aR^2$	0.94		0.93		0.97		0.91	
^{b}RSD	3.12		2.88		0.20		$2.0x10^5$	

a – values have been adjusted for the number of terms in the model.
b – Residual standard deviation for the model.

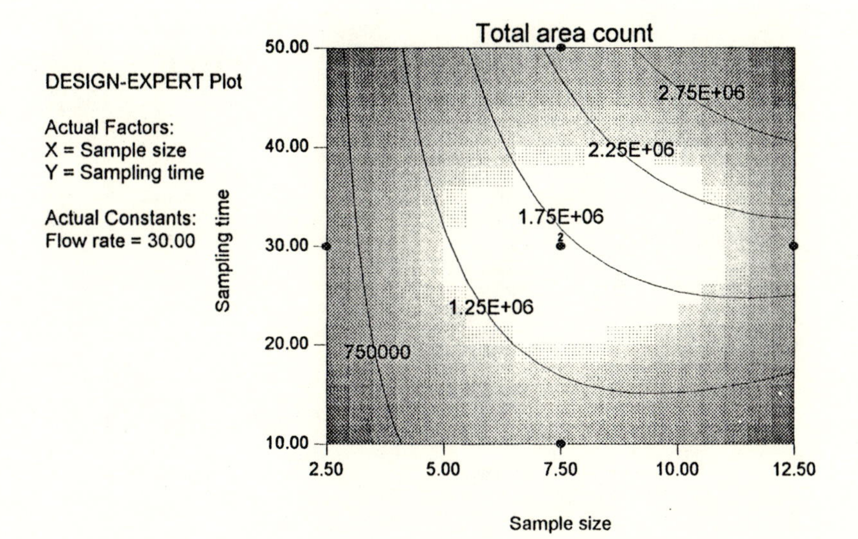

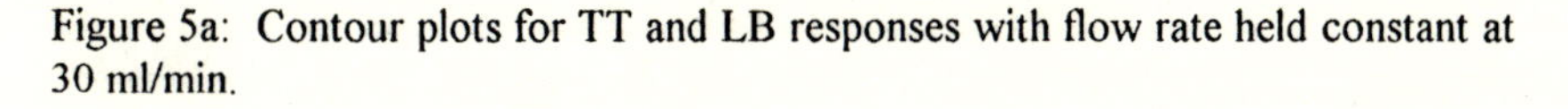

Figure 5a: Contour plots for TT and LB responses with flow rate held constant at 30 ml/min.

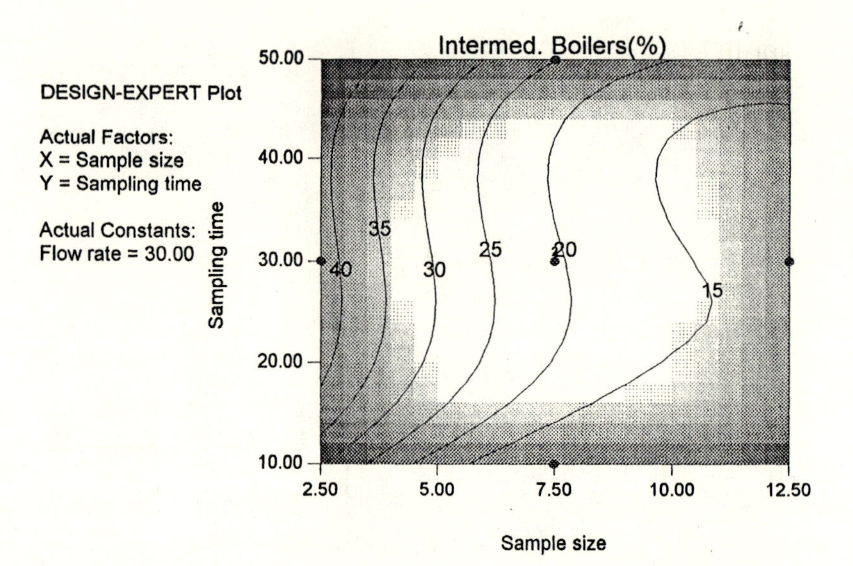

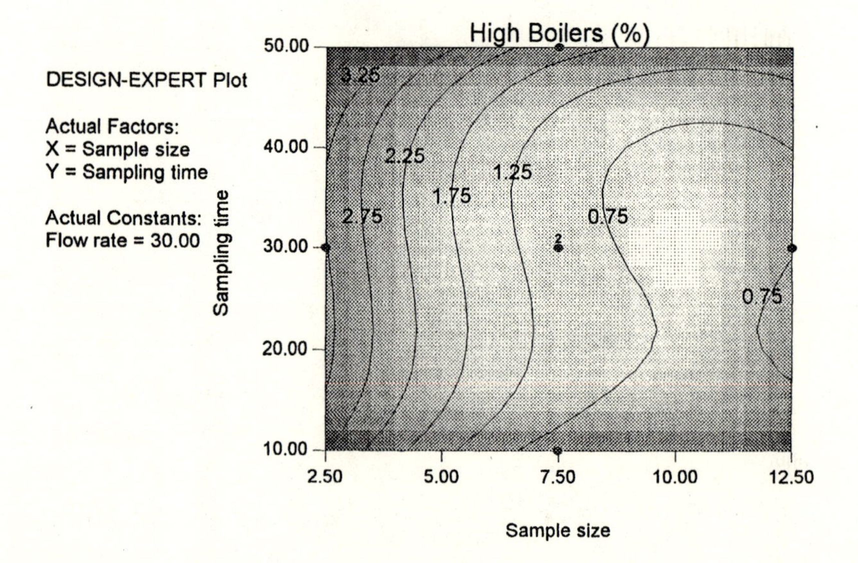

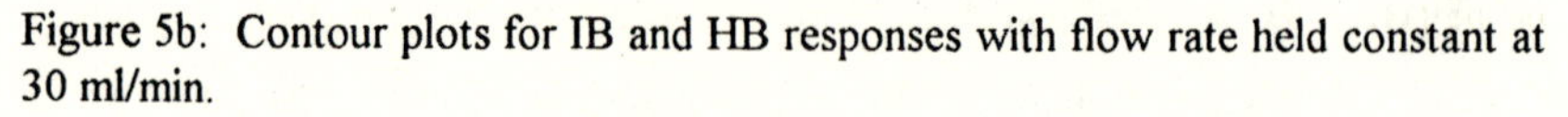
Figure 5b: Contour plots for IB and HB responses with flow rate held constant at 30 ml/min.

Table III. RSM: Optimization of Total Area Counts With Respect to Proportion of Peak-Types.

Allowable Ranges			*Purge-and-trap Parameters*			*Predicted proportions*			*Predicted Areas (TT)*
LB	IB	HB	Time	Size	Flow	LB	IB	HB	
74.1-75.1	23.1-24.1	1.2-2.2	48	7.9	32	75	23.1	1.7	2410000
74.1-75.1	23.1-24.1	1.2-2.2	30	10.7	46.5	75	23.3	2.2	1980000

Table IV. RSM: Confirmation of Predicted Optimal Purge-And-Trap Conditions.

Predicted Parameters				*Product A*			*Product B*	
	Size	Flow		Area	%		Area	%
48	7.9	32	LB	2643770	69.2	LB	2292043	68.8
			IB	1142943	29.9	IB	997404	29.9
			HB	33928	0.9	HB	41979	1.3
			TT	3820641		TT	3331426	
30	10.7	46.5	LB	2631649	69.5	LB	2239154	69.1
			IB	1123249	29.7	IB	951917	29.4
			HB	33252	0.9	HB	51534	1.6
			TT	3788150		TT	3242605	

samples; the model was built in winter and the confirmation samples were run in late spring. A comparison of the confirmation runs indicated that all proportions were similar across products A and B. Additionally, proportions remained identical within samples regardless the level of trapping parameters. This indicated that RSM can be used to optimize the amounts of volatiles sampled for GC analyses, while holding constant the proportions of volatile compounds.

CONCLUSIONS

In developing purge-and-trap methodologies for GC analyses, it is important to evaluate and optimize sampling parameters. When the traditional "one variable at a time" approach was utilized, optimum conditions for analyzing dry pet foods were 30 mL/min for purge gas flow rate, but sample size and sampling time could be estimated only with great difficulty to be 5 g and 30 min, respectively. Additionally, there was distortion in the proportion of volatiles at these sample size and purge time settings. The use of RSM proved effective in this study: The amounts of volatile flavors recovered could be optimized while maintaining the proportions of flavor components isolated. Two sets of conditions (purging time, sample size, and purge gas flow rate) were comparable with respect to quantity of aroma compounds recovered: (1) 48 min, 7.9 grams, and 32 mL/min and (2) 30 min, 10.7 grams, and 46.5 mL/min. This demonstrated that the model could be used to shorten the time of analysis. Furthermore, RSM method allowed the sampling of a much greater amount (2×10^6) of volatiles than the "one variable at a time" method (1×10^6 for 5 g, 30 mL/min, 30 min), which increased the GC detector sensitivity by a factor 2. For samples containing relatively high amounts of volatiles, however, increased concentration can overwhelm detectors and result in poor chromatographic separation. It is the responsibility of the user of this method to adjust the sample size or GC detector sensitivities without sacrificing any chromatographic peak, as some small peaks may be important contributors to the overall aroma of a given food.

References

1. Dicke, M.V.-B.; Posthumus, T.A.; Dom, M.A.; Van Bokhoven, N.B.; DeGroot, A.E. *J. Chem. Ecol.* **1990**, *16*, 381.
2. Sucan, M.K. and Russel, G.F. *Ph.D., Dissertation*, University of California, Davis, Davis, CA, 1995.
3. Sucan, M.K.; Russel, G.F. *J. High Resol. Chromatogr.*, **1997**, *20*, 310.
4. Cessna, A.J.; Kerr, L.A. *J. Chromatography* **1993**, *39*, 185.
5. Withycombe, D. A.; Mookherjee, B. C.; Hruza, A. Isolation of trace volatile constituents of hydrolyzed vegetable protein via porous polymer headspace entrainment; Academic Press, New York, 1978.
6. Hartman, T.G.; Lech, J.; Salinas, J. Rosen, R. T.; Ho, C.-T. In *Flavor Measurement*; Ho, C.-T.; Manley, C. H., Eds.; Marcel Dekker, Inc., New York, 1993, pp. 37.
7. Jordan, E.D.; Hsieh, T.C.-Y.; Fisher, N.H. *Phytochemistry*, **1992**, *31*, 1203.
8. Bodrero, K.O.; Pearson, A.M.; Magee, W.T. *J. Food Sci.* **1981**, *46*, 26.

9. Henika, R.G. *Food Technol.* **1982,** *37*, 41.
10. Pearson, A.M.; Baten, W.D.; Gombel, A.J.; Spooner, M.E. *Food technol.*,**1962,** *16*, 137.
11. Huor, S.S.; Ahmed, E.M.; Rao, P.V.; Cornel, J. A. *J Food Sci.***1980**, *45*, 809.
12. Shaw, J.J. and Ho, C.-T. In *Thermal Degradation of Aromas;* Parliment, T.H.; McGorrin, R.J.; Ho, C.-T., Eds.; ACS Symposium Series 409; American Chemical Society: Washington, D.C.,1989, pp. 217-228
13. Shaw, J. J., Burris, D. and Ho, C.-T. *Perfurmer & Flavorist*, **1990**, 15: 60-66.
14. Izzo, H.V.; Hartman, T.G.; Ho, C.-T. In *Thermally Generated Flavors: Maillard, Microwave and Extrusion Processes*; Parliment, T.H; Morello, M.J.; McGorrin, R.J., Eds.; ACS Symposium Series 543; American Chemical Society: Washington, DC, 1994, pp. 328-333.
15. Derringer, G.; Suich, R. *Journal of Quality Technology* **1980**, *12*, 214.

Chapter 4

The Importance of the Vacuum Headspace Method for the Analysis of Fruit Flavors

Matthias Güntert[1], Gerhard Krammer[2], Horst Sommer[2], and Peter Werkhoff[2]

[1]Flavor Division and [2]Corporate Research, Haarmann and Reimer GmbH, 37603 Holzminden, Germany

The Vacuum Headspace Method is a very gentle work-up procedure especially suited for the analysis of fruits. Comparison of the Simultaneous Distillation-Extraction Method, the Dynamic Headspace Method, and the Vacuum Headspace Method shows clearly the superior performance of the latter with respect to the sensory impression of the resulting extracts. Sensory evaluation of the vacuum headspace extracts of various fruits (e.g., passion fruit, strawberry, pear, peach, raspberry, cherry) leads to very fruit-typical descriptions. Consequently, the qualitative and quantitative flavor patterns of the analyzed fruits represent the genuine fruit flavors very well.

The comparison of the different work-up procedures is explained in detail. Moreover, the volatile constituents of various analyzed fruits are shown and discussed. The importance of individual flavor compounds for the respective fruit flavor is indicated. It is explained how these results can serve as the basis for the composition of a new type of fresh-fruit flavors.

The analysis of food flavors has been dramatically improved over the past 30 years, mainly through the invention of new work-up procedures, modern separation techniques (high resolution gas chromatography on fused silica capillaries, high performance liquid chromatography), and more sensitive spectroscopic methods (nuclear magnetic resonance, mass spectrometry, fourier transform infrared spectroscopy) some of them being available as hyphenated techniques (HRGC-MS, HRGC-FTIR). This continuous development has, over the years, led to results and publications about virtually every food flavor. Thousands of flavor compounds have been identified in the different foods and are collected in libraries (*1*).

Though there have always been groups that have worked this way, a significant trend in the nineties is for flavor research to focus on the few volatiles of a food flavor that are important character impact compounds and that dominate the smell and taste of a food flavor. In order to identify these active flavor compounds and moreover the right qualitative and quantitative compositional picture, the applied work-up procedures have to be selected and screened very carefully. Only if the extract, produced by the applied work-up procedure, shows the sensory properties of the flavor from the analyzed food are the chances good that the analytical results represent a true picture. Flavor analysis of non-thermally treated foods, which are eaten in the raw state (e.g., fruits), needs to be done in a very gentle way. In order to avoid cooked notes and artifacts, fruits have to be worked-up in a way that only the genuine flavor compounds in the right proportions are picked up. Because of the importance of the work-up conditions, we have carried out a comparison between three of the most commonly used procedures: the simultaneous steam distillation-extraction according to Likens-Nickerson (LN), the dynamic headspace (DHS), and the vacuum headspace (VHS). Our goal was to produce an extract that resembles the sensory quality of a freshly picked ripe fruit. Therefore, fruits were first subjected to very thorough sensory evaluation before the application of the work-up procedures. Subsequently, sensory evaluation was run on the extracts in addition to GC and GC/MS analyses. In this way we have elaborated the analysis of numerous fruits. Results of passion fruit, strawberry, peach, pear, raspberry, and cherry are described here.

The analytical results were used as a tool to formulate synthetic (nature-identical) and natural fruit flavors. The main feature of these flavors is their typical strong smell and taste that resembled the freshly picked fruits.

Experimental Procedure

Investigated Fruits. Sensory evaluation, by our flavorists, was run on all fruits. The target was to analyze the fruits at the height of ripeness in their most typical state of taste. The fruits analyzed are listed below.

- Yellow passion fruit (*Passifloraceae*, *Passiflora edulis* var. *flavicarpa*) harvested in 1995 in Columbia (South America)
- Strawberry (*Rosaceae*, *Fragaria x ananassa* var. *Senga sengana*) harvested in 1995 in the environment of Holzminden (Germany)
- Raspberry (*Rosaceae*, *Rubus idaeus*) harvested in 1993
- Pear (*Rosaceae*, *Pirus communis* var. *Bartlett, Williams Christ*) harvested in 1995
- White peach (*Rosaceae*, *Prunus persica* var. *Maria Bianca*) harvested in 1995
- Sweet Cherry (*Rosaceae*, *Prunus avium*) harvested in 1993
- Sour Cherry (*Rosaceae*, *Prunus cerasus* var. *"Schattenmorelle"*) harvested in 1993

Sample Preparation. For the analysis of the fruits four different work-up procedures were employed. These were the Simultaneous Distillation-Extraction Method at ambient pressure (SDE) and the Simultaneous Distillation-Extraction Method under vacuum (SDEV), the Dynamic Headspace Method (DHS), and the Vacuum Headspace Method (VHS). Drawings of the physical principles of these methods are shown in Figure 1.

Simultaneous Distillation-Extraction Method at Ambient Pressure (SDE) and the Simultaneous Distillation-Extraction Method under Vacuum (SDEV). This method is one of the most widely used sample preparation procedures (*2-6*). Water steam volatile flavor compounds from the fruits (between 0.5 kg and 3 kg

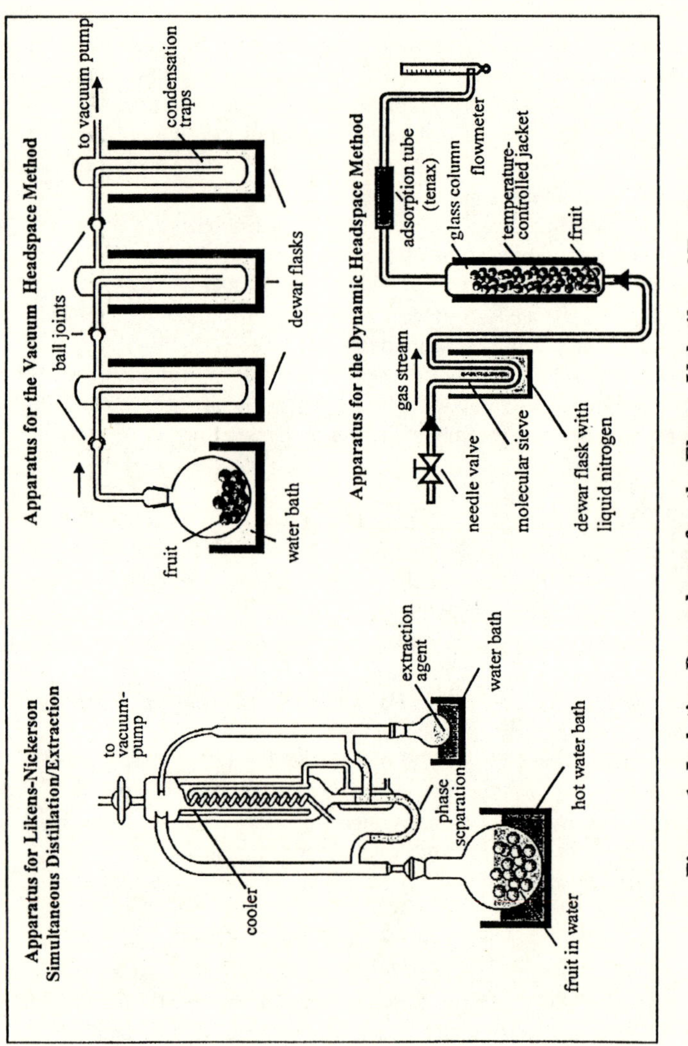

Figure 1: Isolation Procedures for the Flavor Volatiles of Fruits

chopped fruits or fruit pulp depending on the flavor strength) are continuously distilled off and simultaneously extracted from the gas phase using an organic solvent (200 mL pentane/diethyl ether 1:1). The procedure under reduced pressure (80 mbar/50 ^{o}C) was conducted in the same way by using methyl *tert*-butyl ether as an extraction solvent. In this way, the flavor compounds are enriched in the organic phase. Afterwards, the extracts are gently concentrated, using a Vigreux distillation column, to a final volume of about 0.2 mL.

Dynamic Headspace Method (DHS). The physical principle of this sample preparation method is to strip the volatile flavor compounds from the headspace above the foods by a stream of air or inert gas (*7,8*). This flavor isolation procedure has been used successfully by Buttery and co-workers (*9,10*) for the analysis of tomato paste and tortillas .The method has also been used quite often for the analysis of blossoms. Its big advantage is it non-destructive and can be used to isolate volatiles from living nature. In a recent publication Roman Kaiser described the evolution of the analysis of blossom scents in nature (*11*). DHS has also been applied for investigation of fruits on the branch or vine (*12-14*). In our experiments, we used between 1 kg and 3 kg chopped fruits or fruit pulp (depending on the flavor strength), worked at ambient temperature, and used purified (passed through a molecular sieve) helium as a purge gas (150 mL/min flow). The "extracted" flavor compounds were adsorbed onto a polymer carrier material (in our case a glass tube packed with 3.1 g of Tenax TA) and can then be desorbed in different ways (thermal desorption directly into the GC, thermal desorption into a glass desorption trap, desorption with an organic solvent). In our experiments, we ran the adsorption from 5 h to 45 h depending on the fruits. The desorption was carried out with a mixture of pentane/diethyl ether 1:1 (150 mL). Afterwards, the combined extracts were gently concentrated, using a Vigreux distillation column, down to a volume of about 0.2 mL.

Vacuum Headspace Method (VHS). This work-up procedure was first applied by various scientists for the investigation of blossom scents. The first publication appeared in 1986 by Daniel Joulain (*15*). Since then it has become an established method for analyzing blossom scents (*16-20*). More recently, the vacuum headspace technique has been used for flavor analysis of various fruits, such as strawberries, peaches, and cupuaçu (*21-24*). Very recently, Tarantilis and Polissiou reported on the isolation and identification of the flavor compounds of saffron using the vacuum headspace method (*25*). It also has to be mentioned, a similar method for isolating flavor volatiles was described in the literature many years ago under the name "high vacuum distillation" (*26-28*). The vacuum headspace method is, in principle, a vacuum steam distillation that takes place with the water of the respective fruit. The volatile flavor compounds distill off and are condensed in the cooled traps. The contents of the traps are then combined and are cold extracted with an organic solvent. In our experiments we used between 0.5 kg and 5 kg chopped fruits or fruit pulp (depending on the flavor strength) and ran the vacuum headspace from 4 h to 8 h, depending on the fruits (approximately 1-10 mbar). The three cooling traps were cooled with ice-water, dry ice-acetone, and liquid nitrogen. The liquid-liquid extraction was carried out with a mixture of pentane/diethyl ether 1:1 in a rotatory extractor. Afterwards, the extracts were gently concentrated, using a Vigreux distillation column, down to a volume of about 0.2 mL.

The aroma of the flavor concentrates, obtained from fruits by the different isolation methods, were compared with the aroma of fresh fruits by our flavorists. The results are summarized in Table 1.

In order to study the influence of the four different isolation methods on the chemical composition of the fruit volatiles all solvent extracts were subsequently analyzed using completely identical chromatographic and spectroscopic conditions.

Instrumental Analysis. Instrumentation (capillary gas chromatography, spectroscopy) as well as analytical and preparative conditions were described in previous publications (*29,30*) and in a very recent paper from our work group where we described the flavor chemistry of yellow passion fruits in detail (Werkhoff, P.; Güntert, M.; Krammer, G.; Sommer, H.; Kaulen, J. *J.Agric. Food Chem.*, accepted for publication)

Component Identification. Flavor compounds from the fruits investigated were identified by comparison of the compound's mass spectrum and Kovats indices with those of a reference standard. In some cases reference compounds were synthesized in our laboratory. In all cases the respective structures were confirmed by NMR, IR, and MS spectroscopy.

Gas Chromatography-Olfactometry. GC with simultaneous FID and odor port evaluation was carried out using a Carlo Erba Type 5360 Mega Series gas chromatograph. Separations were performed using a 60m x 0.32 mm (ID) capillary column coated with DB-1 (d_f = 1 µm) and DB-WAX (d_f = 0.5 µm) stationary phases. The flow rate of the helium carrier gas was 3-4 mL/min. The column effluent was split 1:50 with a glass-cap-cross (Seekamp, Achim, Germany). The temperature program used was 60 ^{o}C-220 ^{o}C at 3 ^{o}C/min. Injections were made in the on-column as well as in the split/splitless mode. The injector temperature was 220 ^{o}C and the detector temperature was 250 ^{o}C

Results and Discussion

Analysis of the Fruits. The main target of this study was to find the sample preparation method that gives an extract with the sensory impression most typical of the ripe, freshly picked fruit. Subsequently, the qualitative and quantitative flavor patterns, identified in this study, should be the basis for formulating a range of new and unique fruit flavors. Therefore, comparison of the different work-up procedures applied to the various fruits was very important. A second aspect of this study was to compare the dynamic headspace method with the vacuum headspace method. We wanted to determine whether the dynamic headspace method, which has been applied to blossoms, could be successfully applied to fruits. Several reviews on sample preparation and isolation procedures have recently been published (*31-35*). The influence of these different techniques on the resulting flavor extracts was discussed by several scientists (*36-40*). These publications all demonstrate that the isolation procedure is of critical importance for the sensory quality of the resulting flavor extracts. In our study, the results of the different work-up procedures applied on the

Table I: Sensory Evaluation of the Fruit Concentrates Obtained by the Various Flavor Isolation Procedures

Fruit	Vacuum Headspace (VHS)	Dynamic Headspace (DHS)	Simultaneous Distillation-Extraction at ambient pressure (SDE)	Simultaneous Distillation-Extraction under vacuum (SDEV)
Yellow passion fruit	estery, tropical fruit, green, sulfury, juicy, fresh, typical	herbaceous, tropical, weak	sulfury, fruity, tropical fruit, burnt, acidic, somewhat typical	sulfury, tropical, sweet, somewhat typical
White peach	lactoney, fruity, green, white peach skin typical	lactoney, fruity, juicy, overripe, fruit flesh, very volatile	lactoney, fruity, tea note, cooked note	green, fruity, lactoney, cooked, canned
Strawberry	fruity, green, acidic, sweet, earthy, ripe, typical	estery, fruity, sweet, amine-like, weak	green, buttery, weedy, marmelade, cooked	green, cinnamon note, cooked, artificial
Pear	estery, fruity, green, juicy, pear skin, typical	fruity, fatty, rancid, weak	overripe, cooked	green, estery, fatty, sweaty, juicy, typical
Raspberry	fruity, green, woody, ionone note, violet note, sweet, typical	-	-	-
Sweet cherry	acidic, cinnamon note, sweet, cooked note	-	-	-
Sour cherry	fruity, hay note, juicy, typical	-	-	-

different fruits are described in Table I. The sensory descriptions show very clearly that the vacuum headspace extracts deliver the most typical odor compared to the fresh fruits. In the case of yellow passion fruit, white peach, strawberry, and pear we compared the four described isolation procedures. Raspberry and cherry were only analyzed by the vacuum headspace method. Flavor concentrates were subsequently analyzed by capillary gas chromatography, capillary gas chromatography-mass spectrometry, as well as capillary gas chromatography-olfactometry (GC-O).

Other work-up procedures applied in our laboratories for fresh fruits include liquid-liquid extraction and solid phase microextraction (SPME). Both have their merits, but the results were only used for complementary reasons. Extraction with liquid or supercritical carbon dioxide has been used in the recent past. This seems to be another very gentle and promising isolation procedure but it is too early for conclusions. This topic will be discussed in the future elsewhere.

Yellow Passion Fruit. Table II shows a great part of the flavor compounds identified in yellow passion fruit along with their relative GC area percent. This fruit belongs to the best known tropical fruits in the world. It has its origin in South America and has a floral, estery aroma with a distinct tropical, sulfury note. The flavor of yellow passion fruit is quite complex in its composition. There are no real character impact compounds, but the flavor is a delicate balance of different chemical classes, e.g. fruit esters, green compounds, monoterpenes, sulfur compounds, and lactones. The volatile composition of yellow passion fruits was reviewed by Whitfield and Last (*41*) as well as by Shibamoto and Tang (*42*). To date, more than 200 compounds have been reported. Our newest findings on the flavor of the yellow passion fruit were recently published elsewhere (Werkhoff, P.; Güntert, M.; Krammer, G.; Sommer, H.; Kaulen, J. *J.Agric. Food Chem.*, accepted for publication).

Significant differences for individual flavor compounds were observed in extracts obtained by the different isolation procedures. The difference between the two headspace procedures VHS and DHS is very obvious in that the DHS extract is mainly dominated by the more volatile components. A typical example is ethyl butyrate that comprised 23.0% of the DHS extract whilst the VHS extract contained only 4.7%. These findings are typical for the dynamic headspace method (DHS) and represent the physical principle of this procedure. The vacuum headspace method (VHS) "extracts" much more of the higher boiling point components and delivers a much more balanced composition of the total flavor. Most of the higher boiling and more polar lactones, for example, that are very potent and important flavor compounds (e.g., γ-decalactone, *cis*-γ-jasmin lactone, *trans*-γ-jasmin lactone, γ-dodecalactone) could not be detected in our dynamic headspace experiment. Another example is the sulfur-containing flavor compounds, e.g., 3-mercaptohexanol, 3-(methylthio)hexanol, 3-(methylthio)hexyl butyrate, 3-mercaptohexyl hexanoate, and 3-(methylthio)hexyl hexanoate, which are missing in the DHS extract. So, it can be easily explained why the dynamic headspace extract received a poor sensory judgment. On the other hand, the SDEV extract was judged, by sensory evaluation, as being somewhat typical compared to the fresh fruit. The simultaneous distillation-extraction under vacuum (SDEV) seems to be a valuable alternative to the simultaneous distillation-extraction under atmospheric pressure (SDE), which produces thermally induced artifacts and consequently quite atypical extracts. Typical examples for thermal degradation reactions are the monoterpenes

Table II: Flavor Constituents of Yellow Passion Fruits

Flavor Compound	Area % GC			
	VHS	DHS	SDE	SDEV
2-butanone	0,3	1,3	0,3	-
ethyl propanoate	< 0,1	< 0,1	0,1	< 0,1
propyl acetate	< 0,1	< 0,1	-	< 0,1
2- and 3-pentanone	< 0,1	0,1	0,1	0,1
methyl butyrate	< 0,1	< 0,1	< 0,1	< 0,1
isobutyl acetate	< 0,1	0,1	< 0,1	< 0,1
2-butanol	< 0,1	0,1	< 0,1	< 0,1
α-pinene	< 0,1	0,3	< 0,1	< 0,1
ethyl butyrate	4,7	23,0	7,9	9,1
2*E*-butenal	-	-	< 0,1	< 0,1
2,3-pentanedione	< 0,1	< 0,1	< 0,1	< 0,1
ethyl 2-methylbutyrate	< 0,1	< 0,1	< 0,1	< 0,1
butyl acetate	0,1	0,6	< 0,1	0,1
hexanal	0,1	0,1	0,1	0,2
2,6,6-trimethyl-2-vinyltetrahydropyrane	-	-	0,4	-
3-pentanol	< 0,1	< 0,1	-	< 0,1
2-pentanol	< 0,1	-	-	< 0,1
β-pinene	-	< 0,1	< 0,1	< 0,1
2- and 3-methylbutyl acetate	< 0,1	0,3	< 0,1	< 0,1
propyl butyrate	< 0,1	< 0,1	< 0,1	< 0,1
ethyl pentanoate	< 0,1	< 0,1	< 0,1	< 0,1
4-methyl-3-penten-2-one	< 0,1	-	< 0,1	< 0,1
3-hexenal	-	-	< 0,1	< 0,1
butanol	0,2	0,1	0,2	0,3
3*E*-hexenal	< 0,1	-	< 0,1	< 0,1
3-heptanone	< 0,1	< 0,1	< 0,1	< 0,1
Δ-3-carene	< 0,1	< 0,1	-	< 0,1
myrcene	0,3	4,4	0,7	1,2
isobutyl butyrate	< 0,1	< 0,1	-	< 0,1
α-phellandrene	-	< 0,1	< 0,1	0,1
α-terpinene	-	0,1	< 0,1	0,2
2-heptanone	< 0,1	0,1	< 0,1	< 0,1
methyl hexanoate	< 0,1	< 0,1	< 0,1	-
limonene	0,2	1,8	0,4	0,7
3-methyl-2-butenal	< 0,1	-	-	< 0,1
2- and 3-methylbutanol	0,2	< 0,1	0,1	0,2
trans-anhydrolinalool oxide	-	-	0,3	-
β-phellandrene	-	< 0,1	-	0,2
1,8-cineole	< 0,1	< 0,1	< 0,1	< 0,1
2*E*-hexenal	0,1	-	0,4	0,4
ethyl hexanoate	5,2	7,2	2,0	5,8
cis-anhydrolinalool oxide	-	-	0,2	-
cis-β-ocimene	-	< 0,1	< 0,1	-
γ-terpinene	0,1	0,3	0,1	0,2
trans-β-ocimene	0,3	4,1	0,7	1,0
hexyl acetate	2,0	7,9	0,7	1,6
ethyl 3*E*-hexenoate	< 0,1	< 0,1	< 0,1	0,1
3*Z*-hexenyl acetate	1,6	4,2	0,5	1,3

Continued on next page.

hexanol	8,0	3,3	2,5	5,4
2-methyl-3-(2*H*)-furanone	-	-	< 0,1	-
hexyl butyrate	8,9	15,5	2,9	8,7
α-angelica lactone	-	-	< 0,1	-
trans-linalool oxide (f)	0,2	-	1,2	0,4
ethyl (methylthio) acetate	< 0,1	-	-	-
3*E*-hexenylbutyrate	0,3	0,3	0,1	0,2
furfural	< 0,1	< 0,1	2,3	< 0,1
3*Z*-hexenylbutyrate	2,1	2,2	0,5	1,3
nerol oxide	-	-	< 0,1	-
cis-linalool oxide (f)	< 0,1	-	0,6	0,1
theaspirane A	< 0,1	-	< 0,1	< 0,1
benzaldehyde	1,9	< 0,1	1,9	2,9
ethyl 3-hydroxybutanoate	1,6	< 0,1	0,2	< 0,1
cis-2-methyl-4-propyl-1,3-oxathiane	< 0,1	-	< 0,1	< 0,1
theaspirane B	< 0,1	-	-	< 0,1
linalool	0,5	0,3	4,6	4,9
trans-2-methyl-4-propyl-1,3-oxathiane	< 0,1	< 0,1	< 0,1	< 0,1
octanol	2,3	0,4	0,3	0,8
ethyl 3-(methylthio) propanoate	< 0,1	< 0,1	< 0,1	< 0,1
5-methylfurfural	-	-	0,1	-
hexyl hexanoate	26,7	7,6	10,3	24,4
phenylacetaldehyde	-	-	0,8	< 0,1
furfuryl alcohol	-	-	0,1	-
3*Z*-hexenyl hexanoate	4,7	1,2	1,6	3,7
(*E*)-ocimenol	-	-	0,4	-
ethyl 3-hydroxyhexanoate	0,2	-	0,1	0,1
α-terpineol	0,5	< 0,1	2,0	1,7
3-mercaptohexyl acetate	-	-	< 0,1	-
ethyl 3-(methylthio)-2*E*-propenoate	< 0,1	-	< 0,1	0,1
hexyl octanoate	0,3	< 0,1	0,4	0,6
β-damascenone	< 0,1	-	< 0,1	< 0,1
7,8-dihydro-β-ionone	0,2	-	0,1	0,2
3-mercaptohexanol	0,1	-	0,2	0,2
geraniol	0,3	-	0,8	0,8
benzyl alcohol	2,8	< 0,1	0,3	0,2
3-(methylthio)hexanol	-	-	< 0,1	-
3-(methylthio)hexyl butyrate	< 0,1	-	-	< 0,1
β-ionone	0,1	-	0,1	0,2
2,5-dimethyl-4-hydroxy-3(2*H*)-furanone	0,1	-	-	0,1
3-phenyl-1-propanol	0,2	-	-	-
3-mercaptohexyl hexanoate	< 0,1	-	-	0,1
ethyl cinnamate	0,8	-	0,3	0,3
3-(methylthio)hexyl hexanoate	< 0,1	-	-	< 0,1
3*E*,5*E*-pseudoionone	0,2	-	< 0,1	< 0,1
γ-decalactone	0,2	-	0,1	< 0,1
eugenol	< 0,1	-	0,1	-
cis-γ-jasmin lactone	< 0,1	-	-	-
trans-γ-jasmin lactone	< 0,1	-	-	-
γ-dodecalactone	< 0,1	-	0,1	< 0,1
hexadecanoic acid	-	-	14,6	-

2,6,6-trimethyl-2-vinyltetrahydropyrane, the *trans*- and *cis*-anhydrolinalool oxides, nerol oxide, and *trans*-ocimenol as well as some furans like 2-methyl-3(2*H*)-furanone, furfural, 5-methylfurfural, and furfuryl alcohol. All of the compounds mentioned were only identified in the simultaneous distillation-extraction under atmospheric pressure (SDE).

White Peach. Another fruit analyzed in our study was white peach. A great part of the volatile flavor composition (DHS, VHS, and SDEV experiments) is shown in Table III. Peach seems to have its origin in China, but nowadays it is mainly grown in Europe and the United States. It has a very distinct flavor that is mainly dominated by fruity, green, and lactoney notes. The volatile composition was reviewed by J. Crouzet et al. (*43*). Again it is very obvious from our results that the dynamic headspace method (DHS) does not lead to a well balanced flavor extract. Concentrations of the more volatile components are rather over-represented, e.g., hexyl acetate was 33.1% of the DHS extract compared to 13.3% in the VHS extract. On the other hand, lactones that are very important from a sensory standpoint are not represented in the DHS extract at all, e.g., γ-heptalactone, γ-octalactone, δ-octalactone, γ-nonalactone, γ-decalactone, *trans*-marmelolactone, *cis*-γ-jasmin lactone, *cis*-marmelolactone, *trans*-γ-jasmin lactone, δ-decalactone, jasmin-δ-lactone, γ-dodecalactone, (*Z*)-6-dodecene-1,4-olide, and δ-tetradecalactone. Our experiments show that the composition of the VHS extract is somewhat similar to the SDEV extract. Again as in the case with passion fruit the simultaneous distillation-extraction under vacuum (SDEV) is the closest alternative to the vacuum headspace method (VHS). This is supported by the sensory descriptions, though slight cooked and canned notes were detected in the SDEV extract. The sensory differences between DHS and VHS become even bigger while the differences between VHS and SDEV become smaller, since the flavor of white peach is mainly dominated by higher molecular weight and polar compounds. Narain et al. (*44*) used the dynamic headspace technique to analyze the flavor volatiles of a promising new peach cultivar under development. Typically enough, they found mainly the more volatile and less polar compounds in the extract. The lactone with the highest boiling point they identified was γ-octalactone in small amounts.

Strawberry. A further fruit investigated in our laboratories was strawberry, one of the best known fruits of all. Strawberry belongs to the genus *Fragaria (Rosaceae)*. The plant has its origin probably in the Himalayas and Southeast Asia. Nowadays there is a wide variety of different variants grown mostly in Europe, Asia, and North America. The composition was reviewed by Honkanen and Hirvi (*45*). Fischer and Hammerschmidt (*21*) investigated the flavor of fresh strawberries with the "Closed Loop Stripping" technique and the vacuum distillation. The vacuum distillate (closely comparable with our vacuum headspace distillate) showed a very fresh, typical strawberry flavor, which was analyzed further by Aroma Extract Dilution Analysis (AEDA). Recently, Ulrich et al. gave an interesting overview about the flavor differences in various strawberry cultivars (*46,47*). Strawberry has a very delicate flavor dominated mainly by acidic, green, fruity, and sweet notes.

The composition of the volatile flavor compounds (DHS and VHS experiments) is in part shown in Table IV. In this case, it is very obvious that only the VHS

Table III: Flavor Constituents of White Peaches

Flavor Compound	Area % GC		
	VHS	DHS	SDEV
propyl acetate	<0,1	0,4	-
isobutyl acetate	<0,1	0,7	<0,1
2-butanol	<0,1	0,8	<0,1
propanol	0,2	8,2	<0,1
ethyl 3-methyl butyrate	<0,1	0,4	-
butyl acetate	<0,1	0,3	<0,1
hexanal	<0,1	<0,1	0,5
iso-butanol	0,1	0,8	<0,1
2- and/or 3-methylbutyl acetate	0,1	4,2	<0,1
2E-hexenal	0,1	<0,1	2,0
hexyl acetate	13,3	33,1	9,3
3*E*-hexenyl acetate	0,1	0,2	0,1
3*Z*-hexenyl acetate	3,3	5,0	3,0
2*E*-hexenyl acetate	25,4	38,5	15,5
hexanol	1,0	0,1	4,7
3*Z*-hexenol	<0,1	<0,1	0,5
nonanal	<0,1	-	0,8
2*E*-hexenol	2,9	0,3	12,0
theaspirane A	<0,1	<0,1	0,1
benzaldehyde	3,5	0,1	7,2
theaspirane B	<0,1	<0,1	0,1
linalool	<0,1	-	0,1
methyl-4*Z*-decenoate	<0,1	-	0,3
γ-hexalactone	1,4	<0,1	0,6
γ-heptalactone	<0,1	-	<0,1
β-damascenone	<0,1	-	0,1
7,8-dihydro-β-ionone	0,2	-	0,2
geraniol	0,1	-	0,1
geranyl acetone	<0,1	-	0,1
benzyl alcohol	0,6	-	0,1
γ-octalactone	0,6	-	<0,1
β-ionone	0,1	-	0,2
β-ionol	0,1	-	0,2
δ-octalactone	<0,1	-	<0,1
7,8-dihydro-β-ionol	0,4	-	0,5
5,6-epoxyionone	0,2	-	-
3,4-dehydro-β-ionol	<0,1	-	<0,1
γ-nonalactone	0,1	-	0,2
γ-decalactone	17,1	-	12,4
trans-marmelolactone	<0,1	-	<0,1
6-amyl-α-pyrone and *cis*-γ-jasmin lactone	4,0	-	2,0
cis-marmelolactone	<0,1	-	<0,1
trans-γ-jasmin lactone	<0,1	-	<0,1
δ-decalactone	1,9	-	0,8
jasmin-δ-lactone	0,2	-	0,2
γ-dodecalactone	0,3	-	0,9
6*Z*-dodecen-1,4-olide	<0,1	-	<0,1
δ-tetradecalactone	-	-	<0,1

Table IV: Flavor Constituents of Strawberries

Flavor Compound	Area % GC		
	VHS	DHS	SDE
ethyl propionate	0,5	0,9	<0,1
methyl butyrate	0,5	1,1	<0,1
ethyl butyrate	5,2	11,7	<0,1
ethyl 2-methyl butyrate	0,7	1,7	<0,1
butyl acetate	0,2	0,5	<0,1
2- and/or 3-methylbutyl acetate	0,5	1,3	<0,1
ethyl crotonate	0,3	0,4	<0,1
methyl hexanoate	1,9	3,9	0,7
2- and/or 3-methylbutanol	1,5	0,2	<0,1
ethyl hexanoate	31,8	52,9	0,7
hexyl acetate	1,2	3,7	0,6
3*Z*-hexenyl acetate	0,2	0,3	<0,1
2*E*-hexenyl acetate	2,5	5,4	1,7
hexanol	0,5	0,1	0,4
2*E*-hexenol	0,8	0,1	1,0
ethyl octanoate	0,6	0,5	<0,1
acetic acid	0,4	<0,1	1,6
furfural	-	-	3,3
linalool	1,1	0,1	0,5
isobutyric acid	1,5	<0,1	1,1
2,5-dimethyl-4-methoxy-3(2*H*)-furanone	4,0	0,1	2,0
butyric acid	10,0	<0,1	11,3
2-methylbutyric acid	7,3	<0,1	11,6
α-terpineol	-	-	0,2
benzyl acetate	0,2	<0,1	<0,1
nerol	-	-	<0,1
4-methylpentanoic acid	0,2	-	0,4
octyl hexanoate	<0,1	<0,1	0,2
geraniol	-	-	0,1
hexanoic acid	9,8	<0,1	26,3
2,5-dimethyl-4-hydroxy-3(2*H*)-furanone	0,2	-	<0,1
E-nerolidol	1,5	<0,1	2,0
octanoic acid	0,3	<0,1	1,2
ethyl-*E*-cinnamate	0,3	<0,1	<0,1
γ-decalactone	4,5	<0,1	7,2
γ-jasmin lactone	0,2	-	0,1
δ-decalactone	0,2	-	-
γ-dodecalactone	0,4	-	1,0
p-vinylphenol	-	-	0,1
dodecanoic acid	-	-	0,4
tetradecanoic acid	-	-	0,7
hexadecanoic acid	-	-	4,4

experiment leads to satisfactory results concerning the sensory comparison of the different work-up procedures. In the DHS concentrate the lower molecular weight fruit esters (e.g., ethyl butyrate 11.7%, ethyl hexanoate 52.9%, hexyl acetate 5.4%) are substantially over represented while the rather important acids (isobutyric acid <0.1%, butyric acid <0.1%, 2-methylbutyric acid <0.1%, hexanoic acid <0.1) were only detected in traces. In addition the polar furanones 2,5-dimethyl-4-methoxy-3(2*H*)-furanone and 2,5-dimethyl-4-hydroxy-3(2*H*)-furanone which are important flavor compounds of strawberries were almost not detected.

Figure 2 shows a comparison of the resulting chromatograms of two of the four applied work-up procedures (VHS and DHS). Visually, it is very obvious that the DHS extract is dominated by the high volatile compounds while the VHS extract shows a more balanced composition of lower and higher molecular weight flavor compounds.

Figure 3 shows a GC-O chromatogram from the strawberries worked-up by the vacuum headspace method. It can be seen quite nicely how the strawberry flavor is composed of acidic, green, sweet (caramellic), and fruit notes.

Pear. Pear belongs to the genus *Pirus communis (Rosaceae)*. The flavor of Bartlett pears is mainly dominated by estery, fruity, and green notes. Shioto (*48*) reported on the differences between two cultivars of pears: La France and Bartlett. The flavor composition was reviewed by Paillard (*49*). A very interesting work about pear flavor has recently been published by Suwanagul (*50*). The results of our own experiments are shown in Table V. The flavor of pears consists mainly of methyl and ethyl esters of saturated or unsaturated fatty acids with one, two, or three double bonds. Other compounds are alcohols, aldehydes, acetates, a few sulfur volatiles, as well as a few norisoprenoids. A still relatively new compound is 1,3-octanediol. So far, it has only been reported in apple flavor and French cider (*51-53*). In a very recent publication its presence in pear flavor was described (*54*). It is interesting to note that despite its polarity the relative amount in our VHS extract (1.8%) is quite high compared to the DHS (not detected) and the SDEV extracts (<0.1%).

Comparison of the three work-up procedures vacuum headspace (VHS), dynamic headspace (DHS), and simultaneous distillation-extraction under vacuum (SDEV) shows very similar results as already described for passion fruit, peach, and strawberry.

Raspberry. Raspberry is economically the most important species in the genus *Rubus*. Its delicate flavor is mainly dominated by fruity, woody, sweet, and ionone notes. The flavor composition was reviewed by Honkanen and Hirvi (*45*). Our own results are shown in Table VI. We only applied the vacuum headspace method (VHS) for our studies of this fruit. Of sensory importance for the raspberry flavor are the C-13 norisoprenoids. Cycloionone (6,7,8,8a-tetrahydro-2,5,5,8a-tetramethyl-5*H*-1-benzopyrane) I is an interesting flavor compound published for the first time here in raspberries. So far it was only found in cognac (*55*). It is a potent flavor compound with fruity, sweet, floral, and woody notes.

Interestingly, we were not able to find 4-(4-hydroxyphenyl)butan-2-one, the so-called raspberry ketone in our studies. On the other hand, 2,5-dimethyl-4-hydroxy-3(2*H*)furanone and its methyl ether 2,5-dimethyl-4-methoxy-3(2*H*)furanone were found in small amounts.

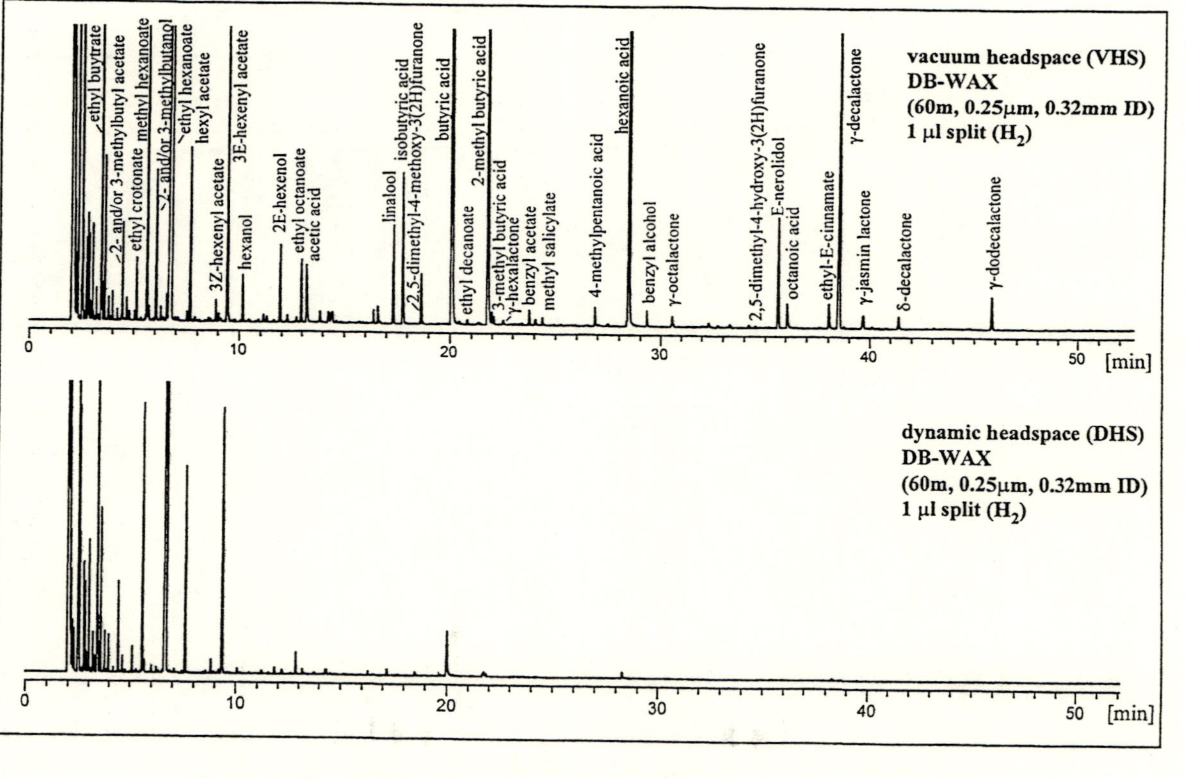

Figure 2: Capillary Gas Chromatograms of two Strawberry Concentrates

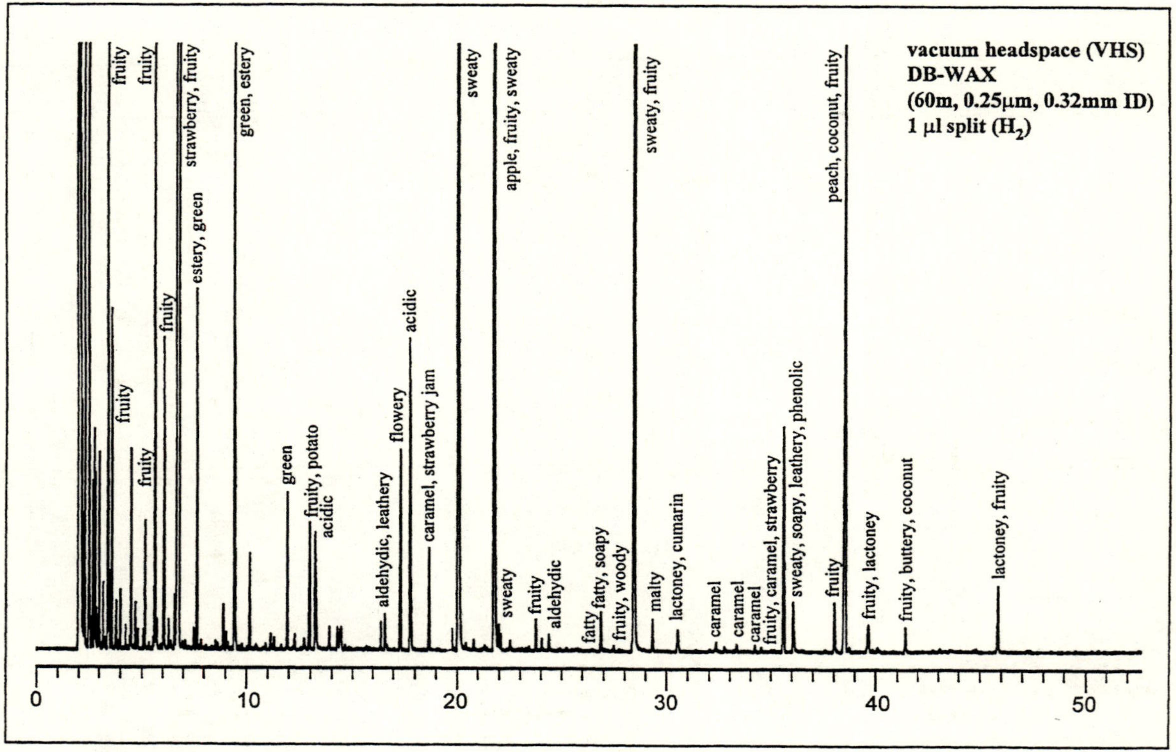

Figure 3: Capillary Gas Chromatogram (GC-Olfactometry) of a Strawberry Concentrate

Table V: Flavor Constituents of Pears

Flavor Compound	Area % GC		
	VHS	DHS	SDEV
propyl acetate	1,3	6,7	1,3
propanol	0,1	<0,1	3,0
butyl acetate	10,6	39,8	7,5
hexanal	0,1	<0,1	0,2
isobutanol	1,0	0,2	1,3
butanol	8,5	4,6	7,6
pentyl acetate	0,6	1,3	0,4
2- and/or 3-methylbutanol	2,3	1,4	0,7
hexyl acetate	14,3	22,9	8,3
3-hydroxy-2-butanone (acetoin)	0,2	0,1	0,2
2*E*-hexenyl acetate	0,3	0,2	0,2
hexanol	5,0	2,9	1,5
2*E*-hexenol	0,2	<0,1	<0,1
ethyl octanoate	0,2	<0,1	0,1
trans-linalool oxide (f)	<0,1	-	<0,1
methional	<0,1	-	<0,1
1-octen-3-ol	<0,1	-	<0,1
methyl-3*Z*-octenoate	<0,1	-	<0,1
cis-linalool oxide (f)	<0,1	-	<0,1
ethyl-3*Z*-octenoate	<0,1	<0,1	<0,1
methyl-2*E*-octenoate	<0,1	<0,1	<0,1
ethyl-2*E*-octenoate	0,2	<0,1	<0,1
octanol	0,3	<0,1	<0,1
ethyl-3(methylthio)propanoate	-	-	<0,1
methyl decanoate	<0,1	<0,1	<0,1
methyl-4*Z*-decenoate	0,2	<0,1	0,2
ethyl decanoate	0,3	<0,1	0,5
2- and/or 3-methylbutyric acid	2,5	<0,1	0,6
methyl-2*E*-decenoate	0,1	<0,1	0,2
3*E*,6*E*-α-farnesene	0,6	<0,1	1,6
ethyl-2*E*-decenoate	0,4	<0,1	0,5
methyl-2*E*,4*Z*-decadienoate	9,5	1,7	9,5
β-damascenone	-	-	<0,1
ethyl-2*E*,4*Z*-decadienoate	19,3	2,5	24,3
methyl-3-hydroxyoctanoate	2,2	<0,1	1,6
ethyl-11-dodecenoate	<0,1	-	0,2
ethyl-3-hydroxyoctanoate	6,0	<0,1	3,8
ethyl-2*E*,4*Z*,7*Z*-decatrienoate	2,5	0,2	1,1
ethyl-3-acetoxyoctanoate	0,2	-	<0,1
propyl-2*E*,4*Z*-decadienoate	0,1	-	0,2
3,4-dehydro-β-ionol	-	-	<0,1
butyl-2*E*,4*Z*-decadienoate	0,1	-	0,2
ethyl-5*Z*-tetradecenoate	0,2	-	1,0
methyl-5*Z*,8*Z*-tetradecadienoate	0,2	-	0,6
ethyl-5*Z*,8*Z*-tetradecadienoate	2,2	-	2,0
1,3-octanediol	1,8	-	<0,1
hexyl-2*E*,4*Z*-decadienoate	<0,1	-	<0,1
3*E*,6*E*-farnesol	0,1	-	0,3

Table VI: Flavor Constituents of Raspberries

Flavor Compound	Area % GC
	VHS
isobutanol	0,9
1-pentene-3-ol	0,1
α-phellandrene	0,7
2-heptanone	0,6
2- and/or 3-methylbutanol	5,9
2*E*-hexenal	0,2
diisopropyl disulfide	<0,1
hexyl acetate	0,1
2-hydroxy-3-butanone (acetoin)	6,0
3*E*-hexenyl acetate	1,0
2-heptanol and 2*Z*-pentenol	0,6
hexanol	2,2
3*E*-hexenol	0,7
3*Z*-hexenol	10,6
2*E*-hexenol	0,2
acetic acid	30,7
6-methyl-5-hepten-2-ol	0,3
theaspirane A	<0,1
propionic acid	0,2
linalool	3,0
trans-2-*p*-menthenol	0,5
isobutyric acid	0,2
2,3-butanediol	0,1
2,5-dimethyl-4-methoxy-3(2*H*)-furanone	<0,1
β-caryophyllene	3,4
1-terpinen-4-ol	0,6
butyric acid and *cis*-2-*p*-menthenol	2,0
cycloionone	<0,1
isovaleric acid	0,7
α-terpineol	0,5
piperitone	0,3
trans-piperitol	0,2
nerol	0,3
7,8-dihydro-β-ionone	0,2
hexanoic acid and geraniol and trans-α-ionone	10,1
benzyl alcohol	0,8
α-ionol	0,5
2-phenylethanol	0,5
γ-octalactone and 8-myrcenol	0,2
β-ionone	2,0
β-ionol	<0,1
3*E*-hexenoic acid	0,3
δ-octalactone and 7,8-dihydo-β-ionol	0,4
dihydrocumine alcohol	0,4
caryophyllene oxide	0,9
2,5-dimethyl-4-hydroxy-3(2H)-furanone	0,1
octanoic acid	0,2
δ-decalactone	0,7

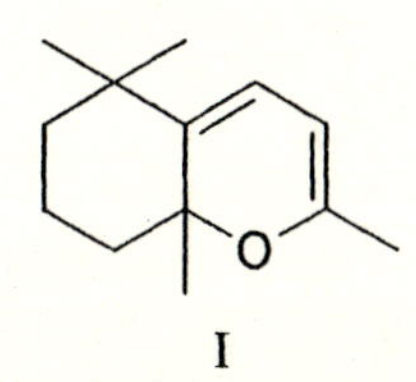

I

Cherry. The sweet cherry, *Prunus avium*, and the sour cherry, *Prunus cerasus*, both originated in Asia. They are now widespread throughout the world. Cherry flavor was reviewed by Crouzet et al. (*43*). The flavor of sweet cherries is less dominated by benzaldehyde than the flavor of sour cherries. This is clearly shown by our results in Table VII. In sweet cherries, benzyl alcohol and the green compounds are far more represented than in sour cherries. Both cherry varieties were only worked-up by the vacuum headspace method (VHS). The sensory description of the sour cherry extract was very typical of the fresh fruit, while the sensory description of the sweet cherry extracts indicates this may be one of the few examples where the vacuum headspace method is not the isolation procedure of choice.

Flower Scents and Flavors. As mentioned before, the dynamic headspace or purge and trap method has been used for the analysis of living blossoms and fruits (*11-14*). The results were the basis for the creation of fragrances and flavors that mimic the natural models. Therefore, it is an interesting question whether flower scents and flavors are comparable with regard to the biological function of blossoms and fruits (Table VIII). The main biological function of blossoms is to attract insects and other smaller animals to assist the plant in reproduction. Their scent is emitted as an aerosol into the air. Humans smell the scents orthonasally. Therefore, blossom scents are certainly intended to be smelled in a living state. Though not very well suited from its physical principle to represent the balanced composition of a scent, the dynamic headspace method is the most well-known method to be applied in living nature. Recently, solid phase microextraction (SPME) was also used for analyses in living nature (*56*).

The main biological function of fruits is the seed carriage and to serve as food for humans and animals. Fruit flesh and accordingly the flavor are often covered by stable shells and skins. In most cases, the actual flavor does not emerge until a freshly picked fruit is eaten. It is based on the interplay of oral sensation (texture), taste (acids, sugars), and the retronasally olfacted flavor. In many fruits an essential part of the flavor (e.g., the green part) develops only after cutting through enzymatic processes. Some fruits (e.g., some varieties of apples) even need to be stored a few weeks until they develop their optimally balanced flavor. Consequently, flavors are intended to be "tasted" in picked rather than in living fruit. Since dynamic headspace is not well suited for the analysis of the balanced fruit flavor composition, the vacuum headspace method (VHS) appears to be the method of choice for the analysis of the flavor of many fresh fruits.

Table VII: Flavor Constituents of Sweet and Sour Cherries

Flavor Compound	Sweet Cherry	Sour Cherry
	Area % GC/VHS	Area % GC/VHS
hexanal	0,1	<0,1
isobutanol	0,2	0,1
3-pentanol	<0,1	<0,1
2-pentanol	<0,1	<0,1
1-pentene-3-ol	0,2	<0,1
2- and 3-methyl butanol	1,2	0,3
2*E*-hexenal	1,0	0,9
3-methyl-3-butenol	2,8	1,0
diisopropyl disulfide	<0,1	0,2
2-hydroxy-3-butanone (acetoin)	0,4	<0,1
2*Z*-pentenol	<0,1	-
3-methyl-2-butenol	3,8	-
3-methyl pentanol	-	1,5
2*E*-hexenyl acetate	-	<0,1
hexanol	4,0	1,7
3*Z*-hexenol	0,3	0,2
2*E*-hexenol	27,4	11,5
acetic acid	<0,1	1,3
heptanol	<0,1	-
benzaldehyde	11,3	41,3
linalool	<0,1	<0,1
octanol	<0,1	-
isobutyric acid	-	0,3
γ-butyrolactone	0,5	0,2
phenylacetaldehyde	0,2	-
2- and 3-methylbutyric acid	0,1	2,0
2-hydroxy-1,8-endo-cineol	<0,1	-
methyl salicylate	<0,1	-
2*E*,4*E*-decadienal	<0,1	-
geraniol	0,2	0,1
hexanoic acid	-	0,3
benzyl alcohol	40,6	20,3
2-phenyl ethanol	0,6	0,1
benzothiazole	0,1	-
2-ethylhexanoic acid	-	0,1
2*E*-hexenoic acid	-	<0,1
octanoic acid	<0,1	-
2*E*-hexenyl benzoate	0,1	<0,1
eugenol	0,3	1,1
nonanoic acid	<0,1	<0,1
anisic alcohol	<0,1	-
cinnamic alcohol	<0,1	-
decanoic acid	<0,1	-
8*E*-hydroxylinalool	<0,1	-
benzoic acid	-	<0,1
dodecanoic acid	<0,1	-
benzyl benzoate	<0,1	-
octadecanol	-	<0,1

Table VIII: Comparison between the Scents of Blossoms and the Flavor of Fruits

Blossoms	Fruits
main biological function is the attraction of insects and other small animals for the reproduction	main biological function is to serve as food for humans and animals and the seed carriage
emission of the scent as an aerosol	flavor is often covered in the fruit flesh by stable shells and skins; in many fruits it develops after cutting through enzymatic processes or even after storage
scent is smelled orthonasally	flavor is "tasted" retronasally
blossom scents are intended to be smelled in a living state	flavors are biologically mainly intended to be "tasted" in the picked fruit
the dynamic headspace method (DHS) is not very well suited to represent the balanced composition of a scent but can be applied in the living nature	the dynamic headspace method (DHS) is not suited to represent the balanced composition; picked fruits are rather analysed by the vacuum headspace method (VHS)

Acknowledgments

The authors would like to thank the entire groups of organic synthesis, chromatography, spectroscopy, flavor research, and flavor development for their valuable and skillful work. A special gratitude goes to our senior flavorist Günter Kindel and to Stefan Trautzsch.

Literature Cited

1. Nijssen, L.M.; Visscher, C.A.; Maarse, H.; Willemsens, L.C.; Boelens, M.H. *Volatile Compounds in Food. Qualitative and Quantitative Data*. Seventh Edition. TNO Nutrition and Food Research Institute, Zeist, The Netherlands, 1996.
2. Likens, S.T.; Nickerson, G.B. *Proc. Am. Soc. Brew. Chem.* **1964**, 5.
3. Schultz, T.H.; Flath, R.A.; Mon, T.R.; Eggling, S.B.; Teranishi, R. *J. Agric. Food Chem.* **1977**, *25*, 446.
4. Godefroot, M.; Sandra, P.; Verzele, M. *J. Chromatogr.* **1981**, *203*, 325.
5. Blanch, G.B.; Tabera, J.; Herraiz, M.; Reglero, G. *J. Chromatogr.* **1993**, *628*, 261.
6. Maignial, L.; Pibarot, P.; Bonetti, G.; Chaintreau, A.; Marion, J.P. *J. Chromatogr.* **1992**, *606*, 87.
7. Werkhoff, P.; Bretschneider, W. *J. Chromatogr.* **1987**, *405*, 87.
8. Werkhoff, P.; Bretschneider, W. *J. Chromatogr.* **1987**, *405*, 99.
9. Buttery, R.; Teranishi, R.; Ling, L.C.; Turnbaugh, J.C. *J. Agric. Food Chem.* **1990**, *38*, 336.
10. Buttery, R.; Ling, L.C.; *J. Agric. Food Chem.* **1995**, *43*, 1878.
11. Kaiser, R. *Perfumer & Flavorist* **1997**, *22*, 7.
12. Mookherjee, B.D.; Trenkle, R.W.; Wilson, R.A.; Zampino, M.; Sands, K.P.; Mussinan, C.J. In *Flavors and Fragrances: A World Perspective;* Lawrence, B.M.; Mookherjee, B.D.; Willis, B.J., Eds.; Elsevier: Amsterdam, New York, Tokyo, 1988, pp 415-424.
13. Trenkle, R.W.; Mookherjee, B.D.; Zampino, M.J.; Wilson, R.A.; Patel, S.M. US 5, 269,169, 1963.
14. Mookherjee, B.D.; Trenkle, R.W.; Wilson, R.A. *Pure & Appl. Chem.* **1990**, *62*, 1357.
15. Joulain, D. "Study of the Fragrance Given off by Certain Springtime Flowers". In *Progress in Essential Oil Research;* Brunke, E.J., Ed.; Walter de Gruyter: Berlin, New York, 1986, pp 57-67.
16. Surburg, H.; Güntert, M. *H&R Contact* **1991**, *51*, 12.
17. Surburg, H.; Güntert, M; Harder, H. In *Bioactive Volatile Components from Plants*; Teranishi, R.; Buttery, G.; Sugisawa, H., Eds.; ACS Symposium Series 525, American Chemical Society: Washington, DC, 1993, pp 168-186.
18. Brunke, E.J.; Hammerschmidt, F.J.; Schmaus, G. *Dragoco Report* **1992**, *39*, 3.
19. Joulain, D. In *Bioactive Volatile Components from Plants*; Teranishi, R.; Buttery, G.; Sugisawa, H., Eds.; ACS Symposium Series 525, American Chemical Society: Washington, DC, 1993, pp 187-204.

20. Brunke, E.J.; Hammerschmidt, F.J.; Schmaus, G. *Rivista Italiana Eppos, Numero speciale* **1993**, 142.
21. Fischer, N.; Hammerschmidt, F.J. *Chem. Mikrobiol. Technol. Lebensm.* **1992**, *14*, 141.
22. Fischer, N. *Dragoco Report* **1993**, 38, 133.
23. Fischer, N.; Hammerschmidt, F.J.; Brunke, E.J In *Progress in Flavour Precursor Studies;* Schreier, P.; Winterhalter, P., Eds.; Allured Publishing Corporation: Carol Stream, Il, 1993, pp 287-294.
24. Fischer, N.; Hammerschmidt, F.J.; Brunke, E. In *Fruit Flavors. Biogenesis, Characterization, and Authentication;* Rouseff, R.L.; Leahy, M.M., Eds.; ACS Symposium Series 596, American Chemical Society: Washington DC, 1995, pp 8-20.
25. Tarantilis, P.A.; Polissiou, M.G. *J. Agric. Food Chem.* **1997**, *45*, 459.
26. Idstein, H.; Schreier, P. *J. Agric. Food Chem.* **1984**, *32*, 383.
27. Idstein, H.; Schreier, P. *Phytochemistry* **1985**, *24*, 2313.
28. Idstein, H.; Schreier, P. *J. Agric. Food Chem.* **1985**, *33*, 138.
29. Güntert, M.; Brüning, J.; Emberger, R.; Köpsel, M.; Kuhn, W.; Thielmann, T.; Werkhoff, P. *J. Agric. Food Chem.* **1990**, *38*, 2027.
30. Güntert, M.; Brüning, J.; Emberger, R.; Hopp, R.; Köpsel, M.; Surburg, H.; Werkhoff, P. In *Flavor Precursors - Thermal and Enzymatic Conversions;* Teranishi, R.; Takeoka, G.R.; Güntert, M., Eds.; ACS Symosium Series 490, American Chemical Society: Washington DC, 1992, pp 140-163.
31. Schreier, P. *Lebensmittelchem. Gerichtl. Chem.* **1987**, *41*, 25.
32. Teranishi, R.; Kint, S. In *Flavor Science. Sensible Principles and Techniques;* Acree, T.E.; Teranishi, R., Eds.; ACS Series, American Chemical Society: Washington, DC, 1993, pp 137-167.
33. Buttery, R.G.; Ling, L.C. In *Biotechnology for Improved Foods and Flavors;* Takeoka, G.R.; Teranishi, R.; Williams, P.J.; Kobayashi, A., Eds.; ACS Series 637, American Chemical Society: Washington, DC, 1996, pp 240-248.
34. Parliment, T.H. In *Techniques for analyzing food aroma;* Marsili, R., Ed.; Marcel Dekker Inc.: New York, 1997, pp 1-26.
35. Wampler, T.P. In *Techniques for analyzing food aroma;* Marsili, R., Ed.; Marcel Dekker Inc.: New York, 1997, pp 27-58.
36. Leahy, M.M.; Reineccius, G.A. In *Analysis of Volatiles*; Schreier, P., Ed.; Walter de Gruyter: Berlin, 1984, pp 19-47.
37. Winterhalter, P.; Lander, V.; Schreier, P. *J. Agric. Food Chem.* **1987**, *35*, 335.
38. Peppard, T.L. *J. Agric. Food Chem.* **1992**, *40*, 257.
39. Misharina, T.A.; Golovnya, R.V.; Beletsky, I.V. In *Trends in Flavour Research*; Maarse, H.; van der Heij, D.G., Eds.; Elsevier: Amsterdam, 1994, pp 117-120.
40. Krumbein, A.; Ulrich, D. In *Flavour Science. Recent Developments*; Taylor, A.J.; Mottram, D.S., Eds.; The Royal Society of Chemistry: Bodmin, UK, 1996, pp 289-292.
41. Whitfield, F.B.; Last, J.H. In *Progress in Essential Oil Research;* Brunke, E.J., Ed.; Walter de Gruyter: Berlin, New York, 1986, pp 3-48.
42. Shibamoto, T.; Tang, C.S. In *Food Flavours. Part C. The Flavours of Fruits*; Morton, I.D.; MacLeod, A.J., Eds.; Elsevier: Amsterdam, 1990, pp 221-280.

43. Crouzet, J.; Etievant, P.; Bayonove, C. In *Food Flavours. Part C. The Flavours of Fruits*; Morton, I.D.; MacLeod, A.J., Eds.; Elsevier: Amsterdam, 1990, pp 43-91.
44. Narain, N.; Hsieh,T.C.-Y.; Johnson, C.E. *J. Food Science* **1990**, *55*, 1303.
45. Honkanen, E.; Hirvi, T. In *Food Flavours. Part C. The Flavours of Fruits*; Morton, I.D.; MacLeod, A.J., Eds.; Elsevier: Amsterdam, 1990, pp 125-193.
46. Ulrich, D.; Hoberg, E.; Rapp, A.; Sandke, G. In: *Flavour Perception – Aroma Evaluation. Proceedings of the 5th Wartburg Aroma Symposium, Eisenach, Germany, March 17-20, 1997*; Kruse, H.P.; Rothe, M., Eds.; Eigenverlag Universität Potsdam: Bergholz-Rehbrücke, 1997, pp 283-293.
47. Ulrich, D.; Rapp, A.; Hoberg, E. *Z. Lebensm. Unters. Forsch.* **1995**, *200*, 217.
48. Shiota, H. *J. Sci. Food Agric.* **1990**, *52*, 421.
49. Paillard, N.M.M. In *Food Flavours. Part C. The Flavours of Fruits*; Morton, I.D.; MacLeod, A.J., Eds.; Elsevier: Amsterdam, 1990, pp 1-41.
50. Suwanagul, A. "Ripening Pear Flavor Volatiles. Identification, Biosynthesis, and Sensory Perception". PhD Thesis, Oregon State University, 1996.
51. Berger, R.G.; Dettweiler, G.R.; Drawert, F. *Dtsch. Lebensm. Rundsch.* **1988**, *84*, 344.
52. Schwab, W.; Scheller, G.; Gerlach, D.; Schreier, P. *Phytochemistry* **1989**, *28*, 157.
53. Beuerle, T.; Schwab, W. In: *Flavour Perception – Aroma Evaluation. Proceedings of the 5th Wartburg Aroma Symposium, Eisenach, Germany, March 17-20, 1997*; Kruse, H.P.; Rothe, M., Eds.; Eigenverlag Universität Potsdam: Bergholz-Rehbrücke, 1997, pp 381-390.
54. Beuerle, T.; Schwab, W. *Z. Lebensm. Unters. Forsch.* **1997**, *205*, 215.
55. Ter Heide, R.; de Valois, P.J.; Visser, J.; Jaegers, P.P.; Timmer, R. In *Analysis of Foods and Beverages - Headspace Techniques*; Charalambous, G., Ed.; Academic Press: New York, San Francisco, London, 1978, pp 249-281.
56. *Spray Technology and Marketing*, **October 1996**, 26.

Chapter 5

Relationship Between Aroma Compounds' Partitioning Constants and Release During Microwave Heating

Deborah D. Roberts and Philippe Pollien

Nestlé Research Center, Vers Chez les Blanc, 1000 Lausanne 26, Switzerland

A method was designed and tested to quantitate aroma compounds eluting from food heated in the microwave. It was readily able to measure aroma compounds present at 4 mg/kg in the meal. The trap and condenser glassware were designed to accommodate the high air and water flow rates, in effect using the distillant nature of microwave heating. SPME was an effective method to quantitate aroma compounds from the trapped fractions. When analyzing frozen spaghetti, aroma compounds were found to elute at different times during heating. The initial release starting time correlated (R^2 = 0.98) with the compounds' air-water partition coefficients. Thus, predictions for which types of compounds will be more readily lost can be made and aroma formulations for microwaveable food can be modified accordingly.

The use of a microwave to reheat foods is highly popular, especially in the U.S. The convenience and quickness of preparation have instigated a range of food products for microwave heating. The mechanisms for heating food in the microwave are different from the conventional oven and can cause differences in the flavor of the prepared food. Due to the differences in the surface moisture and temperature, conventionally cooked foods can form a crust at the surface which reaches high temperatures and acts as a barrier to water and aroma compound evaporation. In microwave food, this crust is usually not formed because of the 100°C maximal temperature and high surface moisture. Thus, there is significant water and aroma migration to the surface and into the air. These losses can cause microwave foods to have less desirable flavor. A recent ACS Symposium Series book (*1*) covers the challenges of flavoring microwave food. In this chapter, a method of aroma trapping is demonstrated that gives information about aroma losses and their relationship to aroma compound properties.

Materials and Methods

Sample Preparation. Buitoni brand spaghetti (1700 g) broken up into 7 cm pieces was immersed in 11.8 L of boiling water and cooked for 8 minutes. After draining and rinsing with cold water for 2 min, the final weight was 3992 g.

The aroma mixture contained (**a-f**): dimethyltrisulfide, ethyl-2-methylbutyrate, 2E-nonenal, 1-octen-3-ol, diacetyl, and 2,3-diethyl-5-methylpyrazine at 3 g/kg dissolved in medium chain triglycerides (MCT), containing approximately 60 % caprylic and 40 % capric acid. The aroma compounds were obtained from commercial sources.

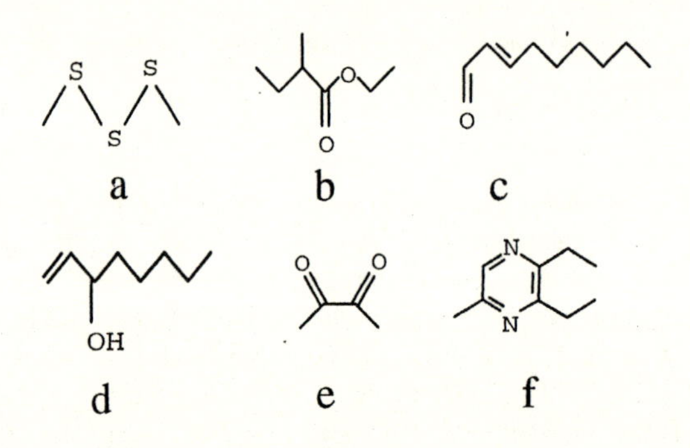

The aroma mixture (0.285 g) was dispersed on top of 200 g of spaghetti in a plastic container and sealed. This gave a final concentration of 4.275 mg/kg in the spaghetti. Samples were frozen at - 25 °C until microwave oven analysis. Eight samples were analysed whose frozen storage times were 2, 3, 5, 11, 13, 14, 16, and 17 days.

Device to Analyze Aroma Released in Microwave. A Panasonic Genius NN 5852 microwave oven with turntable, 23 L volume, 1420 W required power, and 100-900 W output power was used. The actual power output at full power was found to be 711 watts, determined by heating 1 kg water (28 °C) for 1 minute, and measuring the increase in temperature.

The microwave oven was modified by placing a hollow cylinder in the center of the top cover whose dimensions prevent leakage of microwave energy. Specially designed glass vessels, as shown in Figure 1, allowed the trapping of aroma released in the microwave. All glassware was silanized with Sylon CT (5% dimethyldichlorosilane in toluene, Supelco) to avoid adsorption of aroma compounds. The food vessel (19 cm in diameter by 5 cm high) was glued to the center of the turntable to allow reproducible placement of the food. A Teflon washer with Viton o-rings, which connected the food vessel to glass tubing, allowed the food vessel, but not the glass tubing, to turn during microwave heating. The glass tubing was connected to a three-way valve that allowed the air stream to be directed to either the liquid nitrogen trap, or the condenser. The liquid nitrogen trap was placed in this configuration, instead of after the condenser, to avoid a partial trapping of the initial aromas on the condenser. The glass trap was placed in a Dewar of liquid nitrogen and its exit was connected to a deflated plastic bag which prevented oxygen in laboratory air from condensing in the trap. The liquid nitrogen trap and the condenser incorporated serpentine designs which maximized surface area and time for contact of the air flow with the cold surface. The condenser was kept at -5 °C and the flasks, in ice, were rotated to allow multiple fraction collections.

Microwave Analysis Procedure. Immediately before analysis, liquid nitrogen was poured into the liquid nitrogen trap and alcohol coolant was circulated in the condenser. The entire system was filled with nitrogen gas to prevent oxygen condensation in the liquid nitrogen trap. The sample was placed in the food vessel, sealed, and heated for a total of 255 sec. During the first 165 sec, when the temperature was less than 100 °C, the air stream was collected in the liquid nitrogen trap. Then, water began to elute and the air stream was diverted into the condenser. Two consecutive samples were collected (165 - 225 sec, 225 - 255 sec). The liquid nitrogen trap was rinsed with 5 mL of water to recover the trapped aroma. All fractions, which were aroma solutions in water, were weighed. The amounts of aroma compounds in the solutions were determined by SPME. The experiment was repeated eight times in its entirety with separate spaghetti samples.

Solid Phase MicroExtraction (SPME) - GC/FID Quantification. Six external standard curves for quantitation were produced using standard solutions in water, one for each molecule of interest. The most concentrated solution was prepared by dissolving the aroma compounds in water with continued agitation. This solution was diluted to produce five different concentration levels which included the concentration range of the samples. These ranges were (in mg/L) 1-octen-3-ol (0 - 16), diacetyl (0 - 28), ethyl-2-methyl butyrate (0 - 13), 2,3-diethyl-5-methylpyrazine (0 - 19), and 2E-nonenal (0 - 17). As most of the curves were reproducibly slightly curvilinear, a quadratic function was used for regression with R^2 values of at least 0.999.

The standard samples and the microwave elution samples (all flavor solutions in water) were analyzed in the same manner using the following procedure. 3 mL were placed in a 4 mL septum-closed vial and stirred with a stir bar (800 rpm). A carbowax/divinylbenzene fiber (Supelco) was immersed for 10 minutes. A trial with 10 % sodium chloride addition was also performed. The fiber was then desorbed for 5 minutes at 200 °C in the GC injection port containing a 0.75 mm ID liner. During the 5 minutes, the volatiles volatilized almost immediately but the fiber remained for conditioning. Good GC resolution was obtained due to the narrow liner and quick desorption. The GC/FID contained a DBWAX column: 30 m, 0.32 mm, 0.25 μm film thickness, 10 PSI; and was programmed to begin at 50 °C for 3 minutes, heat to 140 °C at 8 °C/min and then to 220 °C at 25 °C/min.

Temperature of Meal. The heating profile of the spaghetti dish was obtained by placing fiberoptic probes at different places in the spaghetti and measuring the temperature during microwave heating, using the same microwave as in the release analysis. Eight measurements were taken from different places in the meal.

Determination of K_{aw}. A method that does not require the use of internal or external standards was used to determine the air-water partition coefficients (K_{aw}) of the compounds studied (*2*). One or two mL of sample were equilibrated with 312 mL of headspace at 30 °C in an enclosed stainless steel sampling cell. After a 30 min. equilibration period, the contents of the headspace were pushed onto a Tenax trap, which was subsequently thermally desorbed using an ATD 400 (Perkin Elmer, Beaconsfield, U.K.). Figure 2 shows how the partition coefficient is calculated.

Results and Discussion

Method Development. In order to study aroma release kinetics during microwave heating, a method was developed based on the distillant properties (*3*). After the attainment of maximal temperature, usually close to 100 °C, a large amount of water evaporates from high moisture food. This leads to dehydrated foods with a high amount of surface moisture. It also causes problems if trying to trap the aroma compounds released. In the case of frozen spaghetti, 15 mL of water evaporated per minute during the last few minutes of heating. This would be too much for a thermally desorbed adsorbent trap. The method developed, therefore, used a series of cold traps especially designed with a large surface area. Liquid nitrogen was used as the coolant for trapping during the first few minutes. When water began to distill, the eluting air-stream was diverted into a condenser in which water and flavor compounds were condensed together. Both traps were very effective in trapping the compounds as traps connected in-series recovered very little. After the experiment, trap 1 was removed from liquid nitrogen and rinsed with water to recover the aroma compounds. Unfortunately, it also trapped oxygen from the air which quickly volatilized as it returned to ambient temperature. This could have resulted in some losses of aroma.

The SPME method was found to be more sensitive than direct GC water injection for compound quantitation. The use of solvents was avoided, allowing analysis of early eluting compounds. The solutions had to be in the concentration range of aroma compound solubility.

One of the key advantages of this method is that it allowed a measurement of released aroma compounds, rather than analysis of the food before and after microwave heating. Another method, developed for popcorn, offered the same advantage (*4*) by using purging gas to strip water and volatiles from food during microwave heating with retention in a cold trap. This trap was then heated to transfer the volatiles to an adsorbent trap, followed by trap heating and GC analysis. The method described here did not require a purging gas and it offered an alternate method (SPME) for quantitation of trapped aroma compounds.

Loss amounts during storage and release. The 8 replicates were sampled with differing storage times, up to 17 days. For most of the aroma compounds, the same results were obtained thoughout this storage period. However, the two compounds with the highest K_{aw} showed a weak relationship between storage time and amount

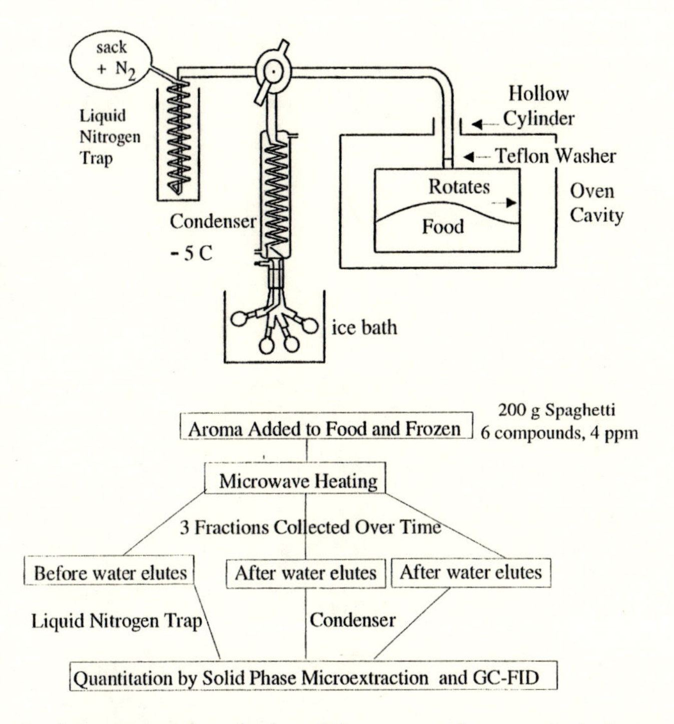

Figure 1. Apparatus and method used for aroma release analysis during microwave heating (Adapted from ref. 3).

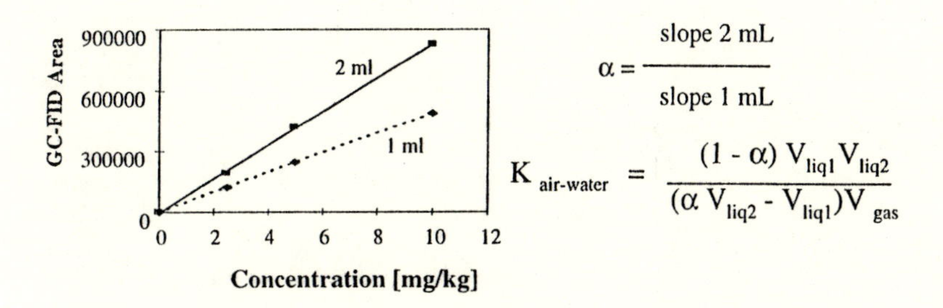

Figure 2. Example data (ethyl-2-methylbutyrate) and calculation for K_{aw}.

released (Figure 3). This was seen for the first trap, where highest amounts of these compounds were collected. The loss during storage for some but not all compounds shows that frozen food can undergo a selective volatilization of compounds, resulting in a product with a different aroma than the freshly prepared counterpart.

In addition, the aroma compounds showed large losses during heating. Respectively for compounds **a-f**: 27, 27, 33, 46, 41, and 44 %. For **a** and **b**, these values were probably larger due to losses from the liquid nitrogen trap. These amounts are similar to what was reported from frozen pancakes (10-56%) (*5*).

Table I. Relationships between the values of K_{aw} [1] (air-water partition coefficient) and the starting release time

Compounds	K_{aw}	K_{aw} RSD	T_o Extrapolated Release starting time, (from Figure 4)	Temperature of meal at T_o
2,3-diethyl-5-methylpyrazine	5.0×10^{-4}	16	62	95
diacetyl	1.1×10^{-3}	4	61	94
1-octen-3-ol	3.1×10^{-3}	1	53	78
2E-nonenal	7.0×10^{-3}	2	49	71
ethyl-2-methylbutyrate	1.5×10^{-2}	1	20	19
dimethyltrisulfide	1.9×10^{-2}	1	the first to be released	

[1] (concentration air/concentration water) at equilibrium

Relationship between estimated release starting time (t_o) and K_{aw}. Figure 4 shows that the aroma compounds were released at different times during microwave heating. The linear relationship between release amount and time of heating indicates that the compounds exhibited a constant release over heating. The point at 225 s of heating was low for ethyl-2-methylbutyrate probably due to losses of this compound in the condenser before the water began to elute. The x-intercept for the release corresponds to the point when the particular aroma compounds began to elute. These points were estimated and are found in Table I. The estimations are more precise for the later eluting compounds than for the earlier eluting compounds. The aroma compounds exhibited different release kinetics, as seen by these estimated times of release commencement, and the differences followed the measured air-water partition coefficients of the compounds. Dimethyltrisulfide and ethyl-2-methylbutyrate were the first two compounds to be released and are also those with the highest air-water partition coefficients. These compounds started to be released early in the heating of the frozen spaghetti, when the average temperature in the meal was still below room temperature. 2E-nonenal was the next compound to elute and also exhibits the next lowest K_{aw}. 1-Octen-3-ol followed in time and also had the next lowest K_{aw}. And lastly, diacetyl and 2,3-diethyl-5-methylpyrazine eluted primarily at the end of heating and had the lowest K_{aw}. These compounds, which have low volatility in water, did not begin their elution until the meal reached its maximum temperature of close to 100 °C. Thus, for a meal that has low fat content, the K_{aw} was found to be an excellent approximation of

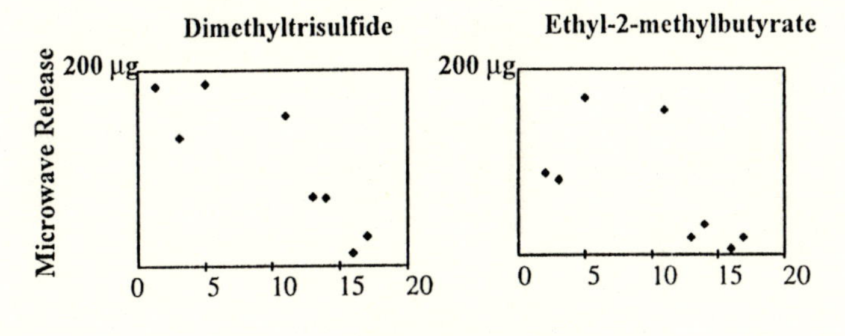

Figure 3. Relationship between release during microwave heating (0 - 165 s) and storage time of spiked frozen spaghetti.

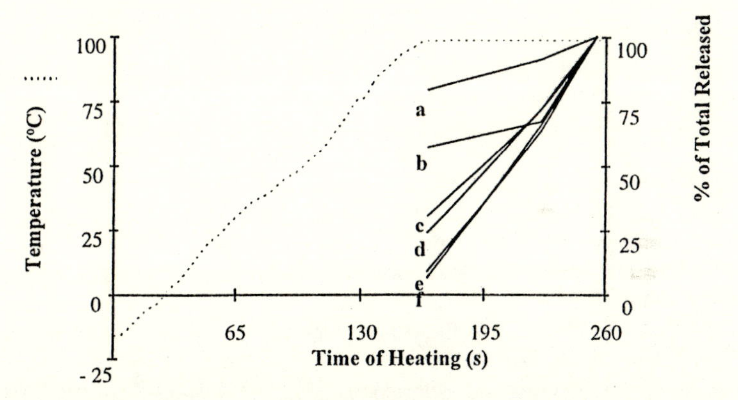

Figure 4. Graph showing the total cumulative release percent for six aroma compounds (**a-f**) and their relationship to the temperature of spaghetti. The line begins at 165 s because the first trap was from 0 - 165 s.

the microwave elution kinetics. A high correlation was found ($R^2 = 0.98$) between the air-water partition coefficient and time of elution commencement (Figure 5) for this low-fat food. Previously, compound parameters such as the air-product partition coefficient (*6*) or the Henry's law constant were used (*7*) to predict losses during microwave heating. This work shows that the partitioning parameters, in addition, predict the kinetics of release.

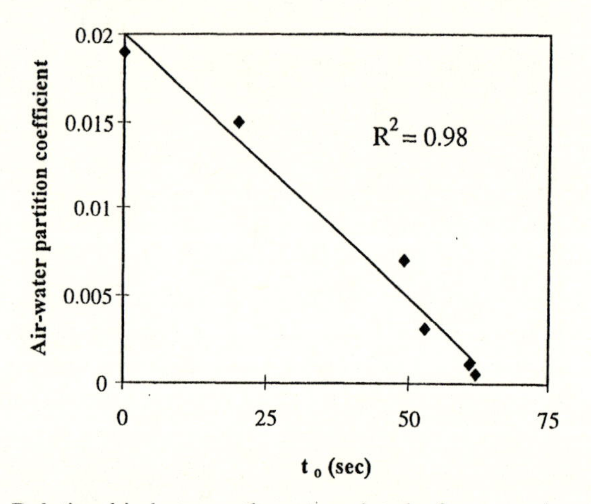

Figure 5. Relationship between the extrapolated release starting time (t_o) and the air-water partition coefficient for compounds (**a-f**).

Literature Cited

1. *Thermally Generated Flavors: Maillard, Microwave, and Extrusion Processes*; Parliment, T.H., Morello, M.J., McGorrin, R.J., Eds.; American Chemical Society: Washington, D.C., 1994;
2. Chaintreau, A.; Grade, A.; Munoz-Box, R. *Anal. Chem.* **1995**, *67*, 3300-3304.
3. Roberts, D.D.; Pollien, P. *J.Agric.Food Chem.* **1997**, *45*, 4388-4392.
4. Risch, S.J.; Keikkila, K.; Williams, R.J. U.S.Patent 5,177,995. Analysis of Migration of a Volatile Substance During Heating with Microwave Energy. 1993.
5. Li, H.C.; Risch, S.J.; Reineccius, G.A. In *Thermally Generated Flavours: Maillard, Microwave, and Extrusion Processes*; Parliment, T.H., Morello, M.J., McGorrin, R.J., Eds.; American Chemical Society: Washington, D.C., 1994; pp 466-475.
6. de Roos, K.B.; Graf, E. *J. Agric. Food Chem.* **1995**, *43*, 2204-2211.
7. Lindstrom, T.R.; Parliment, T.H. In *Thermally Generated Flavours: Maillard, Microwave, and Extrusion Processes*; Parliment, T.H., Morello, M.J., McGorrin, R.J., Eds.; American Chemical Society: Washington, D.C., 1994; pp 405-413.

Chapter 6

Extraction by Headspace Trapping onto Tenax of Novel Thiazoles and 3-Thiazolines in Cooked Beef

J. Stephen Elmore and Donald S. Mottram

Department of Food Science and Technology, The University of Reading, Whiteknights, Reading RG6 6AP, United Kingdom

Aroma extracts of pressure cooked beef were prepared by headspace trapping onto Tenax TA, a commonly used technique, which is rapid, easy to use and provides an extract of similar quality to that perceived by the nose. One drawback of the technique is that it is considered to be less sensitive towards higher boiling compounds, compared with distillation-based methods of extraction. However, thermal desorption of the headspace extracts resulted in gas chromatograms containing alkylthiazoles and alkyl-3-thiazolines with molecular weights in excess of 330 Da.

Purging the headspace above a food onto a trap containing an adsorbent material has been used for the extraction of food aroma for nearly 30 years. Different adsorbent materials have been used over this time but nowadays the most commonly used are the Tenax porous polymers, and Tenax TA has been developed especially for this purpose. Headspace trapping onto an adsorbent gives an extract similar to that which is smelled by a sensory assessor. It is simple to use, sensitive, relatively quick to perform, and there is little artifact formation, compared to distillation based techniques. One apparent disadvantage of the technique, relative to distillation based extractions, is that it is considered to be insensitive towards high molecular weight compounds *(1)*.

This paper describes the identification of thiazoles and 3-thiazolines, with molecular weights 155 to 337, isolated from cooked beef. These compounds were isolated using headspace collection and were analyzed using thermal desorption combined with GC-MS. The results show that headspace trapping is applicable for low volatility aroma compounds.

Experimental

Headspace Trapping of Beef Samples. Two pieces of a steak (50 g each) were placed in separate 100 mL bottles; fitted with air-tight, PTFE-lined screw tops; and cooked at 140 °C in an autoclave for 30 min; after which they were cooled. The pieces of steak were combined and minced twice.

A sample of the minced steak (40 g) was placed in a screw top conical flask (250 mL). A Dreschel head was attached to the flask, using an SVL fitting (Bibby, Stone, UK) (Figure 1). The flask was held in a water bath at 60 °C for 1 h while nitrogen, at 40 mL/min, swept the volatiles onto a glass-lined, stainless steel trap (105 mm × 3 mm i.d.) containing 85 mg Tenax TA (Scientific Glass Engineering Ltd., Ringwood, Australia). A standard (100 ng 1,2-dichlorobenzene in 1 μL hexane) was added to the trap at the end of the collection, and excess solvent and any water retained on the trap were removed by purging the trap with nitrogen at 40 mL/min for 5 min.

Headspace collections were performed in triplicate. Untrimmed and trimmed steaks were examined.

Gas Chromatography–Mass Spectrometry (GC–MS). All analyses were performed on a Hewlett Packard 5972 mass spectrometer, fitted with a HP 5890 Series II gas chromatograph. The mass spectrometer was operated in electron impact mode with an electron energy of 70 eV and an emission current of 50 μA. The mass spectrometer scanned from *m/z* 28 to *m/z* 450 at 1.7 scans/s.

A CHIS injection port (Scientific Glass Engineering Ltd.) was used to thermally desorb the volatiles from the Tenax trap onto the front of a BPX5 fused silica capillary column (50 m × 0.32 mm i.d., 0.5 μm film thickness; Scientific Glass Engineering Ltd.). During the desorption period of 10 min, the oven was held at 0°C. After desorption, the oven was heated at 40 °C/min to 40 °C and held for 2 min before heating at 4 °C/min to 280 °C. Helium at 8 psi was used as the carrier gas, resulting in a flow of 1.75 mL/min at 40 °C. A series of n-alkanes (C_6 - C_{22}) was analyzed, under the same conditions, to obtain linear retention index (LRI) values for the beef aroma components.

Results and Discussion

The collected headspace volatiles gave complex chromatograms, which were dominated by aliphatic aldehydes, alcohols, ketones, and hydrocarbons. Smaller quantities of heterocyclic compounds, such as furans, thiophenes, pyridines, and pyrazines, were also present, as well as alkyl polysulfides. A typical chromatogram is shown in Figure 2. This profile was typical of cooked meat *(2,3)*. Of particular interest were over 60 related compounds, which eluted between 120°C and 270 °C. They all possessed an **M + 2** ion, indicative of sulfur, and an odd molecular mass, indicating an odd number of nitrogen atoms. Each compound had a distinct molecular ion and a characteristic mass spectrum.

The related compounds could be divided into two groups. The compounds in the first of these groups started at MW 155, increasing to MW 239, in units of 14; two

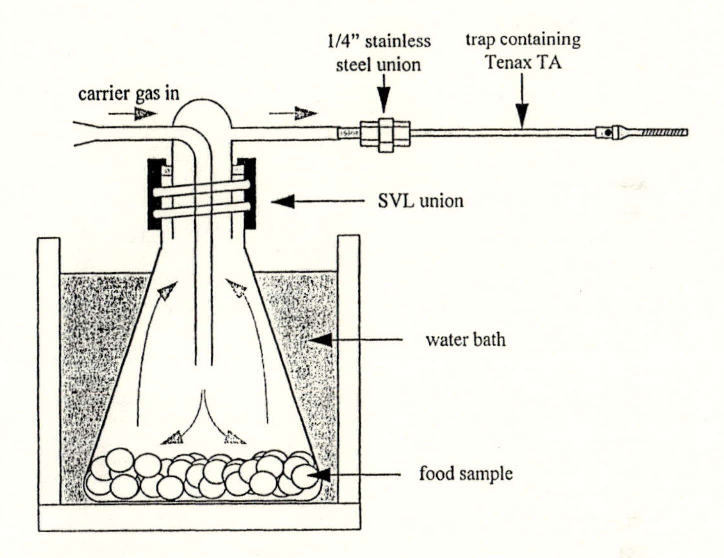

Figure 1. Apparatus used for headspace trapping onto Tenax TA.

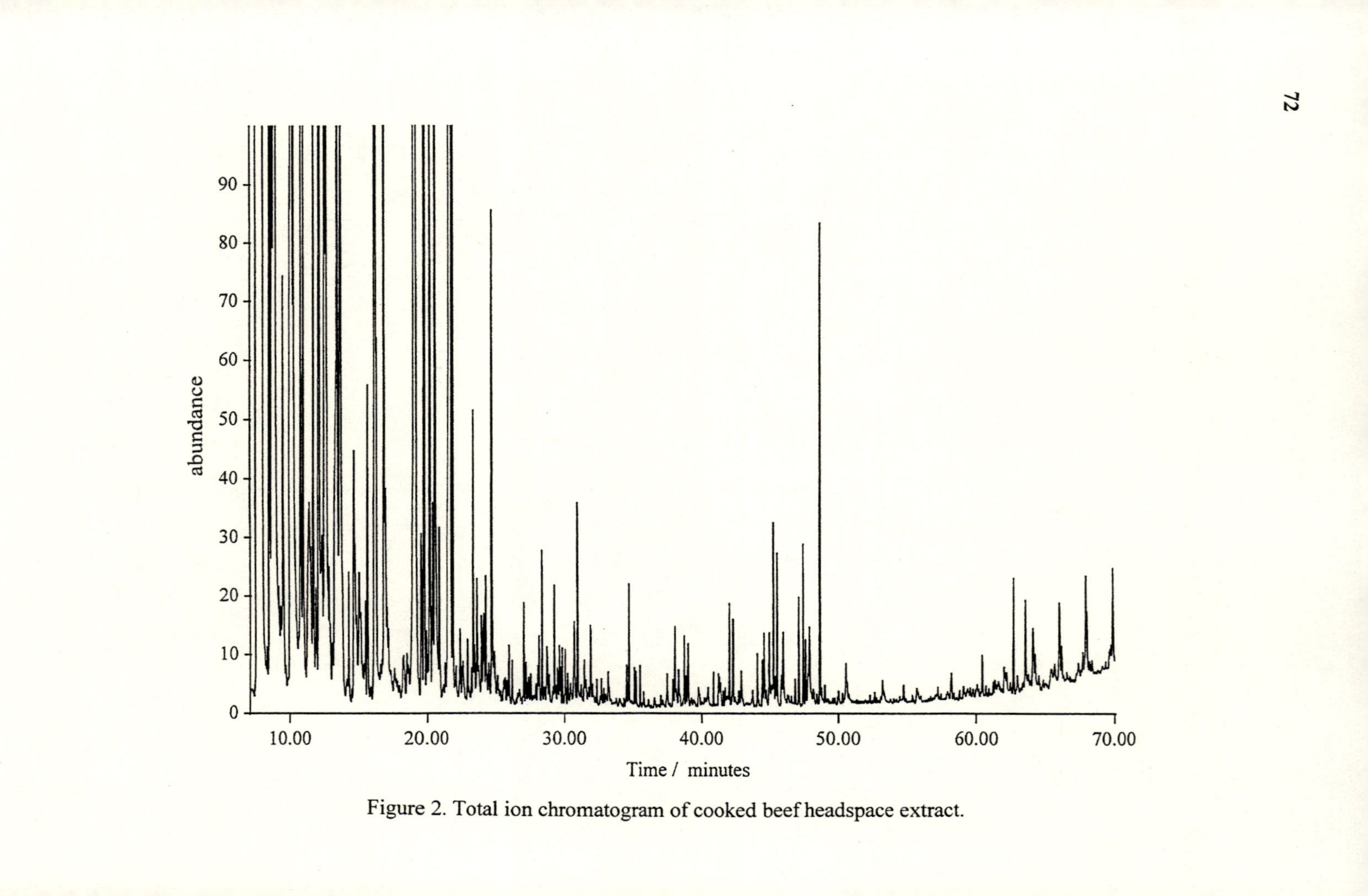

Figure 2. Total ion chromatogram of cooked beef headspace extract.

compounds with molecular weights 323 and 337 were also included in this group. The compounds in the second group started at MW 157 and followed the same pattern, up to MW 255, with additional compounds at MW 311 and MW 325.

Comparison with literature spectra showed that the members of the first group were likely to be alkylthiazoles (**I**), each with an alkyl chain between C_4 and C_9, or C_{15}, in the 2- position and either 4,5-dimethyl, 4-ethyl-5-methyl or 5-ethyl-4-methyl. The dimethylthiazoles gave a base peak of *m/z* 127 and the ethylmethylthiazoles both gave a base peak of *m/z* 141 (Figure 3).

The second group contained five distinctive types of compound: two with base peak at m/z 128 (each present as a pair of peaks), two with base peak at *m/z* 114 (one present as a pair, the other as a singleton), and one with base peak at *m/z* 100 (present as a singleton). Searching the mass spectral databases yielded only one similar compound, 2-isobutyl-4,5-dimethyl-3-thiazoline, identified previously by Werkhoff et al. *(4)*, which had a base peak at *m/z* 114. It was assumed that all of the members of the second group were 3-thiazolines (**II**) and by structure elucidation, tentative identifications were made. The two pairs with base peak at *m/z* 128 were identified as 2-alkyl-4-ethyl-5-methyl-3-thiazolines and 2-alkyl-5-ethyl-4-methyl-3-thiazolines, each present as geometric isomers. The pair with base peak at *m/z* 114 were identified as 2-alkyl-4,5-dimethyl-3-thiazolines, also present as geometric isomers. These identifications were appropriate as their structures corresponded with those of the thiazoles. The singletons at *m/z* 114 were identified as 2-alkyl-4-ethyl-3-thiazolines and the singletons at *m/z* 100 were identified as 2-alkyl-4-methyl-3-thiazolines (Figure 4). Only substitution in the 5-position would result in pairs of geometric isomers.

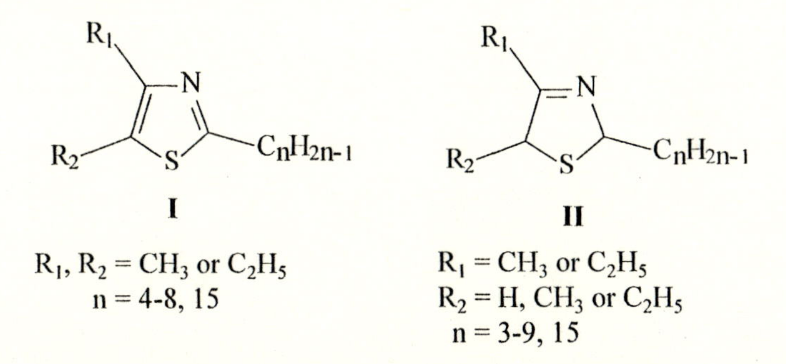

R_1, R_2 = CH_3 or C_2H_5
n = 4-8, 15

R_1 = CH_3 or C_2H_5
R_2 = H, CH_3 or C_2H_5
n = 3-9, 15

In order for a straight alkyl chain of more than 3 carbon atoms to be formed in these thiazoles and 3-thiazolines, a saturated aliphatic aldehyde containing more than 5 carbon atoms must have been present in the cooking process. These aldehydes would be formed from lipid oxidation. The steaks from steers fed on diets supplemented with fish oil contained much higher amounts of C_5 to C_{10} alkanals than the control steaks. These samples also showed the highest concentrations of 3-thiazolines with alkyl chains of 3 to 9 carbon atoms (*5*). The C_{15} sidechain would be formed from a C_{16} aldehyde derived from meat plasmalogens. Plasmalogens are phospholipids containing long-chain alkenyl ether substituents, which hydrolyze to

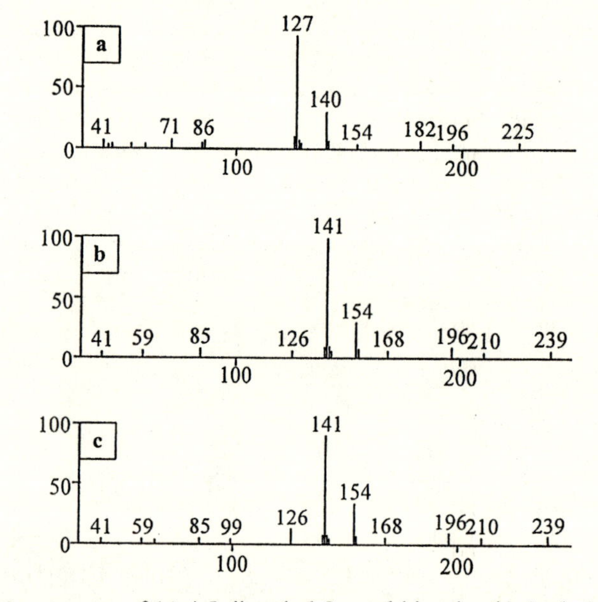

Figure 3. Mass spectra of (a) 4,5-dimethyl-2-octylthiazole, (b) 4-ethyl-5-methyl-2-octylthiazole, (c) 5-ethyl-4-methyl-2-octylthiazole.

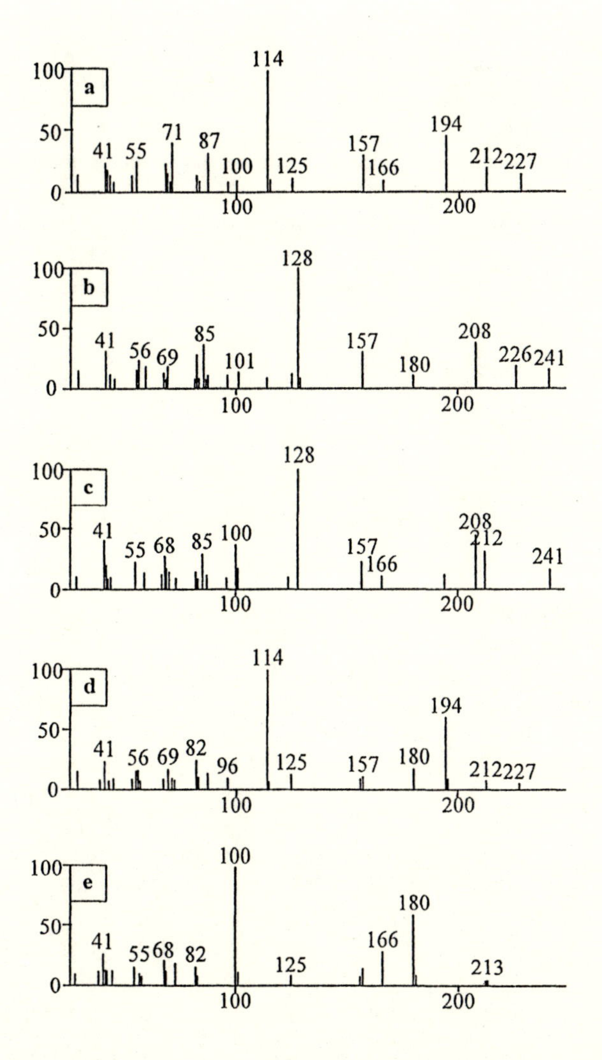

Figure 4. Mass spectra of (a) 4,5-dimethyl-2-octyl-3-thiazoline, (b) 4-ethyl-5-methyl-2-octyl-3-thiazoline, (c) 5-ethyl-4-methyl-2-octyl-3-thiazoline, (d) 4-ethyl-2-octyl-3-thiazoline, (e) 4-methyl-2-octyl-3-thiazoline.

give fatty aldehydes *(6)*. The reproducibility of the C_{15} thiazoles and 3-thiazolines across replicates was poor and did not be correlate with dietary effects.

The identities of many of these thiazoles and 3-thiazolines have been confirmed by comparing their mass spectra and linear retention indices with the products of reaction mixtures *(5,7)*. However, the thiazoles and 3-thiazolines with a C_{15} sidechain in the 2-position have only been tentatively identified and their mass spectra and linear retention indices are reported for the first time in Table I. We proposed that all of the 3-thiazolines and thiazoles were formed from the interaction between an aldehyde (which formed the alkyl chain in the 2-position), ammonia (from cysteine), hydrogen sulfide (from cysteine), and either a hydroxyketone (to form the 3-thiazoline) or a dione (to form the thiazole).The hydroxyketones and diones, which were identified in the beef headspace extracts, were products of the Maillard reaction. This was confirmed by heating aqueous reaction mixtures which contained an alkanal, ammonium sulfide and a hydroxyketone or dione *(7)*.

Table I. New thiazoles and 3-thiazolines tentatively identified in cooked beef.

Compound identity	LRI BPX5	Mass spectrum
4,5-Dimethyl-2-(methyltetradecyl)thiazole	2374	127, 140 (26), 41 (7), 43 (7), 128 (7), 141 (6), **323** (6) 154 (5), 196 (5),
4,5-Dimethyl-2-pentadecylthiazole	2406	127, 140 (30), 126 (11), 43 (10), 128 (8), 182 (8), 41 (7), 141 (6), 196 (6), **323** (6)
4-Ethyl-5-methyl-2-pentadecylthiazole	2447	141, 154 (28), 142 (9), 43 (8), **337** (8), 41 (7), 196 (6), 210 (6), 85 (5), 143 (5), 155 (5)
5-Ethyl-4-methyl-2-pentadecylthiazole	2473	141, 154 (36), 43 (13), 126 (10), 142 (10), 196 (8), 41 (7), **337** (7), 168 (6), 55 (5)
4-Methyl-2-pentadecyl-3-thiazoline	2421	278, 100 (72), 264 (33), 41 (29), 279 (23), 43 (17), 255 (14), 57 (12), 68 (12), 82 (12), 42 (11), 73 (11), 83 (11), **311** (6)
4-Ethyl-2-pentadecyl-3-thiazoline	2497	292, 114 (74), 43 (22), 293 (20), 278 (18), 41 (17), 82 (16), 255 (16), 310 (16), 55 (15), 87 (13), 96 (10), **325** (10)

Conclusion

Headspace trapping onto Tenax TA can be used for the extraction of high-boiling, high molecular weight compounds. 3-Thiazolines and thiazoles of up to 337 Da have been identified in the headspace extracts of cooked beef. These compounds had been formed by the reaction between aldehydes derived from lipid oxidation and Maillard reaction products.

Literature Cited

1. Leahy, M. M.; Reineccius, G. A. In *Analysis of Volatiles: Methods and Applications*; Schrier, P., Ed.; Walter de Gruyter: Berlin, 1984; pp 19-48.
2. Maarse, H.; Visscher, C. A. *Volatile Compounds in Food - Qualitative and Quantitative data*, 6th edn.; TNO-CIVO Food Analysis Institute: Zeist, The Netherlands, 1989.
3. Mottram, D. S. In *Volatile Compounds in Foods and Beverages*; Maarse, H., Ed.; Marcel Dekker: New York, 1991; pp 107-177.
4. Werkhoff, P.; Bretschneider, W.; Emberger, R.; Güntert, M.; Hopp, R.; Köpsel, M. *Chem., Mikrobiol., Technol. Lebensm.* **1991,** *13,* 30-57.
5. Elmore, J. S.; Mottram, D.S.; Enser, M; Wood, J.D. *J. Agric. Food Chem.* **1997**, *45*, 3603-3607.
6. Mottram, D. S. In *Contribution of Low and Non-volatile Materials to the Flavor of Foods*; Pickenhagen, W.; Ho, C.-T.; Spanier, A. M., Eds.; Allured Publishing: Carol Stream, 1996; pp 193-206.
7. Elmore, J. S.; Mottram, D. S. *J. Agric. Food Chem.*, **1997**, *45*, 3595-3602.

Chapter 7

Extraction of Thiol and Disulfide Aroma Compounds from Food Systems

Donald S. Mottram, Ian C. C. Nobrega, and Andrew T. Dodson

Department of Food Science and Technology, The University of Reading, Whiteknights, Reading RG6 6AP, United Kingdom

Thiol-substituted furans and their disulfides are known to have meat-like or roasted, coffee-like aromas at low concentrations. They are widely used as components of flavorings for soups and savory products. During simultaneous steam distillation solvent extraction (SDE) of cooked meat systems containing added bis(2-methyl-3-furanyl) disulfide and bis(2-furanylmethyl) disulfide, considerable loss of these disulfides was observed, compared with an aqueous blank. Significant amounts were converted to the corresponding thiols (2-methyl-3-furanthiol and 2-methylfuranthiol). Methyldithio-derivatives of these thiols were also found. Similar effects were observed using dynamic headspace analysis of the systems, indicating that the conversion of the disulfide to the thiol was not due to the method of volatile extraction. The effect can be explained by interchange redox reactions occurring between the furan disulfides and the sulfhydryl and disulfide groups of proteins.

Thiol-substituted furans, such as 2-methyl-3-furanthiol (**1**) and 2-furanmethanethiol (**2**), and the corresponding disulfides, **3** and **4**, have been shown to have meat-like or roasted, coffee-like aromas at low concentrations. These compounds have exceptionally low odor threshold values, which makes them exceptionally potent aroma compounds. During thermal processing, such compounds may be formed in the Maillard reaction *(1-3)* or from the degradation of thiamin *(4,5)*, and they are produced in significant amounts in heated model systems containing cysteine and ribose, or thiamin, but are only found in very small concentrations in cooked meat.

Their meaty, roast, and savory characteristics have resulted in these thiols and disulfides being widely used as components of flavorings for soups, savory products, and meat substitutes, where they are either added to the flavorings as nature-identical chemicals or as components of reaction-product flavorings.

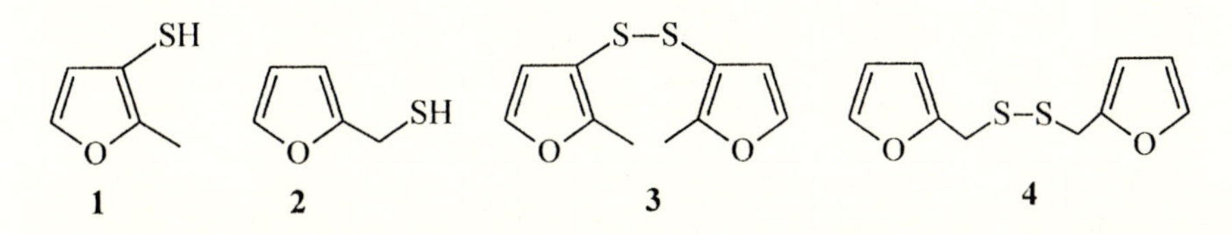

In aqueous solution thiols can be oxidized to the corresponding disulfides, and mixtures of different thiols form mixed disulfides *(6,7)*. It was also shown that, in boiling aqueous solution, disulfides were hydrolyzed to thiols, which were detected using 4-vinylpyridine as a thiol trapping agent *(8)*. In the absence of such a reagent, free thiols were not detected because they re-oxidize to the disulfide, but in a mixture of deuterated and unlabelled disulfides, mixed disulfides were obtained *(8)*. Such changes could result in modification of the sensory properties of foods containing thiols and disulfides. In proteins reduction–oxidation (redox) reactions involving interchange of sulfhydryl and disulfide groups, within the protein or with external thiol groups, are well known *(9)*. This raises the possibility of thiol and disulfide flavor compounds interacting with proteins in foods and causing changes in their relative concentrations. Recently, we demonstrated that when disulfides **3** and **4** were heated in aqueous solution with a protein, such as egg albumin, some of these disulfides were lost but a large proportion of each was converted to the corresponding thiol *(10)*. Interaction with sulfhydryl groups in the protein appeared to be responsible.

This paper reports on the recovery of disulfides and thiols from protein systems using different volatile extraction techniques – simultaneous steam distillation solvent extraction (SDE) and dynamic headspace collection on Tenax porous polymer – and provides further evidence of interaction between proteins and disulfide aroma compounds.

Experimental

Simultaneous Distillation-Extraction (SDE) of cooked beef. Portions (100 g) of meat (beef fillet, *M. Psoas major*) were chopped into small pieces, minced in a blender, transferred to 500 mL bottles, closed with PTFE-lined lids and cooked in an autoclave at 140 °C for 30 min. An ethanolic solution (1 mL) containing 0.5 mg bis(2-methyl-3-furanyl) disulfide (**3**) and 0.5 mg bis(2-furanylmethyl) disulfide (**4**) was added to the cooked minced beef that was blended with 750 mL of distilled water and volatiles were extracted using SDE in a Likens-Nickerson apparatus. Extractions were carried out for two hours with a mixture of redistilled *n*-pentane (27 mL) and diethyl ether (3 mL) as the solvent. After extraction, an internal standard (1,2-dichlorobenzene, 130 μg in 0.1 mL diethyl ether) was added, and the solvent was dried and concentrated to about 0.5 mL by distillation. Blank extractions were carried out using 750 mL water to which both disulfides had been added. Analyses were carried out in triplicate. Quantities of compounds **1 - 4** in SDE extracts were determined by comparison of the peak areas in the total ion chromatogram (TIC) with areas of the compounds in analyses of standard solutions, using 1,2-dichlorobenzene as internal standard. Other compounds were quantitated by comparison of TIC peak areas with those of the internal standard, using a response factor of 1.

Headspace Concentration on Tenax TA. One mL of an ethanolic solution containing 30 µg/mL **3**, 130 µg/mL **4**, and 37 µg/mL methyl decanoate internal standard was added to 10 mL water, and the solution was blended with 100 g cooked beef and left to stand for 0.5 h. A portion (27.5 g) of this mixture was transferred to 250 mL conical flask and volatiles were collected in a glass-lined stainless steel trap (105 mm x 3 mm i.d.) containing 85 mg Tenax TA (SGE Pty Ltd, Ringwood, Australia), as described by Elmore and Mottram *(11)*. During the collection (1 h) the conical flask was maintained at 60 °C, and the volatiles were swept onto the trap using a flow of oxygen-free nitrogen (40 mL/min). At the end of this time, the trap was connected directly to the nitrogen supply for 5 min (40 mL/min) to remove moisture prior to GC-MS analysis. Similar headspace collections were carried out on solutions of egg albumin (6.7 g in 100 mL water) and aqueous blanks, containing the same quantities of disulfides and methyl decanoate as the meat systems. Analyses were carried out in triplicate. The relative amounts of the identified components were estimated by comparison of peak areas with those of the methyl decanoate internal standard.

Gas Chromatography-Mass Spectrometry (CG-MS). Analyses were carried out using a Hewlett-Packard HP5890 gas-chromatograph coupled to a HP5972 mass selective detector. A split/splitless injection was used to introduce 1 µL of each SDE extract on to the GC column. The headspace traps were thermally desorbed, at 250 °C, directly on to the cooled (0 °C) GC column using a CHIS injection port (SGE Ltd). A BPX5 fused silica capillary column (50 m x 0.32 mm; 0.5 µm film thickness; SGE Pty Ltd) was used for all analyses, with helium at 1.6 mL/min (35 cm/s) as carrier gas. The oven was initially held at 50 °C for 2 min and then programmed at 4 °C/min to 250 °C. The interface of the GC to the MS was maintained at 280 °C, and the MS was operated in the electron impact mode with an ionization energy of 70 eV and a scan rate of 1.9 scans/s over the mass range 33-400 amu.

Results and Discussion

The SDE of the aqueous solutions showed that the furan disulfides could be recovered without significant loss and without any formation of the corresponding thiols (Table I). However, a small amount of 2-methyl-3-[(2-furanylmethyl)dithio]furan (**5**) was isolated. This indicated that some hydrolysis of the disulfides had occurred allowing the formation of this mixed disulfide. The headspace analyses of the aqueous blanks containing the disulfides also showed no evidence of furan thiols but, again, small amounts of **5** were isolated. The recovery of disulfides from the aqueous solution by

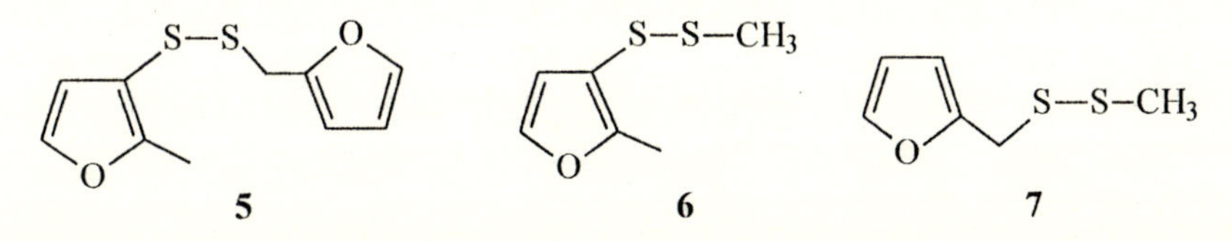

headspace collection depends on the partition of the compounds between the aqueous and vapor phases and, therefore, absolute recoveries cannot be quoted in the same way as for the SDE analyses. However, the absence of any other compounds in the headspace suggests that the disulfides are stable during collection and during thermal desorption on to the GC column. Collected volatiles were desorbed by heating the Tenax trap to 250 °C under a flow of carrier gas in a modified GC injection port. In order to determine whether these conditions caused any oxidation of thiols to disulfides, a mixture of thiols **1** and **2** were added to the trap in ethanolic solution, and the recovery compared with that for a direct injection of the solution. No disulfides were found and complete recovery of the thiols was obtained.

In the presence of meat marked decreases in the concentration of the disulfides were observed in the SDE extracts and significant quantities of the corresponding thiols **1** and **2** were found (Table I). A quantity of the mixed disulfide **5** was also found, together with the methyldithio-derivatives, 2-methyl-3-(methyldithio)furan (**6**) and 2-[(methyldithio)methyl]furan (**7**). Quantitatively, all of the lost disulfides were not recovered as thiols or mixed disulfides, indicating that binding to the protein may have occurred. These results are in agreement with previous data which showed that when disulfides were added to solutions of egg albumin or casein losses of disulfides occurred in SDE analysis with formation of thiols *(10)*.

SDE analysis involves boiling with a large excess of water for several hours and such conditions could promote the hydrolysis of the disulfides and their interaction

Table I. Recovery of thiols and disulfides from cooked meat systems containing added bis(2-furanylmethyl) disulfide and bis(2-methyl-3-furanyl) disulfide.

Compound	Quantity added (μg)	Quantity recovered (μg)		
		Water + Disulfides	Cooked Beef + Disulfides [a]	Cooked Beef alone [a]
bis(2-methyl-3-furanyl) disulfide (**3**)	500	532	126 (27)	– [b]
bis(2-furanylmethyl) disulfide (**4**)	500	437	104 (27)	–
2-methyl-3-furanthiol (**1**)	1 [c]	< 1	302 (35)	1 (0)
2-furanmethanethiol (**2**)	1 [c]	< 1	294 (41)	13 (2)
2-methyl-3-[(2-furanylmethyl)-dithio]furan (**5**)	–	16	39 (6)	–
2-methyl-3-(methyldithio)furan (**6**)	–	–	7 (1)	–
2-[(methyldithio)methyl]furan (**7**)	–	–	3 (0)	–

[a] Mean of triplicate analyses, standard deviations shown in parentheses. [b] not detected.
[c] The disulfides contained very small amounts of the corresponding thiols.

with sulfhydryl groups on the protein. Therefore, headspace collections from both cooked meat and egg albumin systems, with added disulfides, were carried out to determine whether the interactions observed in the SDE systems were an effect of the extraction method. Since complete extraction from the food is not achieved in headspace analysis, the effect of protein on added disulfides was evaluated adding an internal standard (methyl decanoate) to the system, before headspace collection, and amounts of the compounds obtained, relative to this internal standard, were compared with those for aqueous blanks.

The dynamic headspace analyses of the cooked meat showed large losses of the disulfides **3** and **4** compared with the aqueous blank (Figure 1), demonstrating that the effect was not primarily due to the analytical method. Quantities of the corresponding thiols **1** and **2** were found plus relatively large amounts of the methyldithio-compounds **6** and **7**. The latter compounds were found in greater relative concentrations in the headspace analyses than in the SDE. Analysis of a meat sample, without the addition of the disulfides, showed the presence of 2-furanmethanethiol (**2**), its disulfide **4**, and the methyldithio-derivative **7** at levels that would make a contribution to the amounts of these compounds observed in the meat with added disulfides. Even greater losses of disulfide and thiols were noted with the albumin systems, possibly because the protein was in solution allowing easier interaction between protein and the furan disulfides.

The marked effect of the protein on the recovery of disulfides from these systems confirms earlier observations and supports the hypothesis that interaction occurs between the thiol and disulfide flavor compounds with sulfhydryl groups and

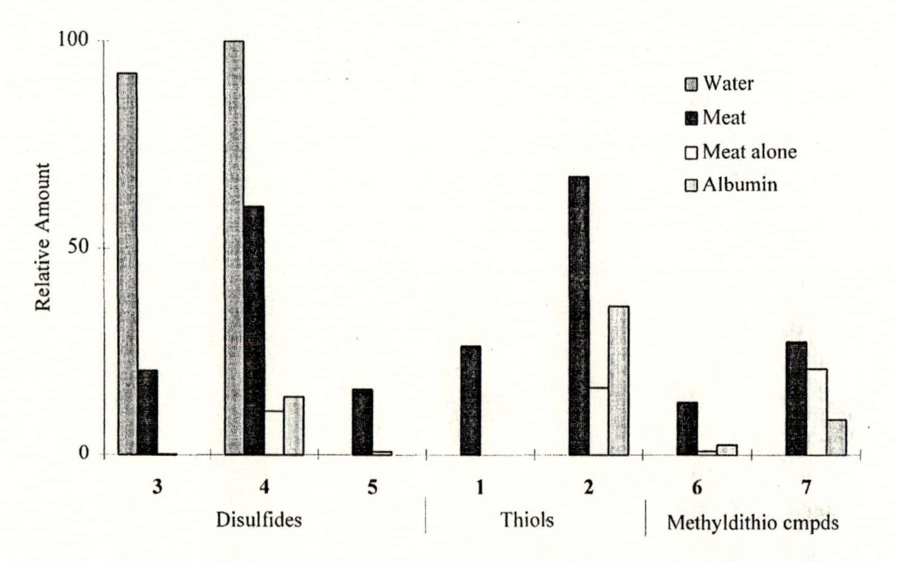

Figure 1. Recovery of Furan Disulfides and their Derivatives in Dynamic Headspace Collections from Cooked Meat and Egg Albumin Systems Compared with an Aqueous Blank.

disulfide bridges, from cysteine and cystine amino acid units in the protein. This results in conversion of the furan disulfides to the corresponding thiols and the formation of new disulfide links between protein and thiol with the associated loss of flavor compound. Such sulfhydryl–disulfide interchange reactions are well known in protein chemistry *(9,12)*. An example of such interchanges is seen in bread-making where sulfhydryl–disulfide interchanges between glutathione and the flour proteins are important in relation to dough rheology *(13,14)*.

The presence of the methyldithio-derivatives indicates that other volatile disulfides, such as dimethyl disulfide, participated in the interchange reactions in the meat and albumin systems. It has been observed that larger amounts of dimethyl disulfide and other methyl disulfides were found in headspace collections of meat volatiles than in SDE *(15)*. This may explain the larger amounts of the methyldithio-compounds **6** and **7** in the headspace systems containing added disulfides. It suggests that the amount of water and the heating conditions during extraction may affect the extent of interchange reactions with the proteins, implying that the relative amounts of thiols and disulfides found in cooked meat may be influenced by cooking conditions. Another interesting result in this present work is the difference between the two disulfides. The overall losses of **3**, and its thiol **1**, were much greater than the losses of disulfide **4** and thiol **2**. It has been shown that thiol **1** is much more easily oxidized than thiol **2** *(7)*, and this may result in greater reactivity between the protein and **1** with correspondingly greater losses.

The odor threshold values of the thiols may differ from those of the corresponding disulfides, e.g., the odor threshold value for the disulfide **3** is reported as 2×10^{-5} μg/kg while that of the corresponding thiol **1** is 5×10^{-3} μg/kg. The aroma characteristics of these compounds may also change with concentration. In general, at low concentrations approaching the odor threshold values, the compounds have pleasant savory or roasted aromas but at higher concentration they become more sulfurous and unpleasant *(16-18)*. Redox-induced changes in the relative concentrations of these thiols and disulfides due to protein–sulfhydryl interchanges could result in significant changes in the sensory properties when the compounds are used in flavorings for food products. The thiol-disulfide exchanges may also contribute to the different aroma characteristics of meat cooked under different conditions.

Literature Cited

1. Farmer, L. J.; Mottram, D. S.; Whitfield, F. B. *J. Sci. Food Agric.* **1989**, *49*, 347-368.
2. Mottram, D. S. In *Thermally Generated Flavors. Maillard, Microwave, and Extrusion Processes*; Parliment, T. H.; Morello, M. J.; McGorrin, R. J., Eds.; ACS Symposium 543; American Chemical Society: Washington DC, 1994; pp 104-126.
3. Hofmann, T.; Schieberle, P. *J. Agric. Food Chem.* **1995**, *43*, 2187-2194.

4. van der Linde, L. M.; van Dort, J. M.; de Valois, P.; Boelens, B.; de Rijke, D. In *Progress in Flavour Research*; Land, D. G.; Nursten, H. E., Eds.; Applied Science: London, 1979; pp 219-224.
5. Werkhoff, P.; Brüning, J.; Emberger, R.; Güntert, M.; Kopsel, M.; Kuhn, W.; Surburg, H. *J. Agric. Food Chem.* **1990**, *38*, 777-791.
6. Mottram, D. S.; Madruga, M. S.; Whitfield, F. B. *J. Agric. Food Chem.* **1995**, *43*, 189-193.
7. Hofmann, T.; Schieberle, P.; Grosch, W. *J. Agric. Food Chem.* **1996**, *44*, 251-255.
8. Guth, H.; Hofmann, T.; Schieberle, P.; Grosch, W. *J. Agric. Food Chem.* **1995**, *43*, 2199-2203.
9. Jocelyn, P. C. *Biochemistry of the SH group*; Academic Press: London, 1972, pp 404.
10. Mottram, D. S.; Szauman-Szumski, C.; Dodson, A. *J. Agric. Food Chem.* **1996**, *44*, 2349-2351.
11. Elmore, J. S.; Mottram, D. S.; Enser, M. B.; Wood, J. D. *J. Agric. Food Chem.* **1997**, *45*, 3603-3607.
12. Whitesides, G. M.; Houk, J.; Patterson, M. A. K. *J. Org. Chem.* **1983**, *48*, 112-115.
13. Chen, X.; Schofield, J. D. *J. Agric. Food Chem.* **1995**, *43*, 2362-2368.
14. Grosch, W. In *Chemistry and Physics of Baking*; Blanchard, J. M. V.; Frazier, P. J.; Galliard, T., Eds.; Royal Society of Chemistry: London, 1986; pp 602-604.
15. Mottram, D. S.; Nobrega, I. N. C.; Dodson, A. T.; Elmore, J. S. In *Flavour Science: Recent Developments*; Taylor, A. J.; Mottram, D. S., Eds.; Royal Society of Chemistry: Cambridge, 1996; pp 413-418.
16. Arctander, S. *Perfume and Flavor Chemicals*; Published by the author: Monclair, NJ, 1969.
17. Tressl, R.; Silwar, R. *J. Agric. Food Chem.* **1981**, *29*, 1078-1082.
18. Fors, S. In *The Maillard Reaction in Foods and Nutrition*; Waller, G. R.; Feather, M. S., Eds.; ACS Symposium Series 215; American Chemical Society: Washington DC, 1983; pp 185-286.

Chapter 8

Isolation of Flavor Compounds from Protein Material

O. E. Mills and A. J. Broome

New Zealand Dairy Research Institute, Palmerston North, New Zealand

The nature of the matrix of spray dried whey protein concentrate or sodium caseinate leads to difficulties in efficiently recovering volatiles for flavor analysis. Although recovery of volatiles is more efficient from aqueous solution the reversible adsorption of flavor compounds to proteins also hinders extraction. Low pressure distillation, solid phase micro extraction (SPME) and purge and trap were compared as methods for extraction of volatile compounds from the two protein products. Low pressure distillation recovered the widest range of compounds while purge and trap was best for the most volatile of compounds. SPME accentuated the extraction of carbonyl compounds.

A study of the volatile components of dairy protein products is an essential part of developing an understanding of the mechanisms of flavor formation. Sodium caseinate and whey protein concentrate (WPC) powders are manufactured by drying from aqueous solution using the spray drying process. These products are then usually presented for analysis in powder form. Although each powder has a low intensity characteristic odor they are tasted after dissolving in water. Volatile analysis is possible in the powder form but better results are obtained from aqueous solution. An explanation of the physical process of spray drying shows why this is so (*1*). During spray drying atomised protein solution is heated by hot air. The surface of the particles begins to dry while water and dissolved volatile compounds diffuse towards the surface of the particle. Drying occurs rapidly at the surface and a semi-permeable membrane begins to form. The rate at which molecules migrate towards the surface depends on their diffusivity which decreases with increasing molecular size. As the porosity of the surface membrane decreases the size of molecules able to escape from the particles decreases until eventually only water is able to escape. This process takes only a few milliseconds and, provided the spray dryer operating parameters are correctly chosen, flavor volatiles are substantially encapsulated in the particle during drying. The semi-permeable membrane restricts release of volatiles from the powder and it is therefore easy to see why the odor of dairy protein powders is low compared with the odor of their solutions.

A further problem is encountered when attempting to extract flavor compounds from protein solutions. This is due to the interaction of flavor compounds with proteins. The nature of this interaction has been studied in mixtures of whey proteins

(*2,3*), with pure β-lactoglobulin (*4,5*) and with pure bovine serum albumin (*6,7*). It has been found that many flavor compounds adsorb reversibly to these proteins and in the studies with pure whey proteins it was deduced that interactions were hydrophobically driven (*4,6*). It was also found that adsorption was accompanied by a conformational change (*4*). Conversely when conformational changes were induced by pH shifts then the extent of binding changed (*6*). Consequently the removal of flavor compounds from a system in which they are in dynamic equilibrium with proteins is going to be influenced by the kinetics of adsorption/desorption processes. Extraction of volatiles from proteins in solution by a competing adsorption process will, therefore, be controlled by the kinetics of the adsorption/desorption and will often be less efficient than extraction from a solution of the dissolved volatiles alone. Extraction of volatiles from the vapor phase above the solution will also be kinetically controlled but will introduce a further stage likely to reduce efficient extraction which is the partitioning of volatiles between the liquid phase and the vapor phase. This is exemplified by the low pressure distillation of volatile compounds from WPC (*8*). Here distillation was carried out by reducing the pressure to allow boiling at 30°C. After each 2 h period of distillation an amount of water equivalent to the volume of distillate recovered during that 2 h period was added to the WPC solution. Even after 8 h volatile compounds were still being distilled. The rate of depletion was dependent on the compound.

The options for the recovery of volatiles from an aqueous protein solution are to recover them from either the liquid phase or the vapor phase above the solution. Although the above arguments favor the recovery of volatiles from aqueous solution no suitable liquid phase extraction methods have been developed for this kind of product. In the work reported here a variety of vapor phase extraction methods will be compared for the analysis of volatile compounds from sodium caseinate and WPC. Methods chosen were low pressure distillation, purge and trap and solid phase microextraction (SPME). Also the use of proteolytic enzymes in conjunction with SPME will be investigated as a means of decreasing the binding of flavor compounds to proteins. Extensive proteolysis was found to decrease the binding of volatile compounds to soybean proteins (*9*).

Experimental

Materials. Protein powders used in this study were WPC containing 80.9 % protein, 7.8 % lactose, 4.07 % fat, 3.27 % ash and 3.65 % moisture and sodium caseinate containing 92 % protein, 0.1 % lactose, 0.8 % fat, 3.6 % ash and 4 % moisture.

Low Pressure Distillation. WPC (500 g) or sodium caseinate (300 g) was dissolved in 3 L distilled water and boiled under reduced pressure at 40°C until approximately 1 L distillate was condensed in the cold trap at -70°C. The distillate was then thawed and continuously liquid/liquid extracted with redistilled diethyl ether. The diethyl ether extract was dried with anhydrous sodium sulphate and concentrated using a Kuderna-Danish apparatus to about 5 mL and finally to 0.1 mL with a gentle stream of nitrogen.

Gas Chromatography/Mass Spectrometry (GCMS). GCMS was carried out using a Fisons MD800 mass spectrometer connected to a Fisons GC 8060 gas chromatograph (Fisons Instruments, Wythenshawe, Manchester, United Kingdom). A 30 m x 0.32 mm i.d. Econocap capillary column with 0.25 μm FFAP phase (Alltech Associates, Deerfield, IL, USA) was used. The column oven was held at 35°C for 5 min then heated to 230°C at 5°C/min. Full scan mass spectra were obtained at 70 eV using a scan range of 35 - 350 M/Z and a rate of 0.5 s/scan. Tentative identifications were made using a computer matching algorithm which compared the spectra generated during the experiment with those in the NIST library of mass spectra.

SPME. WPC and sodium caseinate solutions were prepared at 10 % (w/v). An aliquot (300 mL) was placed in a 500 mL flask fitted with a stopper containing a rubber septum. The flask was submerged in a temperature controlled waterbath at 50°C and the solution stirred with a magnetic bar. SPME was carried out using a 65 μm polydimethylsiloxane-divinylbenzene (PDS/DVB) fiber (Supelco, Inc., Bellefonte, USA). Protein solutions were equilibrated at 50°C for 30 min then the fiber was exposed for 30 min. The fiber was desorbed at 250°C in the injector port of the gas chromatograph.

Purge and Trap. Purge and trap was carried out using a Chrompack CP-4020 TCT (Chrompack, Middelburg, The Netherlands) fitted to the Fison's MD800 GCMS system described above. The trap was packed with Tenax GC 60/80 (Alltech Associates, Deerfield, USA). Volatiles were trapped from the headspace of stirred liquid equilibrated at 50°C. Trapping took place for 10 min at a nitrogen flow rate of 60 ml/min. The trap was backflushed in the purge and trap unit with helium at 50°C for 10 min to remove any condensed water prior to desorption which was carried out at 250°C for 10 min. The desorbed volatiles were cryogenically trapped at -100°C using liquid nitrogen. At the end of the desorption phase the cold trap was rapidly heated to 250°C allowing the volatiles to enter the GC column in a narrow band. GC conditions were the same as those described above.

Enzymatic Treatment. A solution of WPC was prepared as described above for SPME analysis. The solution was equilibrated at 50°C for 30 min and analysed by SPME. Proteolytic enzymes, Alcalase 2.4 L FG, Neutrase 1.5 MC and Flavorzyme 1000 (Novo Nordisk A/S, Bagsvaerd, Denmark) were each added at the rate of 1 % (w/w) on the weight of protein. SPME analysis was again carried out after 5 h at 50°C.

Results and Discussion

Figure 1 shows the total ion current chromatograms of volatiles produced by low pressure distillation of each protein product. Mass spectral data was not collected until the solvent peak had been eluted which was after 0.6 min. Identification of compounds responsible for many of the major peaks is given in Table I. A similarity can be seen between the two chromatograms with a number of relatively volatile compounds being eluted within the first 2.5 min and a series of even carbon numbered fatty acids ranging from acetic acid to tetradecanoic acid, but excluding butyric acid in the sodium caseinate extract. Comparisons of purge and trap with SPME are shown for each protein product in Figures 2 and 3. As there is no solvent used in these methods mass spectral scanning was started immediately on injection. Conditions chosen for purge and trap were a compromise between the use of a sufficiently high temperature that the concentration of volatiles in the headspace was increased and the use of a trapping time that was not too long such that too much water condensed in the trap and interfered with adsorption of volatiles. Purge and trap of both protein products produced a cluster of relatively volatile compounds in the first 3 min similar to low pressure distillation but little at longer retention times. SPME of both protein products showed a more even spread of compounds over 20 min. Results obtained for both extraction methods were very similar for both protein products. The SPME phase used here produced a bias towards the adsorption of aldehydes with the homologous series of hexanal to nonanal being identified. The inability to detect this range of aldehydes in the low pressure distillation extract raises the possibility of oxidation occurring during the preparation of those samples. The disadvantage of purge and trap and SPME as used here was that there was insignificant volatiles with retention times of greater than 20 min. Quantitative comparisons between methods cannot be made even though all these methods rely on extraction from the vapor phase. Low pressure distillation is

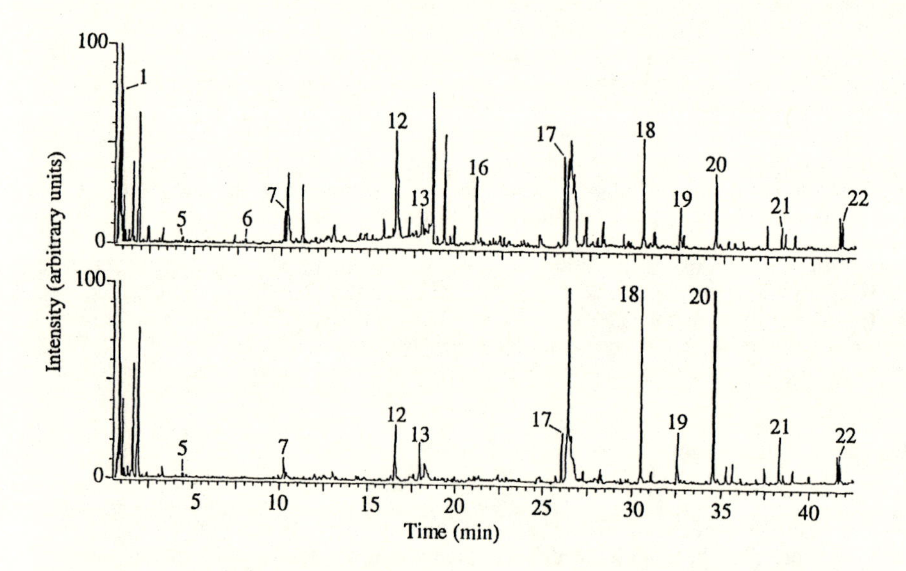

Figure 1. The total ion current chromatograms of volatiles extracted from WPC (top) and sodium caseinate (bottom) by low pressure distillation are shown. Refer to Table I for identifications.

superior for recovery of compounds over a wide boiling point range while purge and trap provides a relatively better recovery of the most volatile compounds. The advantage of purge and trap for the study of the most volatile compounds is that it is

Table I. Identification of compounds in chromatograms

Peak number	*Compound*	*Identification*[1]
1	ethyl acetate	ms
2	2-butanone	ms,rt
3	diacetyl	ms,rt
4	methyl benzene	ms,rt
5	hexanal	ms,rt
6	heptanal	ms,rt
7	styrene	ms
8	octanal	ms,rt
9	dimethyl trisulphide	ms
10	2-nonanone	ms,rt
11	nonanal	ms,rt
12	acetic acid	ms,rt
13	benzaldehyde	ms,rt
14	2-undecanone	ms,rt
15	3,5-octadien-2-one	ms
16	butyric acid	ms,rt
17	hexanoic acid	ms,rt
18	octanoic acid	ms,rt
19	nonanoic acid	ms,rt
20	decanoic acid	ms,rt
21	dodecanoic acid	ms,rt
22	tetradecanoic acid	ms,rt

[1]Identification was by mass spectrometry (ms) and retention time (rt) of the authentic compound or by mass spectrometry alone when the authentic compound was not available.

so much faster to carry out. Other adsorbents can also be used with purge and trap such as Tenax TA, Carbopack and Carbosieve. There may be good reasons for using another phase in that the particular discrimination produced by that phase may offer an advantage.

SPME coupled with GCMS is an extraordinarily simple but sensitive method for the identification of volatile compounds. The PDS/DVB phase used here is specified as being suitable for the detection of polar compounds which would explain the bias towards aldehydes. Other types of SPME phases are also available and may accentuate other classes of compounds.

SPME was also used to monitor the release of volatile compounds during proteolysis. There was no significant increase in volatiles either quantitatively or qualitatively during proteolysis over 5 hours at 50°C. The extent of proteolysis was not determined quantitatively, however, the pH was found to decrease from 7.0 to 5.9 during the incubation period, indicating considerable proteolysis.

Sensory evaluation of the two protein products shows they are clearly different in both their odor and flavor profiles (unpublished data). It is unlikely that

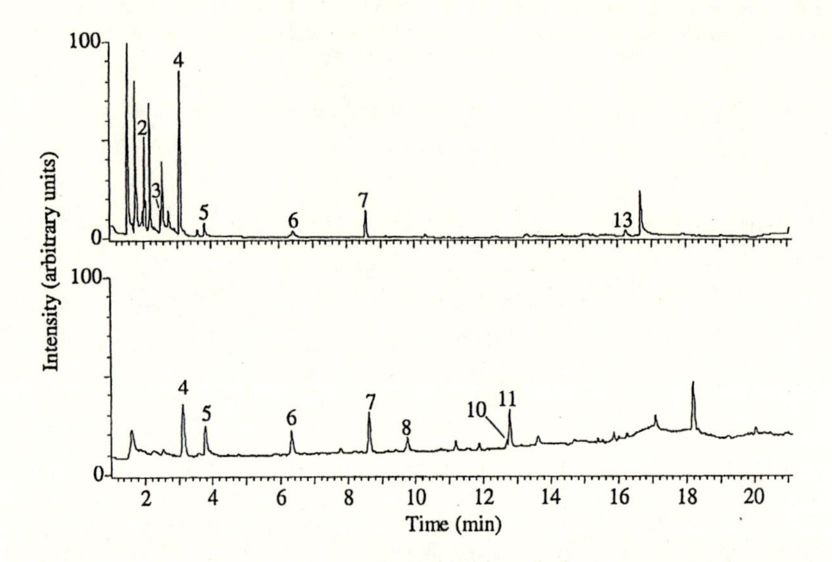

Figure 2. The total ion current chromatograms of volatiles from WPC extracted by purge and trap (top) and SPME (bottom) are shown. Refer to Table I for identifications.

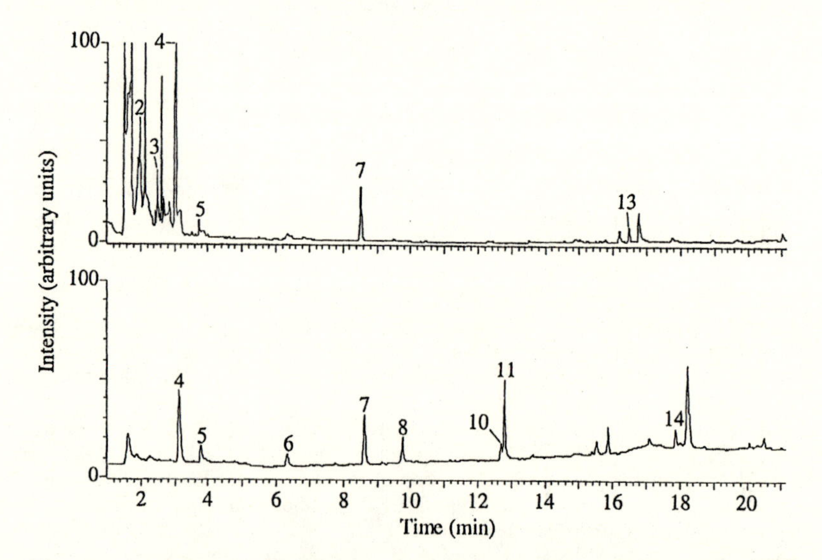

Figure 3. The total ion current chromatograms of volatiles from sodium caseinate extracted by purge and trap (top) and SPME (bottom) are shown. Refer to Table I for identifications.

that such a conclusion would be drawn from an inspection of purge and trap and SPME chromatograms. There are also considerable similarities in the chromatograms produced from low pressure distillation extracts but there are also many more compounds present in these extracts. The sensory differences between the two products would probably be found by employing some form of gas chromatography/olfactometry technique on these extracts.

Literature Cited

1 Bomben, J. L.; Bruin, S.; Thijssen, H. A. C. *Adv. Food Res.* **1973**, *20*, 1.
2 Mills, O. E.; Solms, J. *Lebensm.-Wiss. u. -Technol.***1984**, *17*, 331.
3 Jasinski, E.;Kilara, A. Milchwissenschaft. **1985**, *40*, 596.
4 O'Neill, T. E.; Kinsella, J. E. *J. Agric. Food Chem.* **1987**, *35*, 770.
5 O'Neill, T. E.; Kinsella, J. E. *J. Food Sci.*, **1988**, *53*, 906.
6 Damodaran, S.; Kinsella, J. E. *J. Agric. Food Chem.* **1980**, *28*,567.
7 Damodaran, S.; Kinsella, J. E. *J. Biol. Chem.* **1980**, *255*, 8503.
8 Mills, O. E. In *Food Flavors, Ingredients and Composition;* Charalambous, G., Ed.; Developments in Food Science; Elsevier: Amsterdam, The Netherlands, 1993, Vol. 32; pp 139-149.
9 Arai, S.; Abe, M.; Yaqmasegita, M.; Kato, H.; Fujimaki, M. *Agr. Biol. Chem.* **1971**, *35*, 552.

Chapter 9

Sampling Carbonyl Compounds with Solid Phase Microextraction Utilizing On-Fiber Derivatization

Perry A. Martos and Janusz Pawliszyn

The Guelph–Waterloo Centre for Graduate Work in Chemistry (GWC), Department of Chemistry, University of Waterloo, Waterloo, Ontario N2L 3G1, Canada

Carbonyl compounds can be sampled with solid phase microextraction using on-fiber derivatization with *o*-(2,3,4,5,6-pentafluorobenzyl)-hydroxylamine (PFBHA) adsorbed onto poly(dimethylsiloxane)/divinyl benzene (PDMS/DVB) solid phase microextraction fibers. The products of the reactions are oximes that are thermally very stable; insensitive to light; and can be analyzed by gas chromatography, with FID or other detectors, or HPLC with ultraviolet/visible detection. Loading PFBHA on PDMS/DVB fibers is simple, highly reproducible, and reversible; more than 200 loading, sampling, and analysis steps are possible with one fiber. Formaldehyde was used as a model carbonyl compound to study the viability of the new sampling system. This sampling system provides specificity for carbonyl compounds and requires short sampling times. It has the potential to be an excellent analytical tool for sampling short-lived carbonyl-containing aroma compounds and as a routine sampling method for volatiles from various matrices, such as foods and fragrances.

The sampling and analysis of carbonyl compounds as components of flavor constituents is of particular interest to manufacturers of products containing those flavors. This is due in part to the requirement to control the total cost of the product's aroma constituents by minimizing the amount or number of added analyte(s) required to achieve the desirable overall product aroma. Often, it is the complex arrangement of carbonyl compounds (aldehydes, esters, ketones, and carboxylates) that influences the acceptable aroma impression of a product. The low concentration of desirable aroma compounds can present problems for some analytical methods. Therefore, rapid yet sensitive methods to sample and analyze low concentrations of carbonyl compounds are required; however, the methods should be cost effective as well as provide advantages over standard sampling methods. One of the more popular methods to sample aroma compounds, yet not very rapid nor cost effective, is interfacial cryofocussing of a given volume of headspace from a product at equilibrium. The interface is then heated to load the analytes into a gas chromatograph equipped with a capillary column with subsequent detection using mass spectrometry or other detectors. A sampling system that is ideally suited for this task is solid phase microextraction (SPME).

SPME is a highly desirable sampling method for flavor compounds because it extracts, concentrates, and provides a means to directly deliver the concentrated and extracted analytes into the interface of a GC or HPLC. Figures 1a and 1b present the SPME device. In Figure 1a, the basic SPME holder is presented along with the principle component of the system: a polymer coated onto a solid support inside a needle, which can penetrate a sampling chamber and a GC interface septum. The fibers consist of a fixed liquid film or fixed solid sorbents and are 1 cm long and seven to 100 μm thick, depending on the chosen coating. Once the sampling of the analytes has occurred, they are delivered into the sample introduction interface (Figure 1b). In the case presented, the interface is for a GC septum equipped programmable injector (SPI), but the same mechanism applies for introduction into a split/splitless injector. First, the needle penetrates the SPI septum, then the fiber is introduced into the hot injector, where the analytes rapidly desorb from the fiber, which allows the analytes to be delivered to the head of the column for subsequent separation and detection. Some of the key advantages in using SPME for this purpose are the fiber extracts and concentrates the analytes of interest on one coating, and sample introduction does not involve a solvent. With SPME no cryofocussing is required thus obviating the need for cryotraps, yet sensitivity is not compromised. In addition, more peaks can be resolved per unit time as a result of enhanced separation because the analytes (mass) load onto the head of the capillary column in a much finer band width compared to liquid injection. A more detailed description of SPME can be found in a recently published comprehensive text on the theory and practice of SPME (*1*).

SPME has been used for sampling volatile compounds from fragrance oils and beverages (*2*), truffles (*3*), and fruit juices (*4*); but, the reports have not included analyte specific derivatization, such as presented here. This chapter presents the use of SPME to sample carbonyl compounds with on-fiber derivatization utilizing *o*-(2,3,4,5,6-pentafluorobenzyl)-hydroxylamine (PFBHA) adsorbed onto poly-(dimethylsiloxane)/ divinyl benzene (PDMS/DVB) fibers. Formaldehyde, an ubiquitous carbonyl compound, is used as a model compound to demonstrate the viability of SPME with on-fiber derivatization for sampling carbonyl compounds. Derivatization sampling for carbonyl compounds realizes a number of advantages over techniques not employing derivatization: (1) it provides analyte specificity based on key functional groups, (2) it allows for detection with conventional detectors (for those compounds that do not yield a response or yield a low response), and (3) it can act as a method for confirmation that the compound of interest is present in the sample.

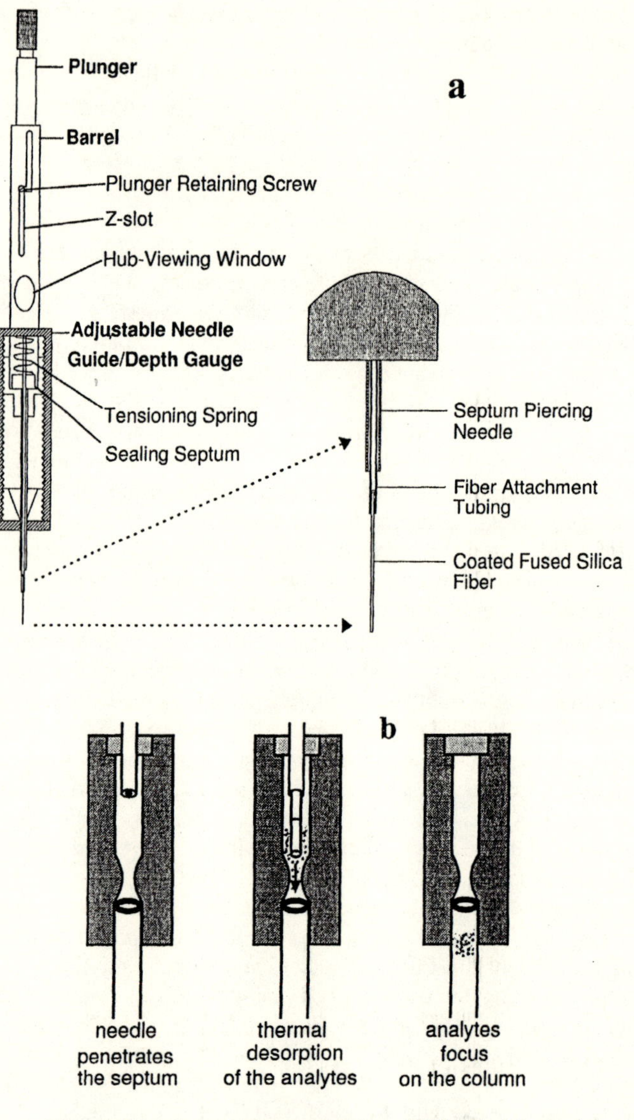

Figure 1: The solid phase microextraction (SPME) (a) device and SPI injector interface (b).

Theory

There are four steps to consider in describing the overall rate of oxime formation on solid sorbent SPME fiber coatings sorbed with PFBHA as a function of the concentration of gaseous carbonyl. Shown below are the steps in the sampling system, where S is the available binding surface of the sorbent,

$$\text{(Step A)}\quad PFBHA + S \xrightarrow{k_1} PFBHA * S \quad \text{(adsorption)}$$

$$PFBHA * S \xrightarrow{k_{-1}} PFBHA + S \quad \text{(desorption)}$$

$$\text{(Step B)}\quad Carbonyl + S \xrightarrow{k_2} Carbonyl * S \quad \text{(adsorption)}$$

$$Carbonyl * S \xrightarrow{k_{-2}} Carbonyl + S \quad \text{(desorption)}$$

$$\text{(Step C)}\quad Carbonyl + PFBHA * S \xrightarrow{K^*} Oxime * S \quad \text{(reaction)}$$

$$\text{(Step D)}\quad Oxime * \text{S} \xrightarrow{k_3} Oxime + S \quad \text{(desorption)}$$

and where $K_A = \frac{k_1}{k_{-1}}$, $K_B = \frac{k_2}{k_{-2}}$ (see below). The first step (Step A) is to load the sorbent with PFBHA. There are two rates to consider in Step A; the rate of adsorption and the rate of desorption. From experimental data, it is understood that following loading the sorbent with PFBHA, its rate of desorption is negligible, i.e., $k_1 >> k_{-1}$. Therefore, it is assumed that loading PFBHA on the sorbent is essentially irreversible. The second step to consider is the possibility that an approaching gaseous carbonyl molecule can bind to unoccupied surface sites (Step B); however, in Step B, the rate of carbonyl adsorption is expected to be small, i.e., $k_2 \approx 0$, due to that fact that almost all sorption sites are occupied by PFBHA. The third step to consider is Step C, where the rate of reaction between sorbed PFBHA and gaseous carbonyl is K^*. It is assumed that the PFBHA aromatic moiety provides the majority of binding affinity to the polymer while the hydroxyl amine moiety is free to react with an approaching carbonyl compound (see Figure 1 for the structure of PFBHA). It is also assumed, from experimental findings, that K* is rate limiting as opposed to the rate of gaseous carbonyl diffusion towards to the sorbent. When the PFBHA loaded sorbent is exposed to gaseous carbonyl, a reaction between the two (Step C) is desired instead of the carbonyl binding to the sorbent (Step B). To achieve this, the experimental sampling conditions are set to favor Step C over Step B by allowing short exposure times (10 to 300 seconds) to high concentrations of gaseous carbonyl and longer exposure times (>300 seconds) for low concentrations. The objective is to minimize the amount of PFBHA consumed from reaction while obtaining a sufficient quantity of the oxime for detection. The last step to consider is Step D. This step is important because if the oxime readily desorbed from the sorbent then there would not be a product to detect. But, from experimental data (at room temperature), the oxime formed has a very strong binding affinity for the sorbent, therefore it is assumed that $k_3 \approx 0$. The presence of oxime on the sorbent, Oxime*S, now occupies a sorption site. It then prevents adsorption of incoming gaseous carbonyl and can reduce the apparent overall rate of reaction. The reduction in reaction rate is possible because as the number of available PFBHA molecules decreases, so does the possibility of reaction between sorbed PFBHA and incoming carbonyl molecules. But small consumptions of PFBHA are expected to have a negligible effect on the overall rate of reaction.

The Langmuir-Rideal theory assumes there is a reaction between an adsorbed molecule and a gas phase molecule on a surface (*5*), which is the case for the new sampling system described herein. Since the sorbent is first exposed to large concentrations of PFBHA, the sites are ostensibly saturated with PFBHA prior to exposure to gaseous carbonyl. The sorbent, therefore, acts as the surface and PFBHA is the adsorbed molecule. Then, the PFBHA loaded sorbent is exposed to gaseous carbonyl. An approaching gaseous carbonyl can therefore either react with PFBHA or displace it due the possibility it has a stronger affinity than PFBHA for the sorbent. Therefore, the possibility that gaseous carbonyl can bind to the sorbent instead of reacting with PFBHA is considered when the former may have larger binding affinity constants than PFBHA.

It is assumed that the loading of PFBHA (Step A) is ostensibly irreversible (based on experimental results). The number of available sites for adsorption of the carbonyl compound of interest are negligible, due to the fact that almost all sorption sites are occupied by PFBHA (Step B); and the rate of desorption of the product (Step D) does not occur to any significant degree at ambient temperature, due to the strong affinity of the PFBHA-oxime for the sorption polymer. Equation 1 shows the relationship between the velocity of reaction for the overall steps implementing the aforementioned assumptions.

$$v = \frac{K^* K_A C_{PFBHA*S} C_{HCHO}}{1 + K_A C_{PFBHA*S} + K_B C_{HCHO}} \tag{1}$$

The equation indicates that when the concentration of the derivatization reagent is extremely large the overall rate of reaction becomes strongly dependent on the concentration of the carbonyl compound and *K**. Equation 1 can be re-written into a form (Equation 2) that can be used to establish a linear relation ship between reaction rate (velocity of the reaction) and the concentration of carbonyl compound.

$$\frac{1}{v} = \frac{1}{C_{HCHO}}\left(\frac{1}{K^* K_A C_{PFBHA*S}} + \frac{1}{K^*}\right) + \frac{K_B}{K^* K_A C_{PFBHA*S}} \tag{2}$$

Equation 2 can be simplified (to Equation 3) by assuming $K^* << K^* K_A C_{PFBHA*S}$.

$$\frac{1}{v} = \frac{1}{C_{HCHO}} \frac{1}{K^*} + \frac{K_B}{K^* K_A C_{PFBHA*S}} \tag{3}$$

A plot of 1/v as a function of $1/C_{HCHO}$ would yield a linear relationship where the slope and y-intercept are defined in equations 2 and 3. The slope in equation 2 contains terms for the rate of reaction between HCHO and sorbed PFBHA as well as the K_A and the surface concentration of PFBHA on the fiber; however, the slope in equation 3 is dependent only on the rate of reaction between PFBHA and carbonyl. This condition is easily satisfied when the loss of PFBHA from the sorbent is negligible (from reaction and/or desorption). The inverse of the slope would yield an apparent first order rate constant when the amount of PFBHA consumed during the reaction is negligible, i.e., $K^* = \frac{\text{weight oxime}}{\text{time}} \frac{1}{C_{HCHO}}$. Therefore, if the velocity of product formation is known, the K^* can be used to quantify unknown carbonyl concentrations without the need for fiber calibration provided that PFBHA is minimally consumed. In addition, with increasing sampling temperature, K^* increases, but conversely, k_3 increases and K_A decreases thus decreasing $Oxime*S$ and $C_{PFBHA*S}$, respectively. This would reduce the apparent rate of product formation. Finally, K^* and/or K_B can vary depending on the carbonyl compound. A more detailed description of this work can be found elsewhere (*6*).

Experimental

Chemicals and Materials. The reagents, *o*-(2,3,4,5,6-pentafluorobenzyl)-hydroxylamine hydrochloride and hexane, were purchased from Sigma-Aldrich (Toronto, Canada). Formaldehyde, 37%, was from Fisher Scientific (Nepean, Canada). All solid phase microextraction fibers, holders, vials, capillary columns (30 m, 0.25 mm i.d., 1.0 μm film thickness), and syringes were purchased from Supelco (Sigma-Aldrich, Toronto).

Standard Gas Concentrations of Formaldehyde. A standard gas generator (Model 491-B, Kin-Tek, Texas City, Texas) was used to generate all the standard gas concentrations of formaldehyde. It was equipped with a mass flow-controlled dilution gas system and temperature-controlled holding zone. National Institute of Standards and Technology (NIST) traceable certified HCHO permeation tubes were from Kin-Tek. Prior to the standard gas generator, the dilution gas (N_2) was scrubbed with a molecular sieve followed by a charcoal scrub.

Headspace Loading of SPME Fibers with PFBHA. Solutions of PFBHA.HCl (17 mg/mL) were prepared in 4 mL Teflon capped vials. PFBHA loading onto SPME fibers occurred by 10 min. headspace extractions of the solutions with the solution stirring at 1800 rpm.

Synthesis of PFBHA-HCHO Oxime and Standard Solutions in Hexane. The oxime formed from the reaction between PFBHA and HCHO was synthesized using a literature method (*7*) which was modified (*6*) and is presented in Figure 2.

Instrumentation and Methods for SPME and Liquid Injections of PFBHA-HCHO Oxime. All GC/FID experiments used a Varian 3400 gas chromatograph equipped with a Septum-equipped Programmable Injector (SPI), maintained at 210 °C for all SPME injections. The column temperature program was 45 °C for 1.00 min, 30 °C/min to 200 °C, then 50 °C/min to 290 °C held for 4.0 min. The column head pressure was set to 26 psi hydrogen (1.7 mL/min, 58 cm/sec). All MS experiments to identify oximes were carried out with a HP-5890 Series II GC coupled to a HP-5970 mass selective detector. All other MS experiments were carried out with a Varian 3400 GC coupled to a Saturn IV ion trap mass spectrometer.

Sampling the HCHO Standard Gas. The HCHO standard gas effluent from the standard generator was directed, with Teflon tubing, into a temperature controlled one liter gas sampling bulb equipped with a sampling port from which the SPME device could be exposed to the gas. The bulb was maintained at 25 °C for all experiments, except the temperature study. All sampling times were measured with a NIST traceable timer.

Headspace Sampling of Leaf and Coffee Grounds Samples. A leaf (approximately three cm^2) from a deciduous plant was placed into a 40 mL vial and left at room temperature for 30 min before the headspace was sampled for two min. with PDMS/DVB fibers sorbed with PFBHA, then analyzed. The lid from a one kg container of fresh coffee grounds was opened and the PFBHA loaded PDMS/DVB fiber was inserted into the headspace for 15 seconds, then analyzed.

Results and Discussion

Selection of a Carbonyl Specific Derivatization Reagent Suitable for Loading onto Chosen SPME Fiber Coatings. Numerous criteria were used for selecting a suitable derivatization reagent specific for carbonyl compounds. The product of the reaction between the derivatization reagent and the carbonyl compound was required to be stable and amenable to analysis with conventional chromatographic equipment and detectors, such as GC/FID. Other criteria included the need for a derivatization reagent which would form a thermally stable product. This would allow for ambient temperature storage conditions so that samples could be field acquired and then analyzed in the laboratory, or if sampling occurred at above ambient temperatures. Further requirements were that the derivatization reagent should be preferably

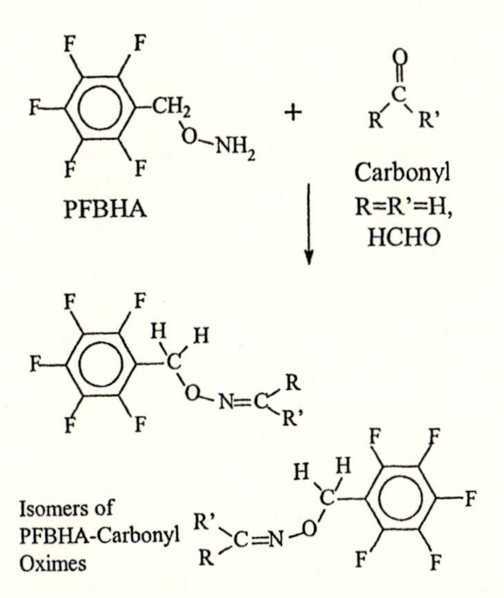

Figure 2: The reaction products formed from the reaction between PFBHA and carbonyl compounds. When R=R'=H, the carbonyl is formaldehyde (HCHO).

non-toxic, insensitive to light, heat and oxygen. The commercially available derivatization reagent PFBHA was found to satisfy all of the aforementioned criteria.

A number of different SPME fiber coatings were examined to establish one, or more, that would provide the highest loading and stability of PFBHA and oxime retention. Of the fibers tested, only PDMS/DVB (65 µm) was selected as the best fiber for this application because of its high mass loading of PFBHA, low release of sorbed PFBHA (at room temperature), reproducibility, and general rugged nature.

Exposure of PFBHA Loaded PDMS/DVB Fibers to HCHO. Figure 3 shows the typical GC/MS data obtained following exposure of PFBHA loaded PDMS/DVB fibers to approximately 650 ppbv HCHO for 10 minutes. Figure 3a shows the PDMS/DVB blank, Figure 3b shows the chromatogram from the PDMS/DVB fiber sorbed with PFBHA, and Figure 3c shows the chromatogram with the PFBHA-HCHO oxime. Table I compares the mass spectrum of this oxime to that reported in the literature (*7*). Therefore, the identity and retention time of the PFBHA-HCHO oxime was confirmed.

Table I: Mass spectral information of PFBHA-HCHO oxime. Literature data (*7*) for PFBHA-HCHO oxime compared to those obtained from the synthesized oxime. The m/e of other ions and their percent of base peak (m/e 181) are also presented.

	M/e	181	195	161	117	182	99	167	93
Literature	%	100	11	10	9	7	7	5	5
Observed	%	100	12	10	9.9	6.7	9.6	5.4	6.3

SPME Fiber Calibration Curve Data for HCHO with GC/FID. Plots of the amount of HCHO-oxime formed during sampling times of 10 and 300 seconds for various concentrations of HCHO are given in Figure 4. The figure shows two curves for the dependence of the amount of product formed on the gaseous concentration of HCHO at 25 °C. The data show that a larger mass of product is formed during longer sampling times than during shorter sampling times, i.e., 300 vs. 10 sec. The slopes of the two curves indicate that sampling for 300 sec results in approximately sixteen times greater sensitivity compared to sampling for 10 sec, which is consistent with the concept that increased exposure time will result in an increased amount of product formed on the fiber. Therefore, depending on the required analytical sensitivity the sampling times can range from 10 to more than 300 sec, which is a function of detector response as well as the amount of product formed. Longer sampling times will result in a larger amount of product on the fiber, but this will be limited by the amount of derivatization reagent available on the fiber. With detectors that are more sensitive than FID, such as a mass spectrometer, shorter sampling times may be used without compromising sensitivity.

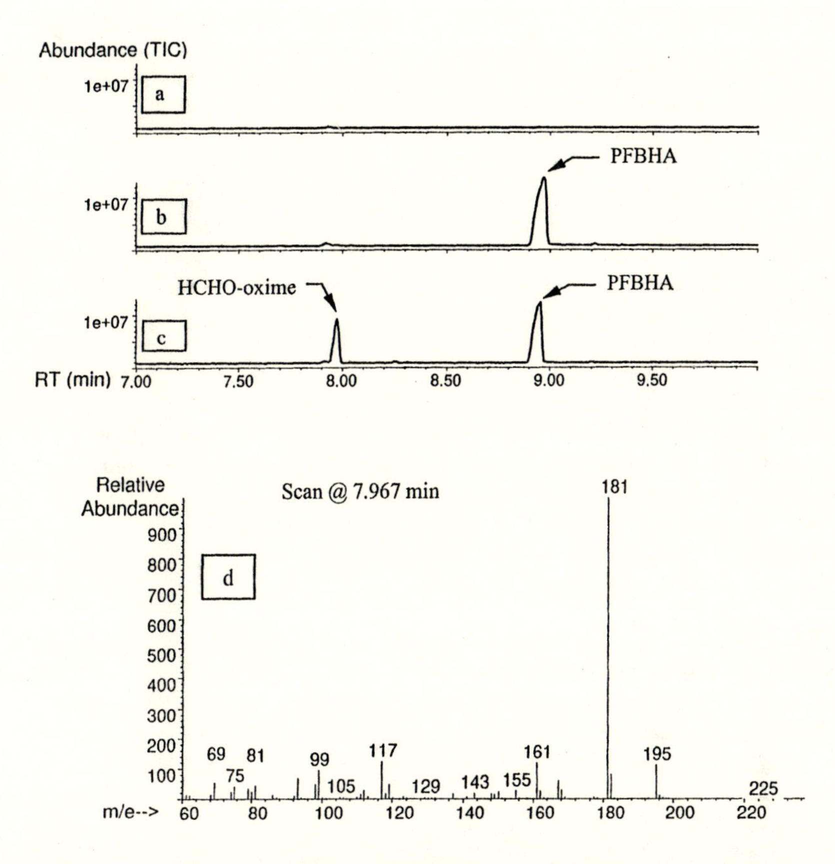

Figure 3: GC/MSD of PFBHA loaded PDMS/DVB fibers (65 μm) with and without exposure to 650 ppbv HCHO. (a). Fiber blank. (b). PFBHA loaded on PDMS/DVB. (c). PFBHA loaded PDMS/DVB fibers followed by exposure to 650 ppbv for 10 minutes. (d). Mass spectrum of the peak at 8 minutes.

Temperature Dependence. Figure 5 shows the dependence of the amount of PFBHA-HCHO oxime formed as a function of the inverse of temperature for a given concentration of HCHO. The plot shows that increasing temperature results in a corresponding decrease in the amount of HCHO oxime formed during a fixed sampling time. It should be noted a temperature variation of ± 5 °C from 25 °C does not significantly affect the amount of HCHO oxime formed. This potentially reduces the requirement to correct for sampling temperature; however, the option to correct for sampling temperature is available if required.

Dynamic vs. Static Sampling. The question as to whether there was a difference between sampling a constant source of standard HCHO (from the standard gas generator) and sampling the standard gas concentration under static conditions gas was addressed. First, one concentration of HCHO standard gas was dynamically generated and sampled with PFBHA sorbed onto PDMS/DVB fibers. Second, the same HCHO standard gas concentration was sampled under static conditions with the same PDMS/DVB fibers sorbed with fresh PFBHA. The data indicated there was no difference between the two sampling methods, i.e., the amount of PFBHA-HCHO oxime formed for a given sampling time were identical for each sampling type. This is presumably due to the fact that K^* is rate limiting. This finding will assist in exploiting the sampling system for use as a real-time or static gas sampling device.

Headspace Sampling of Leaf and Coffee Grounds Samples. This sampling device is not limited to its use as a rapid and sensitive sampler for formaldehyde; it can, also, be used to sample a variety of carbonyl compounds present in complex samples as exemplified in Figures 6 and 7. Figure 6 presents the GC/FID chromatogram obtained from the sampling of the headspace from a small section of leaf. The chromatogram shows the derivatization reagent blank run and, immediately behind it, the chromatogram from the PFBHA derivatization of the headspace carbonyl compounds. The presence of these oxime peaks indicates the leaf headspace had a number of carbonyl compounds. Therefore, it is possible to sample airborne carbonyl compounds, such as those of a plant, in the laboratory or while directly at the site of the plant.

Figure 7 shows GC/MS chromatograms obtained from headspace sampling of coffee grounds. Figure 7a shows the chromatogram of the headspace volatiles sampled with a PDMS/DVB fiber not loaded with PFBHA. Many of the compounds in Figure 7a were identified, according to the NIST library, as carbonyl compounds. Figure 7b shows a chromatogram of the headspace volatiles sampled with a PDMS/DVB fiber sorbed with PFBHA; the large number of peaks to the right of PFBHA represent reaction products between PFBHA and headspace carbonyl compounds. Comparison of Figures 7a and 7b indicates carbonyl compounds can be selectively sampled using this new sampling system.

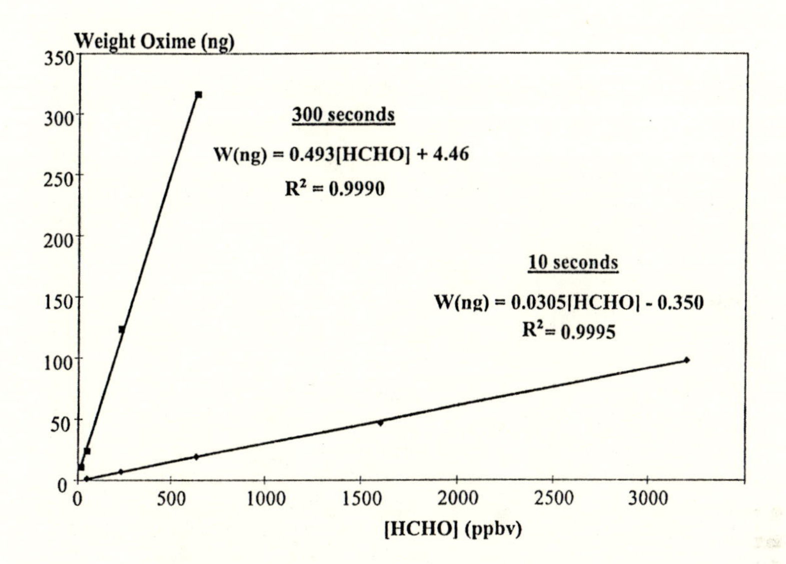

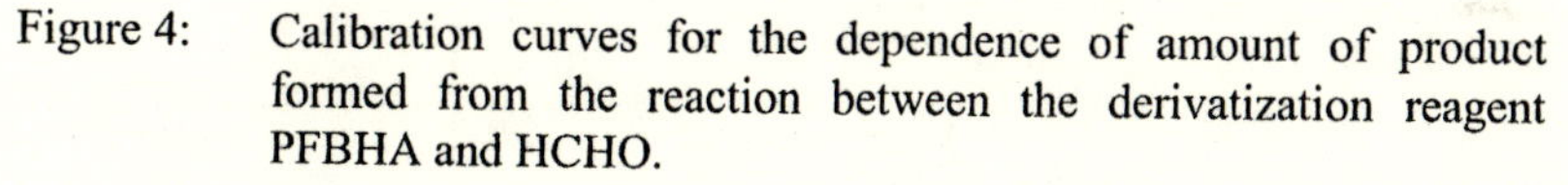

Figure 4: Calibration curves for the dependence of amount of product formed from the reaction between the derivatization reagent PFBHA and HCHO.

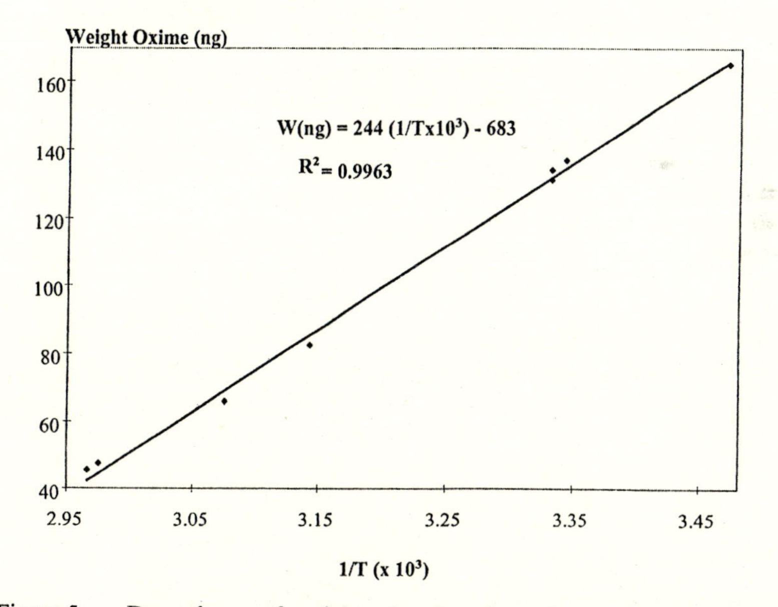

Figure 5: Dependence of weight of oxime formed as a function of the inverse of temperature (Kelvin). The lowest temperature studied was 15 °C and the highest was approximately 60 °C.

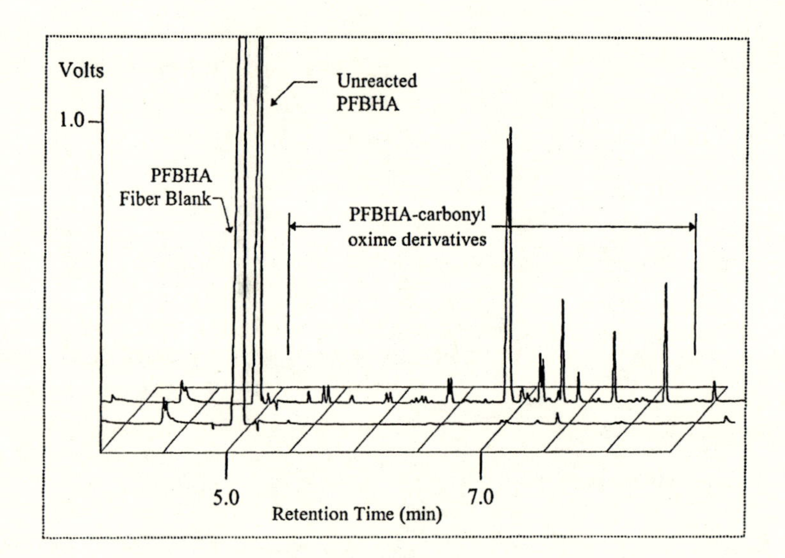

Figure 6: GC/FID chromatograms following the exposure of PFBHA loaded PDMS/DVB fibers (65 μm) to the headspace of approximately 3 cm^2 deciduous leaf.

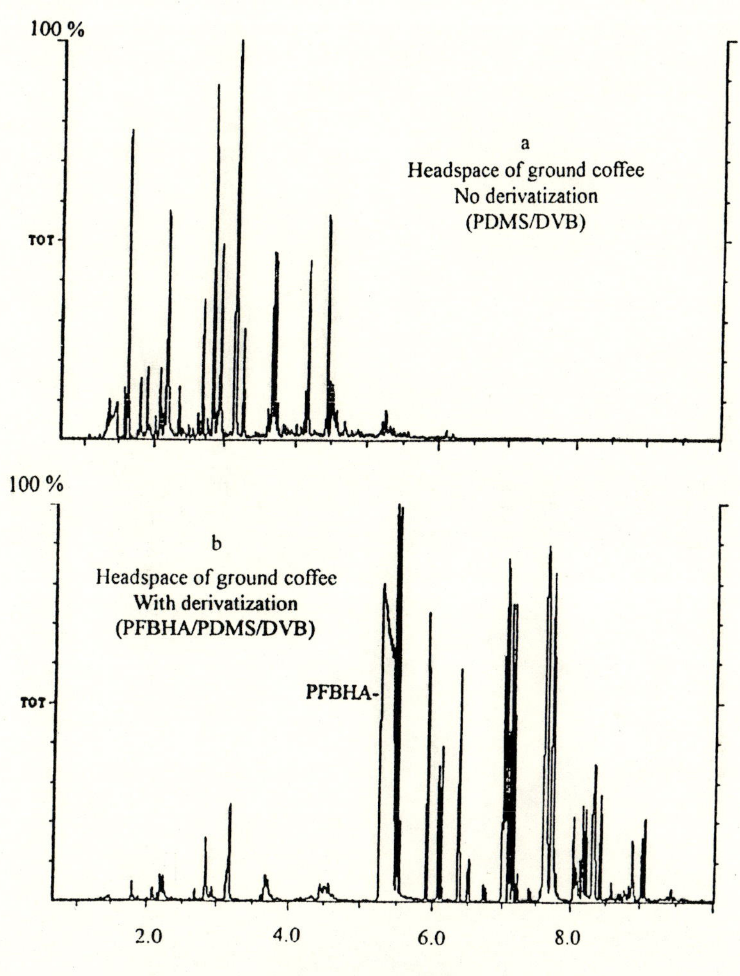

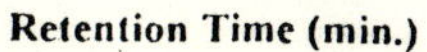

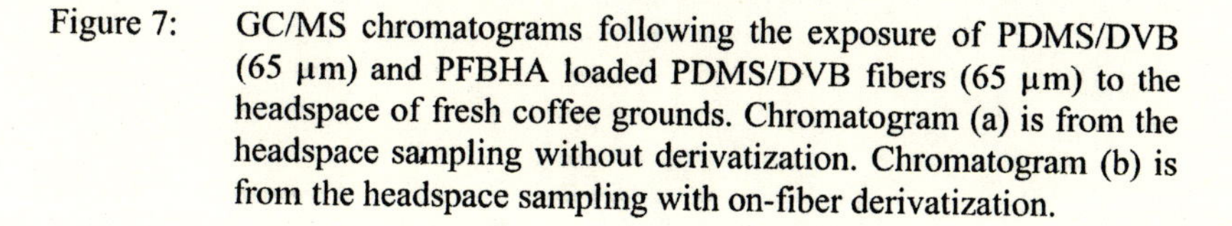

Figure 7: GC/MS chromatograms following the exposure of PDMS/DVB (65 μm) and PFBHA loaded PDMS/DVB fibers (65 μm) to the headspace of fresh coffee grounds. Chromatogram (a) is from the headspace sampling without derivatization. Chromatogram (b) is from the headspace sampling with on-fiber derivatization.

Conclusions

The use of SPME with on-fiber derivatization to sample gaseous carbonyl compounds is extremely promising for measurement of reactive carbonyl compounds. Also, it can be used for the sampling of carbonyl compounds from remote sites, such as field sites where carbonyl compounds are transient or easily degraded, unless immediately sampled and stabilized by derivatization. In addition,
sampling low to high concentrations of carbonyl containing volatiles from fragrances and flavors can be made significantly more convenient with a sampling device that is specific for the analytes of interest.

Acknowledgments

Financial support for this work was provided by NSERC, Supelco, and Varian.

Literature Cited

1. Pawliszyn, J.; *Solid Phase Microextraction-Theory and Practice*, Wiley-VCH, **1997**
2. Yang, X.; Peppard, T. *J. Agric. Food Chem.* **1994**, *42*, pp. 1925
3. Pelusio, F.; Nilsson, T.; Montanarella, L.; Tilio, R.; Larsen, B.; Facchetti, S.; Madsen, J. *J. Agric. Food Chem.* **1995** *43*, pp. 2138
4. Steffen, A.; Pawliszyn, J.; *J. Agric. Food Chem.* **1996** *68*, pp. 3008
5. Laidler, J.L.; *Chemical Kinetics*, McGraw-Hill, Inc., **1965**, pp. 256-320
6. Martos, P.A.; Air Sampling With Solid Phase Microextraction, Dissertation, University of Waterloo, Waterloo, Ontario, Canada, **1997**, pp. 92-128
7. Cancilla, D.A.; Chou, C.; Barthel, R.; Que Hee, S.S., *J. of AOAC International*, **1992** *75*, pp. 842

Chapter 10

Comparison of Volatile Analysis of Lipid-Containing and Meat Matrices by Solid Phase Micro- and Supercritical Fluid-Extraction

Janet M. Snyder, Jerry W. King, and Zhouyao Zhang

Food Quality and Safety Research Unit, National Center for Agricultural Utilization Research, Agriculture Research Service, U.S. Department of Agriculture, Peoria, IL 61604

Contamination and degradation of lipid moieties result in the formation of volatile compounds that affect the flavor and safety of food products. A wide variety of analytical techniques have been developed to determine the concentration of volatile flavor components in foods, such as vacuum distillation, headspace analysis, supercritical fluid extraction, and solid phase microextraction. Previously in our laboratory, volatile compounds from oxidized vegetable oil and fire/smoke damaged meat samples were analyzed by dynamic headspace analysis and supercritical fluid extraction (SFE). In this study, solid phase micro extraction (SPME) methods were also investigated to determine the concentration and identification of compounds from these samples. In applying SPME, different fiber types and analysis conditions were evaluated.

Solid phase microextraction (SPME) has recently been successfully utilized for analyzing many food substances and flavors (*1-4*) on a qualitative basis; however, quantitative studies are still limited. Recently Bartelt described quantitation of solutes by SPME and the difficulties that occur when doing quantitative determinations of headspace volatile for different classes of compounds (*5*). He found that the available fibers are not consistently responsive to all compounds, and equilibrium between the headspace and matrix for several compounds could not be attained at the conditions reported.

Various analytical methods for volatile components from lipids have been reported (*6-10*). Each of these methods have complexities, such as thermal degradation and/or instability of the components formed, that should be considered in developing the analysis method. In previous studies, we have shown that supercritical fluid extraction (SFE) of the volatile compounds has provided a means to quantitatively determine the concentration of lipid oxidation products (*8*). In a similar study, we have utilized SFE to determine aromatic hydrocarbons and polycyclic aromatic hydrocarbons (PAH) as marker compounds formed from the exposure of meats to smoke or fire conditions (*10*). However, the application of SPME to volatile compounds formed from lipid-containing samples has not been fully investigated (*11*). In this study, we have analyzed both a series of oxidized oils and meat samples by a SPME method followed by gas chromatoghraphy/mass spectrometry (GC/MS), and we have determined the effectiveness of this method compared to SFE and traditional purge and trap analysis.

Experimental methods

Samples. Canola oil, corn oil, soybean oil and sunflower oil were stored at 60°C in a forced draft oven until the peroxide values for canola, corn, and soybean were approximately 54, a value consistent with appreciable accelerated oxidation of the seed oils. The final peroxide value for sunflower oil under similar conditions was found to be 80 (*8*). Samples of each oil were also removed when the peroxide values were 2 and 18. Meat samples were obtained from the Food Safety and Inspection Service Laboratory in St. Louis, MO and were kept in a freezer at -45 °C. The samples included a smoked chicken product, ham, and corned beef, which were suspected of being exposed to a fire in an underground storage cavern (*10*). 50 Gram portions of the meat were removed from the original 300 g samples to obtain representative samples; the meat was ground and immediately frozen at -45 °C until analysis.

Standard solutions. Solutions of standard compounds including hydrocarbons and aldehydes, were prepared in concentrations from 1 ppb to 400 ppm to develop response curves for the major volatile compounds resulting from the oxidized oils. Also, solutions of 1 ppb to 400 ppb were prepared for aromatic hydrocarbons to determine the concentration of expected contaminants in the fire-exposed meats. All calibration solutions were formulated in a highly-stable hydrogenated soybean oil with a low volatile profile. The R^2 values of the calibration curves were 0.99 indicating a high degree of linearity (*12*). Dodecane at 1 ppm was added to each sample before analysis as an internal standard.

SPME analysis. In this study, three coated SPME fibers were evaluated: a 100 μm polydimethylsiloxane (PDMS), a 7 μm polydimethylsiloxane, and a 85 μm polyacrylate fiber (Supelco, Inc.; Bellefonte, PA). One-half gram samples, with 1ppm dodecane added as the internal standard, were placed into clear 10 mL vials from Supelco having teflon/silicone septa. Extraction conditions were varied to determine the optimal experimental parameters. Solutions of pentane, hexanal, nonanal, naphthalene and dodecane were preheated to 60 °C from 5 to 30 min using 5 min increments to provide different headspace concentrations. The preheating times were then plotted against peak area from the mass spectral data to assess the time required to reach equilibrium in the vial headspace. For example, as shown for the data for nonanal and dodecane plotted in Figure 1, the preheating time necessary for thermal equilibration was approximately 20 min. Similarly, pentane and hexanal reached equilibrium within 5 min, while naphthalene reached thermal equilibrium in 20 min. The time the SPME fiber was exposed to the headspace of each standard varied from 5 min to 45 min to establish the best extraction time. Care was taken to determine that the equilibration time was sufficient for all analytes studied (*12*). The data from naphthalene and dodecane was also plotted against area from mass spectral data (Figure 2), and 30 min was determined to be the optimal extraction time. The optimum extraction time for the aldehydes used in this study was 20 min; however, a 30 min time was used for all samples.

Gas chromatography/mass spectrometry. SPME injections into the GC/MS system were made using a Varian 8200 Autosampler (Walnut Creek, CA). After the 30 min extraction time, the volatile compounds were desorbed for 1 min into the injector of a Varian Model 3600 GC equipped with a DB-5 capillary column (30 m, 0.25mm i.d., 0.25 μ film thickness) (J&W Scientific, Folsom, CA). The temperature of the column was maintained at 40 °C for 1 min during desorption then ramped at 5 °C/min to 220 °C. The injections were splitless with the injector temperature being held at 220 °C. The GC was interfaced with a Varian Saturn 4D Ion Trap MS/MS (Walnut Creek, CA) for detection and quantitation of the solutes. Mass spectral data were compiled using the electron impact mode.

Results and Discussion

The 100 μm PDMS coated fiber has been previously demonstrated to be a suitable fiber for detecting volatile compounds (*5,13*). The 100 μm PDMS, the 85 μm polyacrylate and the 7 μm PDMS fibers were all used on a mixture of nine target compounds the first seven at a concentration of 10 ppm and dodecane at 1 ppm and hexadecane at 0.5 ppm concentrations (Figure 3). It is apparent from inspecting Figure 3 that the best overall response to the standard mixture is provided by using the 100μm PDMS fiber.

Bartelt has determined the calibration factors (K) for 71 analytes and determined that K was considerably greater for the higher molecular weight hydrocarbons than for aldehydes (*5*). Consequently, the area data from the mass spectral data for dodecane (1ppm) and hexadecane (0.5 ppm) are in much greater proportion to the other compounds measured at 10 ppm concentration levels, except for the compound 2-pentylfuran. The areas for all compounds, except for pentanal, were largest using the 100 μm PDMS fibers; the area for pentanal, a traditional indicator of oil oxidation, was highest with the polyacrylate fiber. The areas of the compounds with the 7 μm PDMS fiber were the smallest except for dodecane and hexadecane which tend to absorb preferentially on the non-polar PDMS fiber. The polyacrylate fiber tends to absorb the more polar analytes (*13*) and was found not to be as effective for the samples that we were studying (Figure 3).

The concentrations of several volatile components from four oxidized vegetable oils were measured using the 100 μm PDMS-coated fiber (Table 1). The concentration of volatiles increased for all compounds as the peroxide values increased during storage. Sunflower oil with 70% linoleic acid oxidizes the most rapidly; and the concentrations of hexanal and decadienal, oxidation products formed from linoleic acid, are greatest for sunflower oil. Nonanal was most prominent during the accelerated storage of canola oil. This is due to the fact that canola oil contains more than 60% oleic acid, the precursor for nonanal formation.

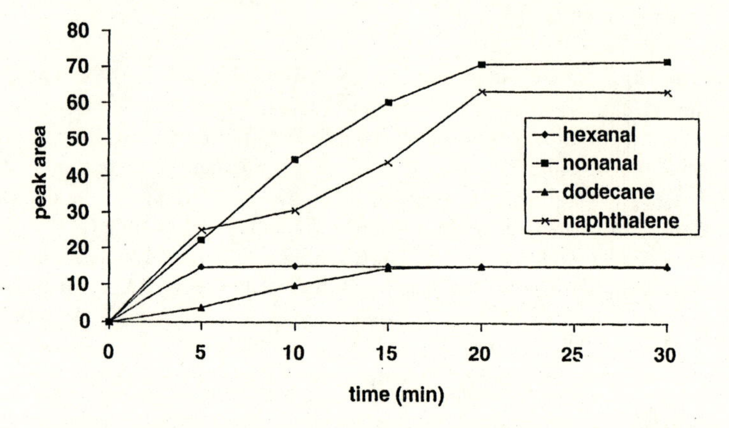

Figure 1. Effect of sample preheat time on equilibration of volatiles.

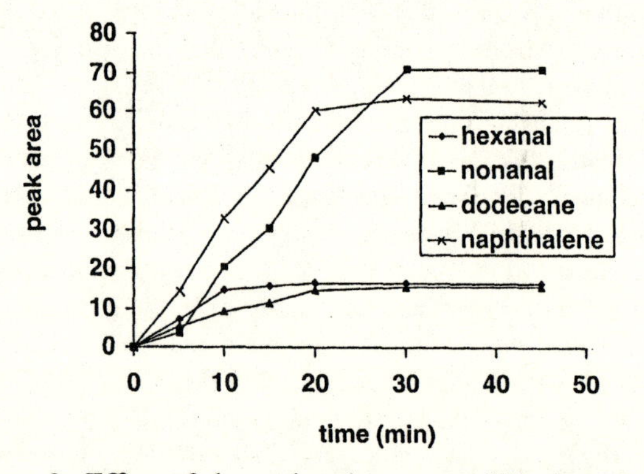

Figure 2. Effect of absorption time on equilibration of solute.

Table I. Volatile Concentration (ppm) in Oxidized Vegetable Oils by SPME Analysis

	Corn oil			Canola oil		
	PV=2	PV=19	PV=52	PV=2	PV=19	PV=58
pentane	0.09	0.44	0.96	0.35	0.65	0.67
pentanal	0.08	0.37	0.70	0.81	1.48	4.54
2-pentenal	0.20	0.30	0.54	0.31	0.27	0.28
hexanal	0.27	2.10	45.17	0.43	7.39	56.74
2-heptenal	0.19	0.48	1.28	0.27	0.37	0.47
2-pentylfuran	0.35	0.31	0.16	0.81	0.34	0.22
octanal	0.36	0.47	8.74	0.79	0.44	0.29
nonanal	1.20	7.13	12.65	0.10	0.91	2.17
2,4-decadienal	0.26	1.03	16.83	0.63	0.71	18.59

	Soybean oil			Sunflower oil			
	PV=2	PV=18	PV=54	PV=3	PV=18	PV=55	PV=81
pentane	0.06	0.44	1.03	0.23	0.34	3.23	5.36
pentanal	0.05	0.98	1.36	0.65	0.82	1.35	1.65
2-pentenal	0.08	0.16	0.37	0.11	0.21	0.32	0.48
hexanal	1.58	7.07	80.59	0.38	8.28	89.41	296.39
2-heptenal	0.96	1.49	15.50	0.14	0.27	2.17	5.49
2-pentylfuran	0.13	0.50	0.14	0.27	1.37	0.46	1.27
octanal	0.06	0.33	0.46	0.24	0.41	0.46	0.97
nonanal	0.73	1.28	2.48	0.09	0.87	1.16	2.01
2,4-decadienal	0.16	5.98	28.17	0.13	8.53	36.90	49.69

PV = peroxide value (meq/kg) a measure of oxidation in oils (8)

Values obtained for the compounds from oils that were highly oxidized and analyzed by an SFE method are shown in Table 2 (*8*). The data from SPME analysis tends to follow some of the same trends inherent in the oxidized oil data determined by SFE (*8*). The concentration of hexanal from the sunflower oil with a peroxide value of 80 was much lower when determined by SPME (296.39 ppm, Table I) than when determined by SFE (365.92ppm, Table II). However, 2-pentyl furan in all oils is up-to-10 fold higher in concentration as determined by SPME analysis relative to the SFE data.

Table II. Volatile Concentration of Oxidized Vegetable Oils by SFE Analysis (8)

	Canola oil	Corn oil	Soybean oil	Sunflower oil
	PV = 53	PV = 53	PV = 60	PV = 82
pentane	0.67	0.22	0.31	0.88
pentanal	2.12	0.90	0.89	1.23
hexanal	52.63	69.93	81.36	365.92
2-heptenal	1.32	2.89	6.97	10.90
2-pentyfuran	0.09	0.08	0.03	0.10
octanal	20.36	1.42	1.32	1.87
nonanal	26.98	0.94	4.51	5.42
2,4-decadienal	16.98	22.80	27.03	30.54

PV = peroxide value (meq/kg) a measure of oxidation in oils

Figure 4 compares the SPME results with those from SFE for the major vegetable oil volatiles produced from the four highly-oxidized oils used in this study. Overall these results show that the pattern of oxidation products formed and detected are very consistent using either SPME and SFE for extraction. Although there are subtle differences between the results from the two techniques, it appears in most cases that the two techniques agree within an order of magnitude for the major volatiles detected. These results indicate, that either SPME or SFE can be used with confidence to monitor the degradation products produced upon aging the oil matrices.

Zhang and Pawliszyn (*14*) demonstrated the SPME technique is highly sensitive to polycyclic aromatic hydrocarbons (PAH) found in environmental samples. Previously we have examined meat samples that were exposed to fire or smoke by an SFE method and found ppb levels of these compounds (*10*). Therefore, SPME was applied to analyze specifically for aromatic compounds in both fire-exposed and control meat samples.

Three of the previously analyzed meat products and their analysis by SPME are listed in Table III. The SPME method proved effective in detecting the aromatic hydrocarbons found previously via SFE in the three meat matrices. The values for both the aromatic hydrocarbons and naphthalene by SPME tend to follow the same trend as that reported by SFE (Table IV) (*10*), with the exception of the values of naphthalene for corned beef. The SPME technique was able to measure lower concentrations than found by the SFE method, especially in the control samples. Also, methylnaphthalene, previously reported by purge and trap headspace analysis to be present in the fire exposed samples (*15*), was not found by the SFE method (*10*). However, using SPME, 1-methylnaphthalene was identified and its presence determined as low as 1 ppb in the corned beef sample.

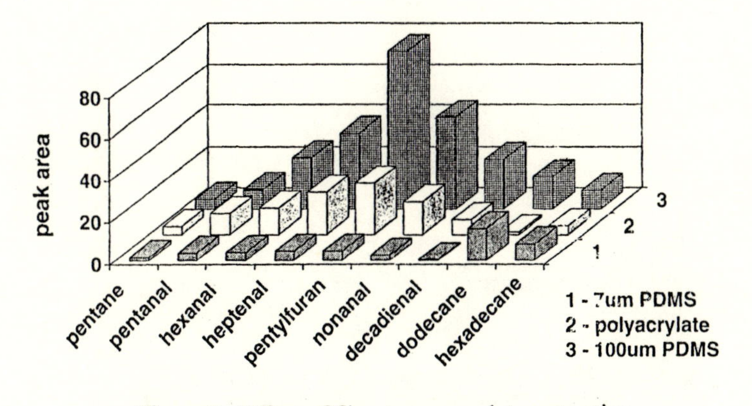

Figure 3. Effect of fiber type on solute extraction.

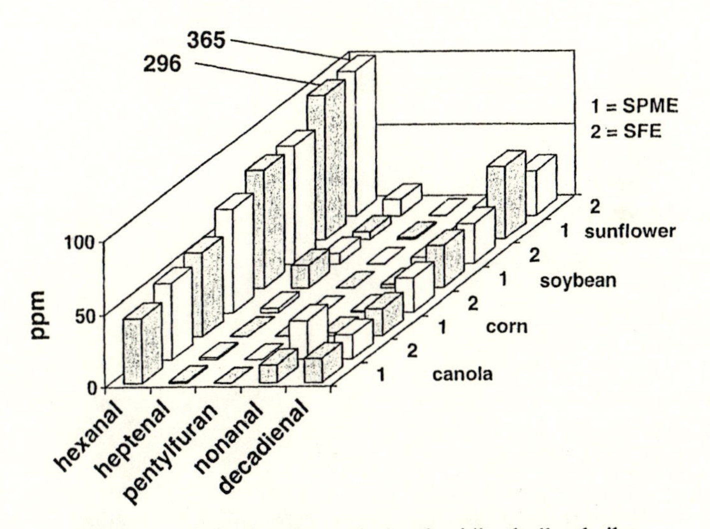

Figure 4. SPME vs SFE analysis of oxidized oil volatiles.

Table III. Aromatic Hydrocarbons (ppb) in Meats Exposed to Fire/Smoke by SPME

	Smoked Chicken		Ham		Corned Beef	
	control	fire*	control	fire*	control	fire*
benzene	7	51	8	44	1	3
toluene	30	98	1	46	38	49
xylene	4	100	4	13	3	7
ethylbenzene	5	14	4	16	4	12
butylbenzene	15	41	22	33	2	5
naphthalene	14	50	3	13	3	2
1-methylnaphthalene	12	23	20	11	2	1

* fire = samples exposed to fire/smoke.

Table IV. Aromatic Hydrocarbons (ppb) in Meats Exposed to Fire/Smoke by SFE

	Smoked Chicken		Ham		Corned Beef	
	control	fire*	control	fire*	control	fire*
benzene	2	43	8	37	2	4
toluene	29	329	1	80	22	52
xylene	26	164	6	15	1	19
ethylbenzene	11	250	2	41	4	31
naphthalene	10	39	5	21	0	7
1-methylnaphthalene	N.D.[1]					

* fire = samples exposed to fire/smoke.
[1] N.D. = not detected

In summary, SPME using the 100 μm PDMS fiber has been shown to efficiently extract and measure volatile compounds in lipid-containing matrices at levels equivalent to those found by our previously-described SFE method. The SPME technique, as with the SFE method, uses moderate extraction temperatures that do not degrade lipid moieties or produce artifacts due to the analytical technique. Therefore, this method can be used in place of the traditional purge and trap method that uses higher temperatures or longer collection times for the extraction and determination of volatile compounds in lipids. In addition, both the SPME and SFE techniques are environmentally benign, use minimal quantities of solvent, and can complement one another for true analysis of analytes (*16*).

However, sample preparation time is shorter with SPME thus SPME is a simpler process than SFE; therefore more samples can be analyzed by SPME than with SFE.

Acknowledgement The authors thank Robert Bartelt (USDA/ARS/NCAUR) for his assistance with this study.

Literature Cited

1. Yang, X.; Peppard, T. *J. Agric. Food Chem.* **1994**, *42*, 1925-1930.
2. Steffan, A.; Pawliszyn, J. *J. Agric. Food Chem.* **1996**, *44*, 2187-2193.
3. Field, J. A.; Nickerson, G.; James, D.D.; Heider, C.J. *J. Agric. Food Chem.* **1996**, *44*, 1768-1772.
4. Chin, H.W.; Bernhard, R.A.; Rosenberg, M. *J. Food Sci.* **1996**, *61*, 1118-1122.
5. Bartelt, R. J., *Anal. Chem.* **1997**, *69*, 364-372.
6. Warner, K.; Frankel, E.N.; Mounts, T.L. *J. Am. Oil Chem.* Soc. **1989**, *66*, 558-564.
7. Ulberth, F., Roubicek, D. *Food Chem.* **1993**, *46*, 137-141.
8. Snyder, J.M. *J. Food Lipids* **1995**, *2* ,25-33.
9. Pinnel, V.; Vandegans, J. *J. High Resolut. Chromatogr.* **1996**, *19*, 263-266.
10. Snyder, J.M.; King, J.W.; Nam, K. *J. Sci. Food Agric.* **1996**, *72*, 25-30.
11. Shirley, R.E; Woolley, C.L.; Mindrup, R.F. *Food Test. and Anal.* **1995**, *1*, 29-42.
12. Pawliszyn, J. *Solid Phase Microextraction:Theory and Practice* **1997** (Wiley-VCH, Inc., USA).
13. Yang, X.; Peppard, T. *LC-GC* **1995**, *13*, 882-886.
14. Zhang, Z.; Pawliszyn, J. *J. High Resolut. Chromatogr.* **1993**, *16*, 689-692.
15. Johnston, J.J.; Feldman, S.E.; Ilnicki, L.P. *J. Agric. Food Chem.* **1994**, *42*, 1954-1958.
16. Hageman, K.J.; Mazeas, L.; Grabanski, C.B.; Miller, D.J.; Hawthorne, S.B. *Anal. Chem.* **1996**, *68*, 3892-3898.

Chapter 11

Thermo Desorption as Sample Preparation Technique for Food and Flavor Analysis by Gas Chromatography

M. Rothaupt

Givaudan Roure Research Ltd., Ueberlandstrasse 138, CH-8600 Duebendorf, Switzerland

Sample preparation is one of the important steps in flavor and food analysis. Thermo extraction of volatile substances simplifies sample preparation by separating volatile compounds from non-volatile compounds. A temperature vaporising program allows an additional pre-fractionation of the analytes. Solvent venting supports the chromatography by removing a large amount of solvent. Solvent venting also helps eliminate interfering and problematic ingredients from the sample. Sample preparation by supercritical fluid extraction (SFE) will be compared to thermo extraction; some specific advantages and disadvantages of both techniques will be discussed.

Existing extraction methods like SFE, solid phase micro extraction (SPME), classical liquid/liquid extraction, and accelerated solvent extraction (ASE) give a variety of different sample preparation choices. Sample preparation via thermo extraction is discussed in this paper. This technique has emerged as a powerful technique for the analysis of food and flavors. It offers a convenient and selective alternative for the preparation of samples for gas chromatographic analysis.

Thermo desorption (TD) is the extraction of compounds by their vapor pressure from a non-volatile, complex matrix. In a thermo desorption experiment, the sample is placed into a specifically designed glass tube. This tube is then inserted into the thermo desorption system (TDS), where the temperatures can be varied from -150 °C to +350 °C by the thermostatic controls. Carrier gas flowing over the sample delivers the volatiles into the GC injection liner. The liner can be maintained at the same temperature as the TDS, or it can be filled with adsorbent material and cooled in order to trap the analytes quantitatively. This additional focusing mechanism is called a cooled injection system (CIS). The key components of a typical TDS are shown in Figure 1.

Instrument Configuration

Temperature and carrier gas flow are the most important parameters for thermo desorption. Higher temperatures can be used even when thermally sensitive substances are present. Temperature can be set so that thermally sensitive compounds are desorbed at low temperatures; the desorption temperature can then be increased to facilitate desorption of molecules with low vapor pressure. Flow can be set to such high rates that even compounds with poor volatility, i.e., those with very low vapor pressure, can be transferred to the injector. Desorption of polar compounds with high boiling points, e.g. glycerol (b.p. = 290 °C), is possible with high carrier gas flow.

Addition of a solvent venting (SV) feature can reduce chromatographic interference from solvents while the analytes are retained in the CIS. Solvent venting is useful for liquid or complex matrices where solvents can interfere with the gas chromatographic separation of the compounds of interest (see Figure 2).

The split valve, of the CIS, is open in solvent venting mode. Thermally extracted volatiles enter the liner, and the analytes are trapped in the CIS, typically at a very low temperature, while the solvent is vented through the split outlet. Venting of the solvent is dependant on temperature and carrier gas flow. Flow rates that are too high will result in loss of analytes. In case the trapping temperature cannot be set low enough, an adsorbent (e.g. Tenax, Porapak) may be placed into the liner to trap the analytes. This action allows analytes to be trapped at a higher (e.g. ambient) temperature *(4, 5)*.

A further enhancement to solvent venting is solvent venting with stop flow. In stop flow mode the carrier gas flow to the analytical column is suppressed by an open back pressure regulator.

The difference between analyses with and without solvent venting can be significant. The presence of water in a sample can be reduced to a level that the interference to the chromatographic separation is minimised. The CIS temperature has to be kept at temperatures above 0 °C when water is present, because at temperatures below 0 °C the liner will be blocked by ice.

The range of applications for TDS can be enlarged to include samples with high water content by using an off-line thermo - extraction instrument with a pre-extraction step. This instrument offers the possibility to vent out all problematic solvents in an off-line venting procedure. The sample is placed into a larger tube, which can also be used for larger sample volumes *(6)*. The extraction takes place by heating up the thermo - extraction tube. The extracted volatiles are then transferred to a TDS tube, which is filled with an adsorbent material. The trapped analytes can then be analysed by the thermo - desorption system as described above.

After the desorption process is completed, the "injection" is carried out by heating the liner to transfer the sample to the analytical column. Mild desorption conditions can be achieved by heating in a suitable temperature gradient. The desorption is also very mild when other compounds are present that can form azeotrops with the analytes. If desorption temperature gradients are too mild, peak broadening may occur and has to be taken into consideration when choosing the cryotrap parameter.

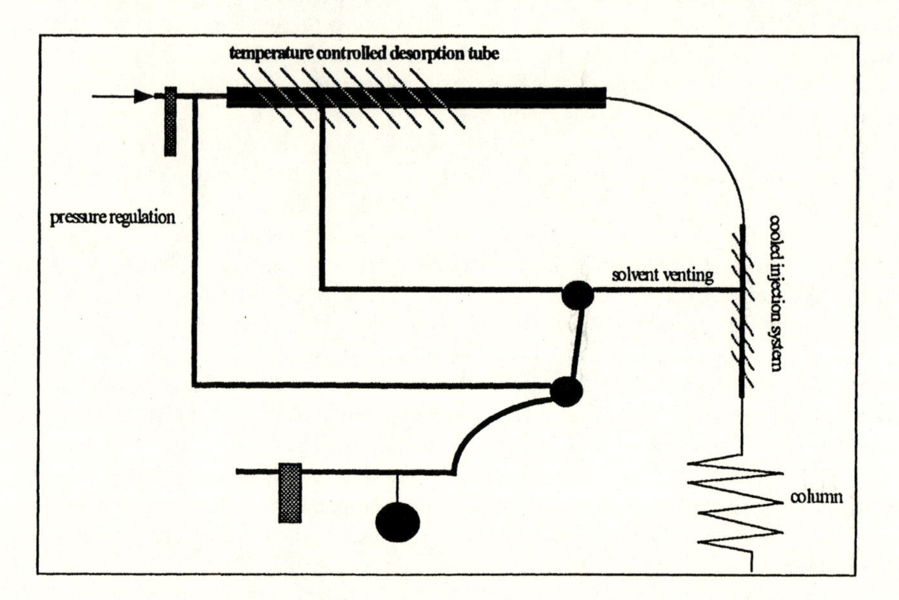

FIGURE 1:. Configuration of a Thermo Desorption Unit

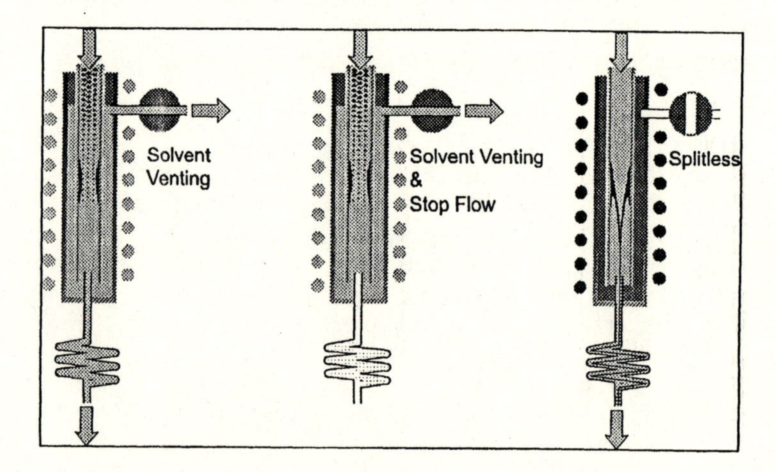

Figure 2. Diagram of a Cooled Injection System (CIS)

Thermal Extraction Applications

The analysis of fragrance in an air freshener stick, which contained a wet soapy matrix, was analysed without any sample preparation except thermo desorption (Figure 3). This analysis would require extensive sample clean up if carried out by classical extraction. Sample preparation consisted of weighing the sample and putting it into the TDS tube. Weighing was done quickly, because recovery rates are influenced by the time of exposure to air. This parameter was found to have a stronger influence on the loss of highly volatile compounds than break through from the CIS.

The fragrance of a shampoo was analysed and results were compared to those from the TD of the shampoo, which used the fragrance. It was observed that most of the fragrance compounds were extracted quantitatively from the shampoo, but a few compounds were hardly recovered. This may have several reasons. Venting time was too long, trap temperature was too high, or incompatibility of the adsorbent material with these molecules. Incompatibility here means that the compounds are not retained on the adsorbent in the trap and are lost through the solvent split, or they are not released from the adsorbent because a strong binding interaction takes place between the compounds and the trapping material.

Isolation of autoxidation products in moistening creams, which contain a considerable amount of water, can be carried using the TDS. In this case, the sample is placed into the TDS tube and thermally desorbed using solvent venting to reduce interference from compounds like water. The desorption temperature (80 °C) was low enough not to desorb the fat, and only the analytes were transferred to the injection system. The low content of these analytes could be overcome by using a "large volume" of oil in order to get enough analytes to be determined.

The analysis of fresh ginger roots by TDS is an example of the moderate conditions that are applied during sample extraction by thermo desorption (Figure 4). Conventional sample preparation for the analysis of essential oils is carried out by steam distillation. This distillation exposes the analytes to high temperatures for quite a long time in the distillation vessel. Degradation of flavor compounds can easily take place in the distillation flask. Thermo extraction can be used to remove most of the water content in a short period of time. This is necessary because the analysis is carried out on a polar (wax) column. The desorption is so mild that thermally sensitive compounds can be detected at much higher concentrations than expected, based on reports in the literature.

Thermo desorption does not require knowledge of the physicochemical properties of the analyte and in most cases leads more quickly to good results. For each single compound extraction efficiency and recovery rate of TDS and SFE have to be determined by spiking the sample with the compound of interest. The amount of spiking material must be in the same range as the compound in the sample.

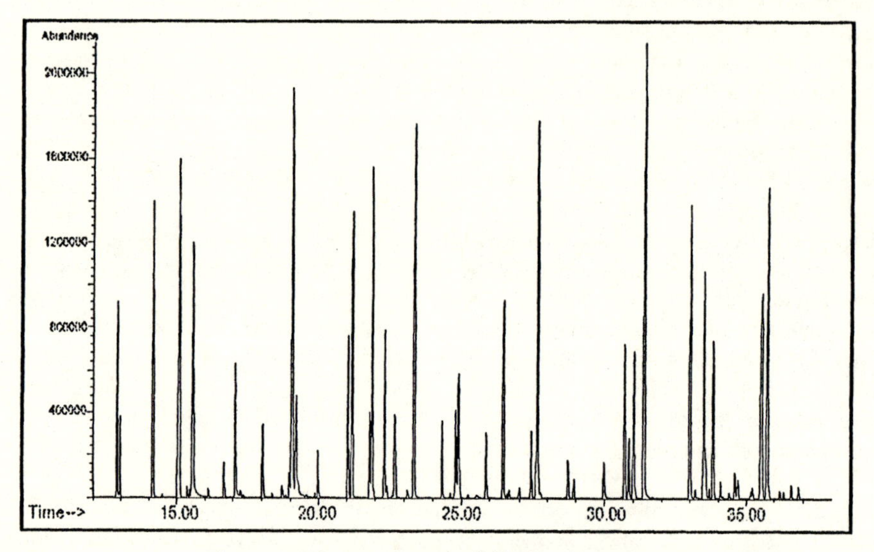

Figure 3. Chromatogram of Air Freshener by TDS

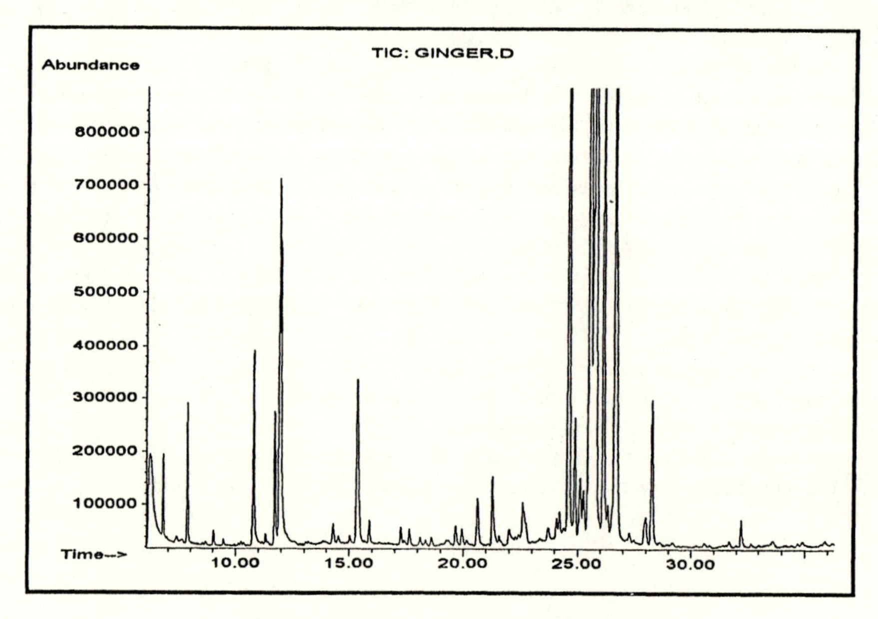

Figure 4. TDS Analysis of essential oil

Comparison to SFE

Supercritical fluid extraction (SFE) is an extraction method comparable to TDS in that it minimises solvent consumption. SFE is particularly effective in extracting hydrophobic compounds *(1, 2)*. Even non-volatile compounds are extractable under very mild conditions. Recoveries are comparable to results obtained with thermo desorption for nonpolar compounds. To extract polar compounds, it is necessary to add a modifier to the supercritical fluid. The modifier is usually the same solvent that is used to rinse the analytes from the traps. Recovery of polar compounds using thermo desorption is also comparable, on a order of magnitude, to SFE. Extracts prepared by SFE are compatible with several separation techniques.

Changes in density of supercritical fluids influence the extraction of nonpolar compounds of different molecular weight. These compounds can be fractionated in a short period of time by the correct choice of density. In addition, hydrophobic compounds can be extracted from complex matrices containing sugar, proteins, and fat in almost quantitative amounts using SFE. To illustrate this point, a chromatogram of a SF extract of cookies is shown in Figure 5.

Although, the recovery of polar molecules is typically poor for SFE, addition of a modifying solvent significantly improves the recovery of these compounds. Addition of a modifying eluent, e.g. acetonitrile or methanol, opens up a new dimension of compounds that can be extracted by SFE. The selection of substances by polarity can be used as an additional fractionation parameter.

Modifiers are typically added in two ways. First, the modifier can be added directly into the extraction thimble. Second, the modifier can be added by a modifier pump that delivers a constant flow rate and improves extraction efficiency and reproducibility.

Modification can also mean addition of an adsorbent to the sample. The adsorbent retains specific substances which make problems for the analysis. For example, extraction of flavors from a fatty matrix can be achieved by adding cellulose powder to the sample. The extraction takes place at low densities. The supercritical carbon dioxide extracts the flavor compounds without any of the problematic sample matrix compounds i.e. fatty acids or triacylglycerols. Analysis of the SF extract can be carried out on any analytical instrument like GC, HPLC, or a capillary electrophoresis. The recovery of terpenes from a hydrophobic matrix isolated by SFE can yield rather high values (80 %).

Conclusion

TDS as a new powerful sample preparation technique was presented and compared to SFE. The decision for either of the presented sample preparation methods is dependant on the compounds which have to be quantified and also on the sample matrix. With these presented techniques a wide range of compounds can be determined in a reliable and reproducible way.

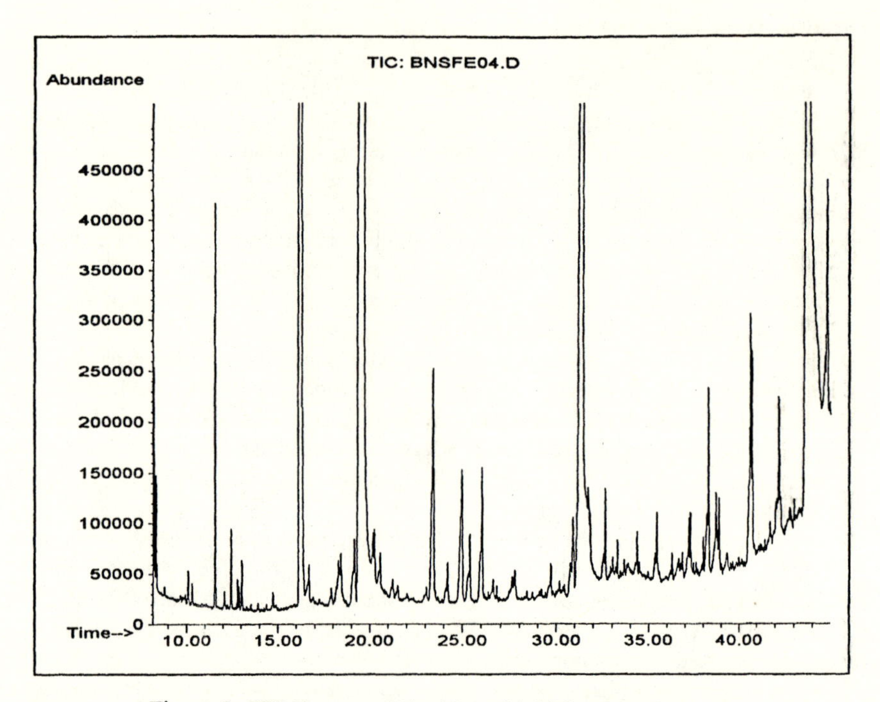

Figure 5. SFE Extract of Cookies with high Water Content

Literature Cited

1. Antinescu G.; Doneanu C.; Radulescu V. *Flavour and Fragrance Journal* **1997**, 12, 173
2. Coleman W. M.; Lawrence B. M. *Flavour and Fragrance Journal* **1997**, *12*, 1.
3. Reverchon E.; Della Porta G.; Gorgoglione D. *Flavour and Fragrance Journal*, **1997**,*12*, 37.
4. Grob K. *Analytical Chemistry*, **1994**, *66 no.: 20*, 1009 - 1018
5. Borgerding A. J., Wilkerson C. W., *Anal.Chem*. **1996**, *68*, 2874.
6. Mol H. G.-J.; Althuizen M.; Janssen H.-G.; Cramers C. A. *J. High Resol. Chromatogr*., **1996**, *19*, 69.

Analytical Characterization

Chapter 12

Analysis of Volatile Maillard Reaction Products by Different Methods

K. Eichner, M. Lange-Aperdannier, and U. Vossmann

Institut für Lebensmittelchemie der Universität Münster, Corrensstrasse 45, D-48149 Münster, Germany

Volatile compounds may serve as markers to reveal different pathways of the Maillard reaction that lead to desirable and undesirable flavor changes. Depending on the conditions and the type of reaction, different classes of volatile aroma and off-flavor compounds, which have different volatility, are formed. Amadori compounds, which are the first detectable Maillard reaction intermediates, may be regarded as flavor precursors; in their presence products of high and lower volatility (Strecker aldehydes, heterocyclic compounds) are formed.
To determine the volatile Maillard reaction products different techniques were applied: headspace gas chromatography, simultaneous distillation/extraction (SDE), and solvent extraction. The formation of artifacts and their possible prevention by application of different methods will be emphasized. The contribution of different components to sensory changes based on their aroma indices will be outlined.

The Maillard reaction, which becomes predominant during thermal treatments in food processing (*1*), may create desirable aroma components like in baking and roasting processes (*1–5*); but, it often causes detrimental sensory changes (*6–7*). The Maillard reaction is principally promoted during drying processes where the product must pass a critical moisture interval.

If we look at the volatile products of tomato powder (Figure 1) we recognize a variety of artifacts formed by chemical reactions that occur during processing and storage. The compounds identified in Figure 1 were isolated by water vapor distillation and extraction with methylene chloride. Similar to a visiting card, the chromatogram gives insight into the quality status or the thermal history of the product. n-Hexanal (peak 1) and 2,4-decadienal (peak 7) represent lipid oxidation products, which impart a rancid off-flavor, and 6-methyl-5-hepten-2-one (peak 2) is

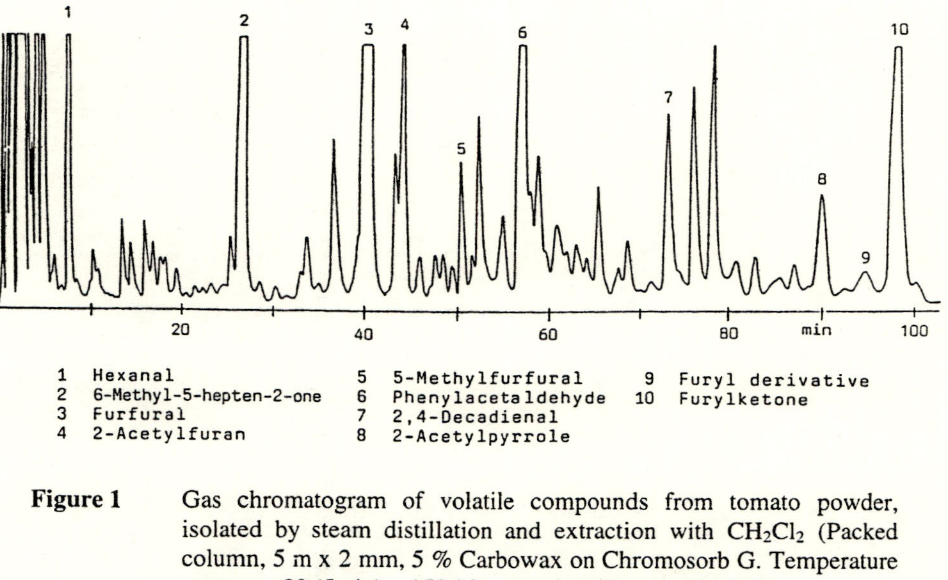

Figure 1 Gas chromatogram of volatile compounds from tomato powder, isolated by steam distillation and extraction with CH_2Cl_2 (Packed column, 5 m x 2 mm, 5 % Carbowax on Chromosorb G. Temperature program: 80 (5min) – 150 °C with a heating rate of 1 °C/min.

formed by oxidation of the tomato dyestuff lycopene. All other products are formed by the Maillard reaction. Peaks 3, 4, 5, 8, 9, and 10 represent sugar degradation products; and peak 6 is the Strecker degradation product of the amino acid phenylalanine.

The relative concentrations of 6-methyl-5-hepten-2-one (a lycopene oxidation product) and some characteristic Maillard reaction products isolated from tomato powder, dried via different thermal treatments, are shown in Table I. To facilitate comparison, concentrations of compounds in the freeze-dried product were set equal to one. It is clear that the concentrations of all Maillard products increase with the intensity of heat during drying; whereas, 6-methyl-5-hepten-2-one, a marker compound for oxidative changes, is predominant in the freeze-dried product that has a more open structure and is more accessible to oxygen during storage.

Reaction Pathways of the Maillard Reaction

To estimate their mode of formation, the reaction pathway for formation of volatile Maillard reaction products must be known. Figure 2 shows the main decomposition pathways for Amadori compounds, intermediates of the Maillard reaction. 1,2-Enolization leads to the 3-deoxyosone; whereas, the 1-deoxyosone is formed via the 2,3-enolization. It is well known that cyclization reactions of the 3-deoxyosone and the 1-deoxyosone lead to hydroxymethylfurfural (HMF) and 5-hydroxy-5,6-dihydromaltol (DHM), respectively (*1*). Therefore, HMF and DHM represent marker compounds for the 1,2- and 2,3-enolization pathways.

A summary of characteristic heterocyclic reaction products of the 1,2- and 2,3-enolization of Amadori products is given in Table II. Hence, it follows that Amadori compounds are important precursors of a variety of heterocyclic compounds, and they can be used as marker compounds for evaluating thermal processes.

Deoxyosones can also form short chain compounds, such as methylglyoxal, hydroxyacetone and diacetyl, via scission reactions. Most of these are dicarbonyl compounds, and they, as well as the deoxyosones, can react with amino acids by the Strecker degradation.

The Strecker degradation of amino acids encompasses the condensation of an amino acid with a dicarbonyl compound to form the Schiff base that following decarboxylation splits into an α-aminoketone and the so-called Strecker aldehyde. At roasting temperatures, α-aminoketones form pyrazines, which are important components of coffee and cocoa aroma.

Model roasting experiments with fructose-alanine demonstrated that Amadori compounds are important precursors of roast aroma components like methylpyrazines (*5*). On the other side, Amadori compounds have been shown to be precursors of the Strecker aldehydes, which typically contribute to the undesirable flavors produced by heat processing of foods (*6,8*). Therefore, the concentration and distribution of Amadori compounds must be known (*5*) in order to evaluate the aroma formation potential of heated foods.

Table I. Volatile Products of Tomato Powder Dried in Different Ways

Peak-Nr.	Compound	C_r (F)	C_r (C)	C_r (H)
2	6-Methyl-5-hepten-2-one	1	0.31	0.38
3	Furfural	1	2.4	3.5
4	2-Acetylfuran	1	3.7	10.3
5	5-Methyl-2-furfural	1	7.1	9.5
6	Phenylacetaldehyde	1	3.2	4.0
8	2-Acetylpyrrole	1	6.7	12.4

C_r = relative concentration (F = freeze-dried; C = cold-spray-dried; H = hot-spray-dried)

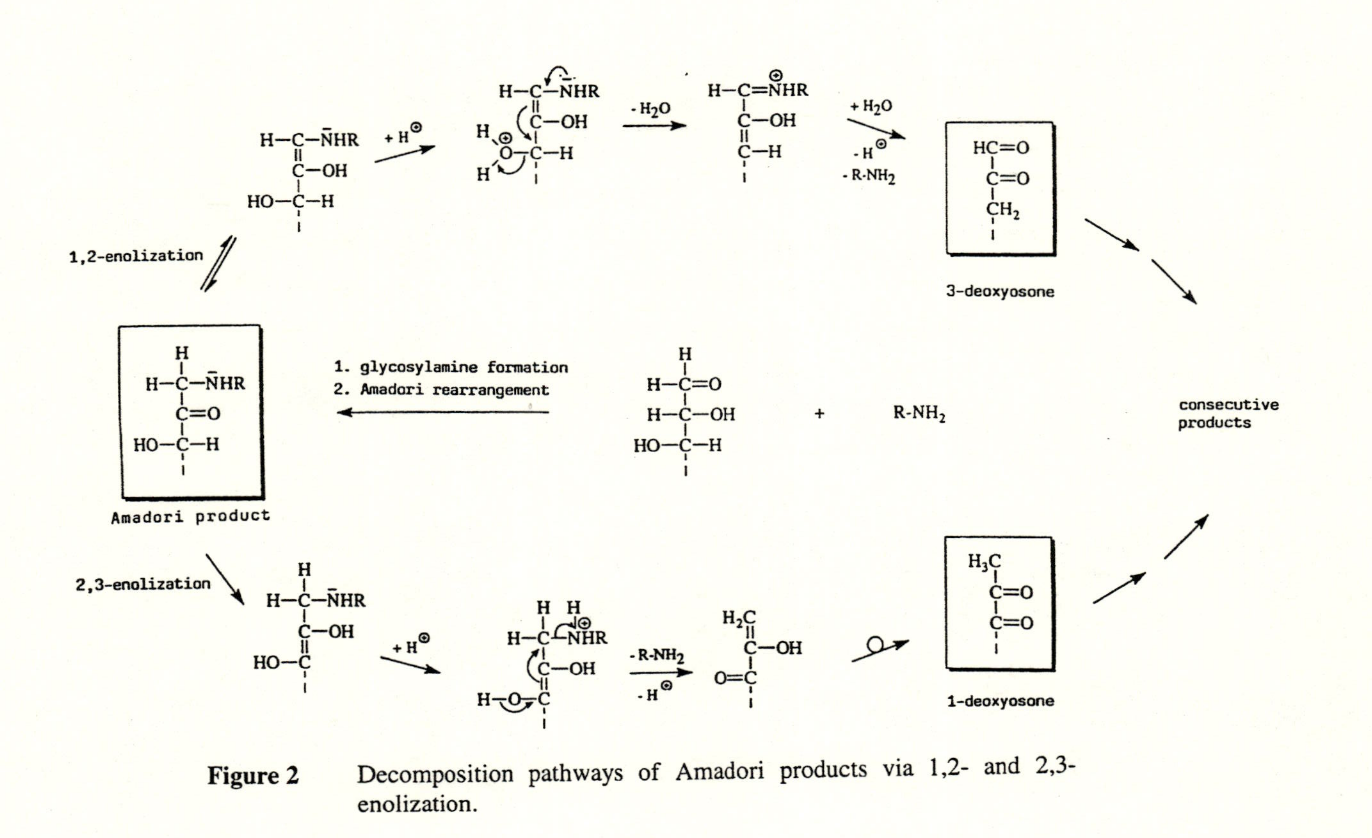

Figure 2 Decomposition pathways of Amadori products via 1,2- and 2,3-enolization.

Table II. Formation of Furans, Pyrroles, Pyrones and Furanones by Decomposition of Amadori Compounds via

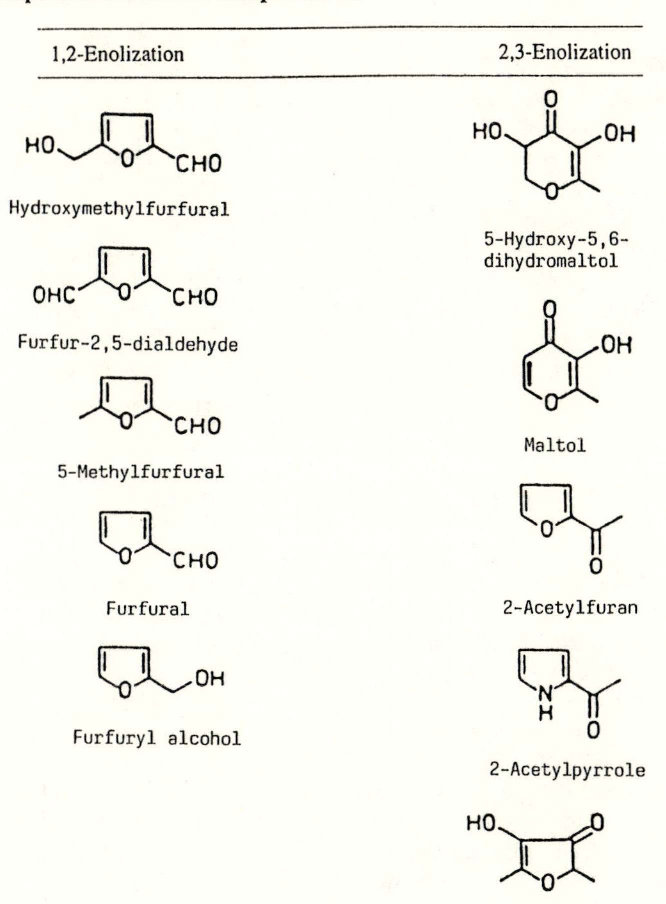

Occurrence of Volatile Maillard Reaction Products and Their Precursors in Foods and Model Systems

Formation of HMF and DHM by Decomposition of Amadori Products. Amadori compounds have been detected in a variety of heated foods like dried vegetables, kilned malt, and fermented cocoa beans (*5, 8*). Figure 3 shows a standard gas chromatogram of Amadori compounds that had been transformed into their oximes and converted to the trimethylsilyl ethers (*8*). Using this same gas chromatography method the decomposition pathway of Amadori compounds can be followed. Decomposition of fructose-glycine under acidic conditions (pH 3) forms the 3-deoxyosone, which subsequently forms HMF by cyclization. On the other side, DHM, a characteristic marker compound for the 2,3-enolization pathway, was also formed when tomato paste (pH 4.2) was drum dried. Figure 4 shows the gas chromatographic separation of HMF and DHM in drum dried tomato flakes. Samples were prepared for analysis by extracting with water, precipitating the pectins, extracting the water phase with methylene chloride, evaporating the solvent, and reacting the residue to form the oximes and trimethylsilyl ethers (*8*). Gas chromatography was carried out on a quartz glass capillary (30 m x 0.26 mm) covered with DB-5 (0.1 µm) programmed from 100 °C at 4 °C/min to 240 °C. HMF and DHM appear as characteristic double peaks: corresponding to the syn- and anti-form of the oximes (*8*). Table III demonstrates that drum dried tomato flakes contain relatively high portions of DHM compared to HMF, and both compounds increase after further heating. After heating tomato paste for 1 hr at 90 °C the portion of DHM is much lower. The 2,3-enolization pathway seems to predominate when heating foods having higher pH values; only DHM could be analyzed after heating freeze-dried carrot powder or cocoa.

Analysis of Volatile Maillard Reaction Products. The possible methods for analytical determination of volatile Maillard reaction products depend on the volatilities of the individual compounds. Highly volatile compounds like Strecker aldehydes can be analyzed by headspace gas chromatography. Figure 5 shows the headspace gas chromatogram of tomato paste. Of primary interest is 2- and 3-methylbutanal formed by Strecker degradation of the amino acids isoleucine an leucine. These Strecker aldehydes have very low sensory threshold values, and they provide a significant contribution to the undesirable sensory changes of heated foods from plant origin.

Samples were prepared for headspace analysis by mixing 3.0 g of tomato paste or 0.5 g of drum-dried tomato flakes with 1.0 – 3.0 mL water and 1 mL standard solution. The samples were homogenized and heated for 20 min at 90 °C in septum vessels in order to achieve equilibrium of the volatile compounds between the gas and liquid phases. Following equilibration, 2.5 ml of the head-space gas was withdrawn and injected into a gas chromatograph equiped with a PTV injector. The gas chromatographic separation was carried out on a glass capillary column (60 m x 0.32 mm) covered with Stabilwax (0.1 µm). The GC temperature program was 40 – 220 °C: with a heating rate of 4 °C/min and holding time of 20 min.

In Table IV the concentrations of Strecker aldehydes and dimethyl sulfide, a decomposition product of S-methyl-methionine, in tomato paste and tomato flakes

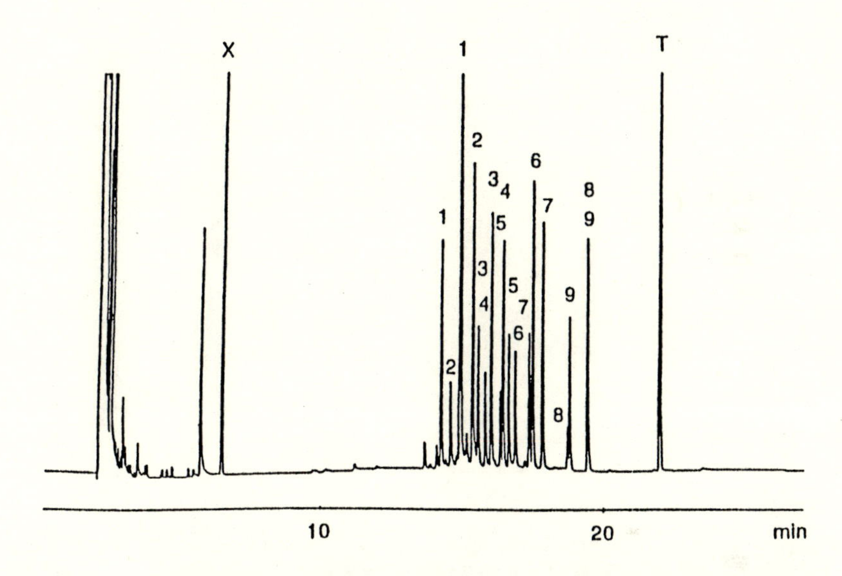

Figure 3 Separation of Amadori compounds by capillary gas chromatography (standard chromatogram) 1 Fru-Ala; 2 Fru-Gly; 3 Fru-Val; 4 Fru-Leu; 5 Fru-Gaba; 6 Fru-Ser; 7 Fru-Thr; 8 Fru-Asp; 9 Fru-Glu (Gaba = γ-aminobutyric acid). Internal standards: X = xylitol; T = trehalose. Gas chromatographic conditions (*8*).

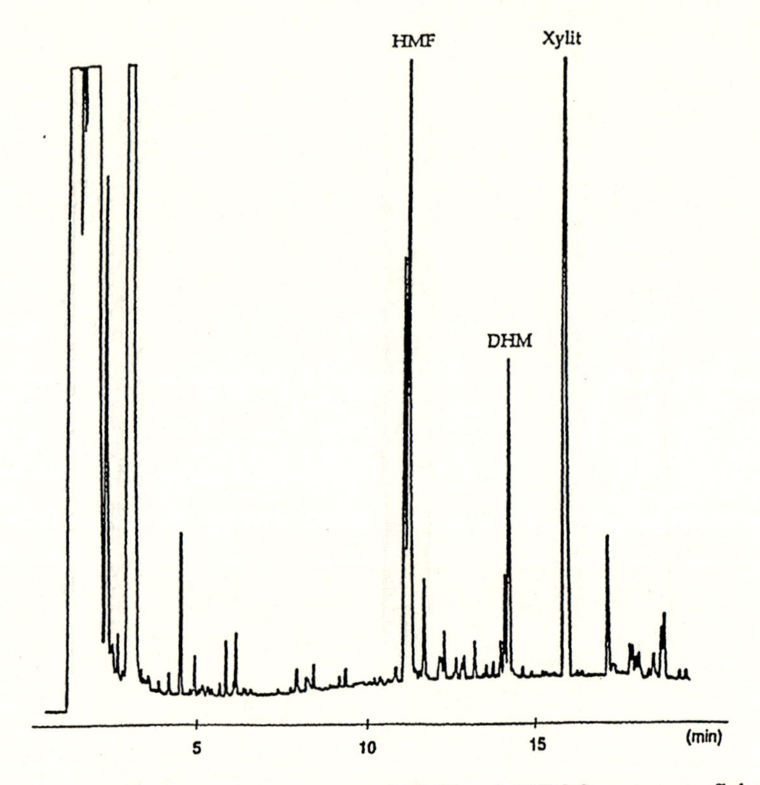

Figure 4 Capillary gas chromatogram of HMF and DHM from tomato flakes after oximation and trimethylsilylation.

Table III. Concentrations of DHM and HMF in Heated Foods (mg/kg DM)

Type of food	Heating time at 90 °C	DHM	HMF
Tomato Paste	0 h	-	14
	1 h	32	128
Freeze-dried tomato powder	0 h	-	14
(a_w = 0.23)	1 h	47	237
Drum-dried tomato flakes	0 h	189	249
	1 h	500	947
Freeze-dried carrot powder	0 h	-	-
(a_w = 0.63)	1 h	19	-
Cocoa	20 Min. 130 °C	142	-

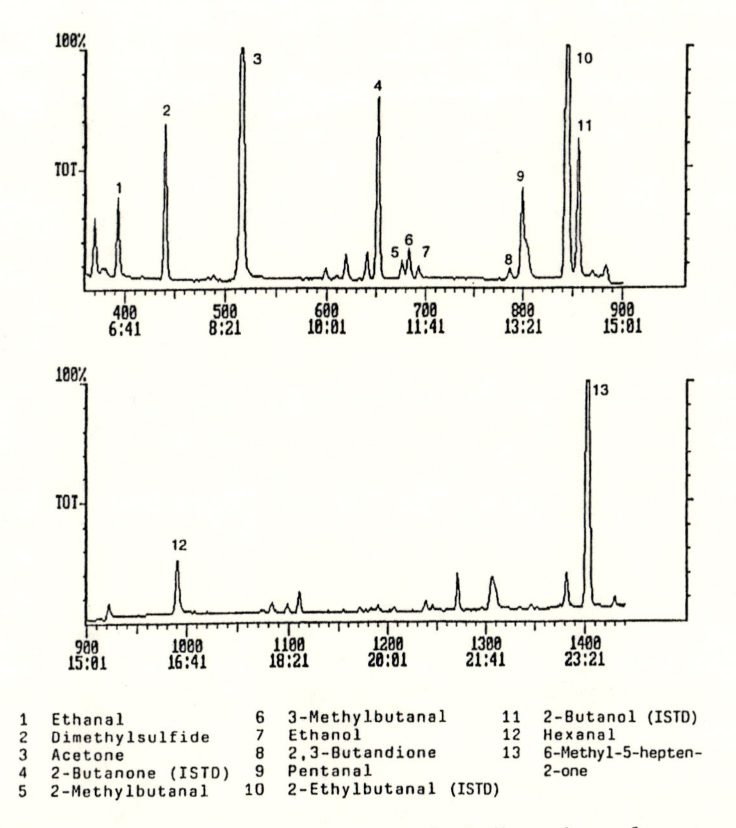

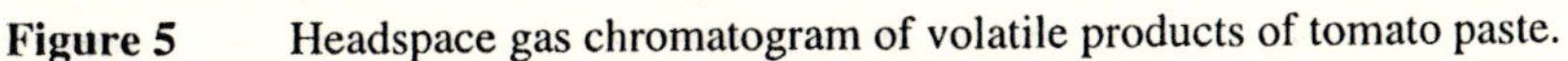

Figure 5 Headspace gas chromatogram of volatile products of tomato paste.

Table IV. Determination of Volatile Compounds in Tomato Paste and Tomato Flakes by Static Head-Space-Gas Chromatography

Compound	Tomato paste (mg/kg DM)	Tomato flakes* (mg/kg DM)	Sensory threshold values (mg/L H_2O)
Acetaldehyde	6.4	75.1	0.015
2-Methylpropanal + Acetone	18.0	81.9	
2-Methylbutanal	0.21	27.6	2×10^{-3}
3-Methylbutanal	0.26	31.2	0.2×10^{-3}
Dimethyl sulfide	10.6	63.9	0.3×10^{-3}

* containing 56 % starch; all data are related to tomato solids

are presented. Their concentrations are all far above their sensory threshold values. It can be seen that the concentrations of Strecker aldehydes are much higher in tomato flakes than in tomato paste. The main reason for this is that tomato flakes contain high proportions of Amadori compounds, which proved to be precursors of Strecker aldehydes (*8,9*).

Figure 6 shows that the initial concentrations of isovaleraldehyde (the sum of 2- and 3-methylbutanal) and its rate of formation during holding of reconstituted tomato powder at elevated temperatures depend on the intensity of heat impact during drying and the resulting concentration of Amadori compounds (9). This means that a portion of the Strecker aldehydes analyzed by head space gas chromatography are formed during the equilibration step, and they must be regarded as artifacts. Their concentrations increase with the amount of Amadori compounds in the dried product and the time of heating.

The gas chromatogram of an aroma extract obtained by simultaneous distillation-extraction (SDE) of tomato flakes, using methylene chloride as the solvent, is shown in Figure 7. Again, a series of important marker compounds of the Maillard reaction can be found. These include furfural, 5-methylfurfural, and 2-formyl-pyrrole (products of the 1,2-enolization of Amadori compounds); 2-acetylfuran and 2-acetylpyrrole (products of the 2,3 enolization of Amadori compounds); and phenylacetaldehyde (the Strecker degradation product of phenylalanine). SDE samples were prepared for analysis by evaporating the solvent. Gas chromatography was performed in the same way as described above for headspace analysis.

Table V again demonstrates that the concentrations of the Maillard products mentioned above are much higher in the tomato flakes than in the tomato paste. The concentration of phenylacetaldehyde is far above its sensory threshold value, so it highly contributes to flavor changes of the product.

Figure 8 shows the gas chromatogram of a methylene chloride extract of tomato flakes. The extraction procedure was performed at room temperature in order to avoid the formation of artifacts; the extract was purified by gel chromatography before evaporation of the solvent. It can be seen in Figure 8 that furaneol, DHM, and HMF can be analyzed when tomato flakes are extracted with methylene chloride; these compounds could not be isolated by SDE due to their low vapor pressure. The gas chromatographic separation was carried out on a glass capillary (60 m x 0.32 mm) covered with Stabilwax (0.25 μm) and programmed 40 °C for 5 min, 5 °C/min to 220 °C, and held 30 min.

Table VI again shows that the concentrations of volatile Maillard products are much higher in tomato flakes than in tomato paste.

In Table VII the results of the described methods for analysis of volatile compounds in tomato flakes are compared. It becomes clear, that the concentrations of Maillard products found by SDE are significantly higher than the values obtained by cold extraction. This indicates that during water vapor distillation artifacts are formed from Amadori compounds present in the tomato flakes. On the other side, the concentrations of 2-acetylpyrrole and 2-formylpyrrole found by extraction are much higher than by SDE, because only a small portion of these compounds evaporate during steam distillation.

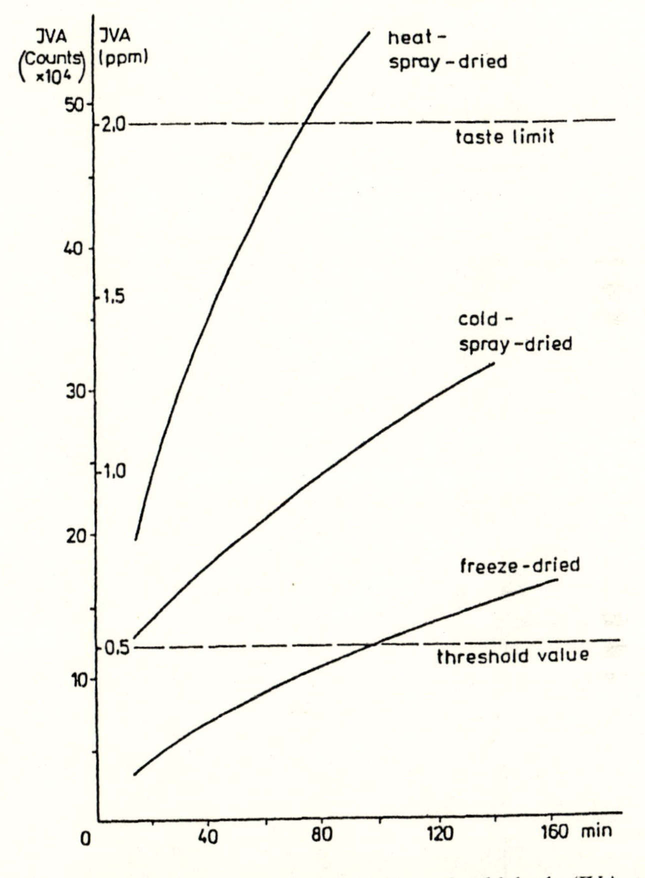

Figure 6 Increase of the concentration of isovaleraldehyde (IVA, sum of 2- and 3-methylbutanal) in freeze-dried and industrially spray-dried tomato powder reconstituted with water (3 ml/g tomato powder) at 70 °C.

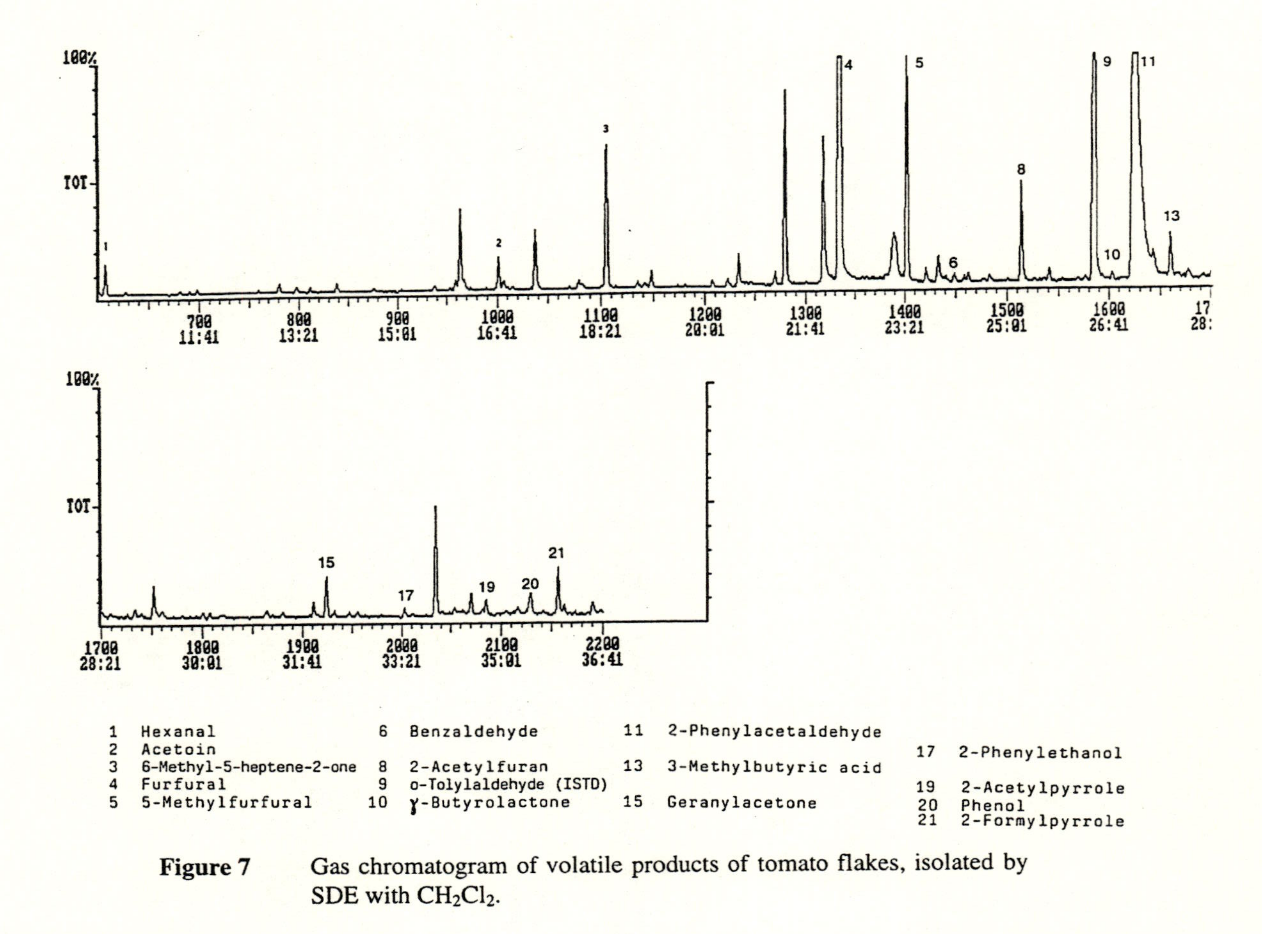

Figure 7 Gas chromatogram of volatile products of tomato flakes, isolated by SDE with CH_2Cl_2.

Table V. Determination of Volatile Compounds in Tomato Paste and Tomato Flakes by SDE with Methylene Chloride

Compound	Tomato paste (mg/kg DM)	Tomato flakes (mg/kg DM)	Sensory threshold values (mg/L H_2O)
Furfural	3.34	51.10	3.0
5-Methylfurfural	0.06	7.95	10*
2-Acetylfuran	n.d.	1.30	110*
Phenylacetaldehyde	2.19	48.0	4×10^{-3}
2-Acetylpyrrole	n.d.	0.31	200*
2-Formylpyrrole	n.d.	0.85	

n.d. = non detectable. * determined in orange juice

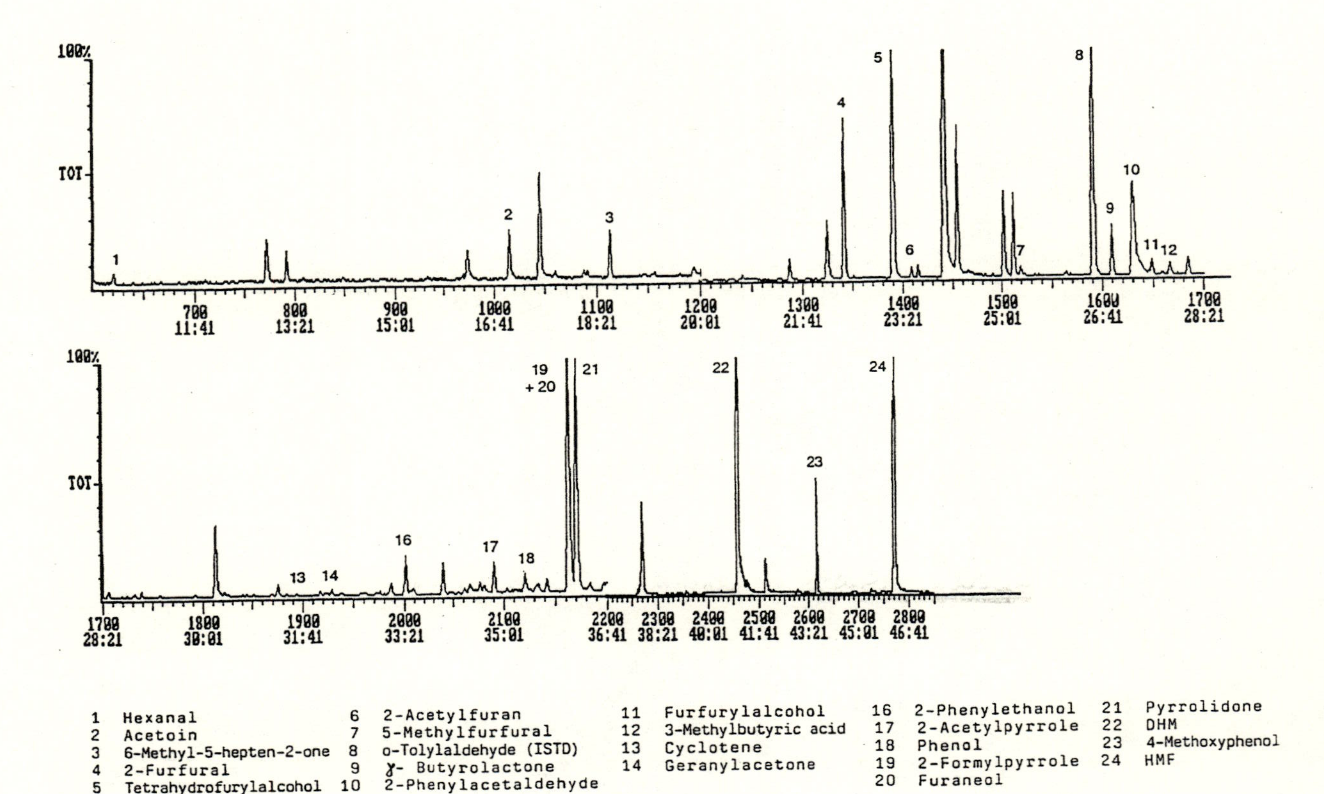

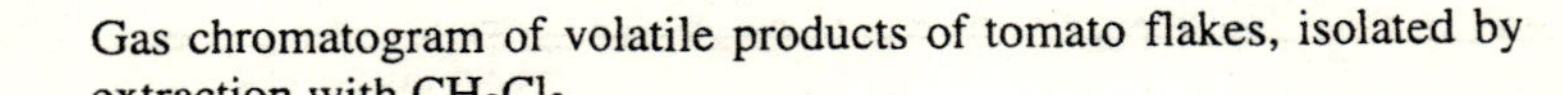

Figure 8 Gas chromatogram of volatile products of tomato flakes, isolated by extraction with CH_2Cl_2.

Table VI. Determination of Volatile Compounds in Tomato Paste and Tomato Flakes by Extraction with Methylene Chloride

Compound	Tomato paste (mg/kg DM)	Tomato flakes (mg/kg DM)	Sensory threshold values (mg/L H_2O)
Furfural	< 0.05	11.1	3.0
5-Methylfurfural	< 0.05	0.5	10*
2-Acetylfuran	< 0.05	0.36	110*
Phenylacetaldehyde	0.3	31.2	4×10^{-3}
2-Acetylpyrrole	< 0.05	2.0	200*
2-Formylpyrrole	< 0.05	16.5	
Furaneol	0.42	8.2	0.03
HMF	14.0	249.0	200*
DHM	< 0.05	189.0	200*

* determined in orange juice

Table VII. Determination of Volatile Compounds in Tomato Flakes (mg/kg DM). Comparison of Methods

Compound	Headspace-GC	SDE	Extraction
3-Methylbutanal	31.2	-	-
Furfural	-	51.10	11.1
5-Methylfurfural	-	7.95	0.5
2-Acetylfuran	-	1.3	0.36
Phenylacetaldehyde	-	48.0	31.2
2-Acetylpyrrole	-	0.31	2.0
2-Formylpyrrole	-	0.85	16.5
Furaneol	-	-	8.2
HMF	-	-	249.0
DHM	-	-	189.0

Conclusions

The rate of formation of volatile Maillard reaction products depends on the concentration of Amadori compounds, precursors of flavor and off-flavor development, formed during heating processes like drying, kilning, and roasting. The composition of the volatiles depends on reaction conditions like temperature and pH that influence the decomposition pathways of the Amadori compounds.

HMF and DHM proved to be useful marker compounds for evaluating the proportions of the Maillard reaction that procede through the 1,2- and 2,3-enolization pathways.

Volatile Maillard reaction products can be analyzed by headspace gas chromatography, simultaneous distillation-extraction (SDE), and solvent extraction. Selection of the isolation procedure depends on the vapor pressure of the compounds to be analyzed. Application of heat during the analytical procedure (head space gas chromatography, SDE) leads to artifacts that depend on the concentration of Amadori compounds present in the analyte.

Literature cited

1. Namiki, M. *Adv. Food Res.* **1988**, *32*, 115 – 184.
2. Baltes, W. *Dtsch. Lebensm. Rdsch.* **1979**, *75*, 2 – 7.
3. Baltes, W. *Lebensmittelchem. Gerichtl. Chem.* **1980**, *34*, 39 – 47.
4. Danehy, J.P. *Adv. Food. Res.* **1986**, *30*, 77 – 138.
5. Eichner, K.; Schnee, R.; Heinzler, M. In *Thermally Generated Flavors*; Parliment, T. H.; Morello, M. J.; Mc Gorrin, R. J., Eds.; ACS Symposium Series 543, 1994; pp. 218 – 227.
6. Sapers, G.M. *J. Food Sci.* **1970**, *35*, 731 – 733.
7. Eichner, K. *Dtsch. Lebensm. Rdsch.* **1973**, *69*, 4 – 12.
8. Eichner, K.; Reutter, M.; Wittmann, R. In *Thermally Generated Flavors*; Parliment, T. H.; Morello M. J.; Mc Gorrin, R. J., Eds.; ACS Symposium Series 543, 1994; pp. 42 – 54.
9. Eichner, K.; Ciner-Doruk, M. In *Maillard Reactions in Food*; Eriksson, C., Ed.; Progress in Food and Nutrition Science 5 (1 – 6), 1981; pp. 115 – 135.

Chapter 13

γ- and δ-Thiolactones: An Interesting Class of Sulfur-Containing Flavor Compounds

Karl-Heinz Engel, Irmgard Roling, and Hans-Georg Schmarr

Technische Universität München, Lehrstuhl für Allgemeine Lebensmitteltechnologie, Am Forum 2, D-85350 Freising-Weihenstephan, Germany

This contribution is dedicated to Dr. Roy Teranishi on the occasion of his 75th birthday.

γ-Thiolactones (5-alkyldihydro-2(3H)-thiophenones) and δ-thiolactones (6-alkyltetrahydro-2H-thiopyran-2-ones) were synthesized by reaction of the corresponding oxygen containing lactones with thiourea and hydrobromic acid. Analytical data (MS, ^{1}H and ^{13}C NMR), chromatographic behavior, GC separation of the enantiomers, and sensory properties of the newly described compounds are presented. The substitution of oxygen by sulfur in the lactone ring induces tropical fruit notes. Odor thresholds and sensory characteristics of the thiolactones vary with ring size and chain length.

Sulfur-containing compounds are known as outstanding contributors to pleasant notes as well as off-flavors of many foods (*1,2*). On one hand, they are formed during thermal processing of foods, such as roasting of coffee (*3*). On the other hand, they may be biogenetically derived and determine the aroma of natural systems, such as some tropical fruits (*4,5*).

Another widely distributed class of substances known to have an impact on many flavors are aliphatic γ- and δ-lactones (*6*). Nectarines (*7*), apricots (*8*), and peaches (*9*) are classic examples of fruits with a lactone-type aroma.

The effect of a substitution of oxygen by sulfur on flavor quality and potency has been shown for many substances (*10*); one of the most impressive examples is the difference between α-terpineol and 1-p-menthene-8-thiol (*11*). The replacement of oxygen in the lactone ring by sulfur, resulting in γ- and δ-thiolactones, has not yet been investigated. The short chain homologue, γ-thiovalerolactone, has been identified after heating model systems with sulfur-containing amino acids (*12,13*); however, systematic data on this type of compound have not been described. In this contribution the analytical and sensory characterization of a series of γ- and δ-thiolactones is presented.

Experimental

Synthesis. Experimental details of the synthesis and purification of the γ- and δ-thiolactones are described elsewhere (*14*).

Capillary Gas Chromatography. (a) Carlo Erba Mega II 8575 (C.E. Instruments, Rodano, Italy) equipped with a flame ionization detector (FID) and a flame photometric detector (FPD, FPD 80, C.E. Instruments). Parallel detection was achieved by dividing the effluent of the column (DB-Wax, J&W, Folsom, CA; 60 m x 0.32 mm i.d., 0.25 µm film thickness) via press-fit splitter and short pieces of deactivated fused silica capillaries to the two detectors. Split injection was performed at 220°C. Column temperature was programmed from 40°C (5 min hold) to 235°C (5 min hold) at 4°C/min. Hydrogen was used as carrier gas at a constant inlet pressure of 105 kPa. (b) Carlo Erba Fractovap 4160 equipped with FID and split injector; column: 30 m x 0.32 mm i.d. fused silica capillary column coated with a dimethylsiloxane stationary phase (PS-255, ABCR, Karlsruhe, Germany) providing a film thickness of 1.0 µm. Column temperature: 40°C (10 min hold) to 300°C (5 min hold) at 4°C/min; carrier gas: hydrogen (inlet pressure 50 kPa).

Separation of the enantiomers of the thiolactones was achieved on a fused silica column (30 m x 0.25 mm i.d.) coated with 25 % of heptakis-(2,3-di-*O*-methyl-6-*O*-*tert.*-butyl dimethylsilyl)-β-cyclodextrin in SE 54 to provide a film thickness of 0.25 µm. Gas chromatograph: Carlo Erba, Mega 5160 with a FID detector and split injection at 210°C. Hydrogen was used as carrier gas at a constant inlet pressure of 100 kPa. To separate the enantiomers of the thiolactones the oven temperature was programmed from 100°C (2 min hold) to 125°C (10 min hold) at a rate of 3 °C/min and then to 205°C at a rate of 1.5°C/min.

Gas Chromatography–Mass Spectrometry (GC–MS). Mass spectral data were acquired on an HP 5890 gas chromatograph coupled to an HP 5970 mass selective detector (Hewlett-Packard, Palo Alto, CA). The mass spectrometer interface temperature was set to 250°C, and the electron energy was 70 eV. The column used for GC-MS was a 50 m x 0.2 mm i.d. FFAP fused silica capillary with a film thickness of 0.33 µm (Hewlett-Packard). Split injection was performed at 230°C, and oven temperature was programmed from 70°C (5 min hold) to 235°C (10 min hold) at 5°C/min. Helium was used as carrier gas at a constant inlet pressure of 175 kPa.

NMR spectroscopy. ^{1}H and ^{13}C NMR spectra were recorded at 22°C using a DRX 500 and an AC 200 spectrometer, respectively, from Bruker, Karlsruhe, Germany. ^{1}H detected experiments were performed using an inverse broadband probehead, and ^{13}C detected experiments were performed using a dual ^{13}C/^{1}H probehead. DEPT and two-dimensional double quantum filtered COSY and HMQC experiments were performed according to standard Bruker software. Chemical shifts were referenced to solvent signals (^{1}H, 7.24 ppm; ^{13}C, 77.0 ppm).

Odor Thresholds. Odor thresholds in water were determined in Teflon squeeze bottles according to the procedure described by Guadagni and Buttery (*15*). Odor thresholds in air were determined by gas chromatography-olfactometry according to the method reported by Schieberle (*16*).

Synthesis and Identity of Compounds

The structures of the thiolactones synthesized in the course of this study are depicted in Figure 1. The reaction sequence starting from the corresponding oxygen containing γ- and δ-lactones and proceeding via isothiouronium bromides as intermediates (*17*) is outlined in Figure 2. The method was applied according to the procedure described for the conversion of γ-butyrolactone by Kharasch and Langford (*18*). The identity of the products was established by means of IR, MS as well as ^{1}H and ^{13}C NMR (*14*).

In analogy to the typical ions m/z 85 and 99, known for γ- and δ-lactones, the mass spectral patterns of the thio-compounds showed the corresponding fragments at m/z 101 and 115, respectively. As examples, the spectra of γ- and δ-thiooctalactone are shown in Figure 3.

^{1}H and ^{13}C NMR spectroscopic data are summarized in Figure 4. The substitution of oxygen by sulfur in the lactone ring results in a marked upfield shift of the signals obtained for the proton as well as for the carbon in position 4 (position 5 for δ-lactones, respectively). The ^{13}C NMR signal assigned to the carbon in position 1 is significantly shifted downfield in the sulfur containing compound. These differences are indicated in Figure 4 for the C_6 and C_{10} homologues.

Gas chromatographic Data

Kovats retention indices (KI) of aliphatic γ- and δ-lactones and the synthesized thio-compounds on a polar and an apolar stationary phase are listed in Table I. These data demonstrate that the chromatographic behavior of the thiolactones is comparable to that of their oxygen containing counterparts.

Table I. Comparison of Retention Indices (KI) for γ- and δ-Lactones and their Sulfur Containing Counterparts

	KI DB-Wax		KI PS-255	
	O-lactones	S-lactones	O-lactones	S-lactones
γ-C_6	1671	1697	0999	1094
γ-C_8	1889	1909	1209	1312
γ-C_{10}	2122	2142	1426	1534
γ-C_{12}	2350	2382	1643	1754
δ-C_6	1762	1791	1038	1124
δ-C_8	1941	2004	1234	1344
δ-C_{10}	2169	2233	1448	1565
δ-C_{12}	2404	2473	1663	1789

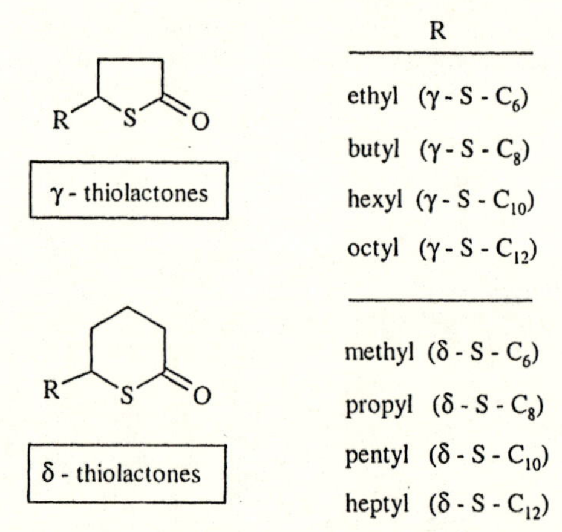

Figure 1. Structures of the synthesized γ-thiolactones (5-alkyldihydro-2(3H)-thiophenones) and δ-thiolactones (6-alkyltetrahydro-2H-thiopyran-2-ones)

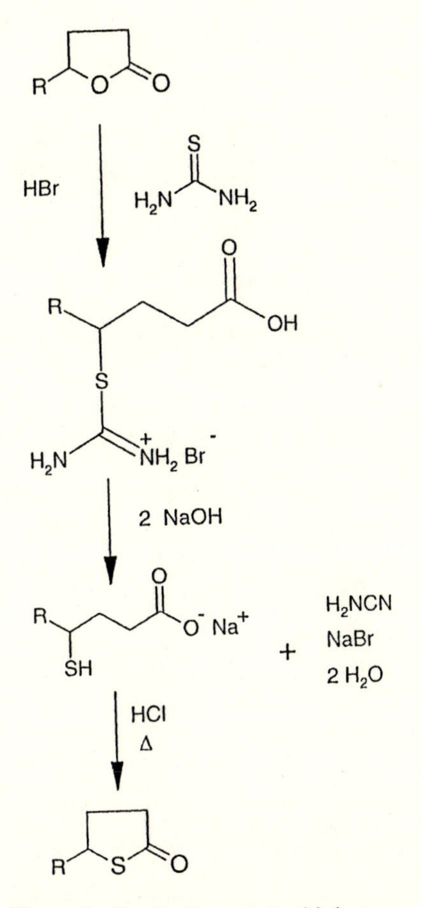

Figure 2. Synthetic route to thiolactones

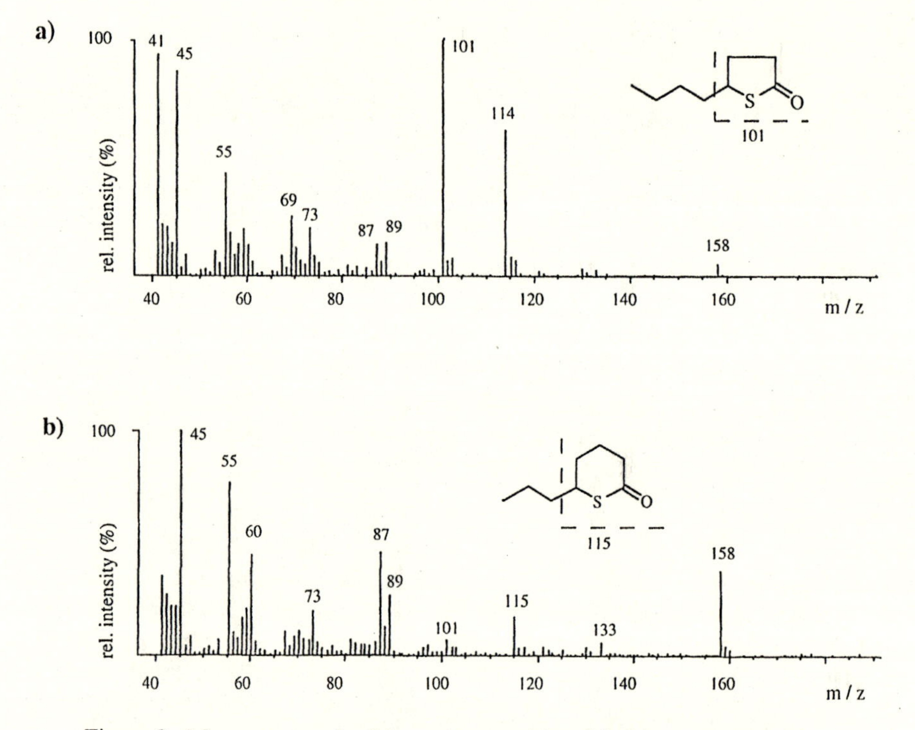

Figure 3. Mass spectra of γ-thiooctalactone (a) and δ-thiooctalactone (b)

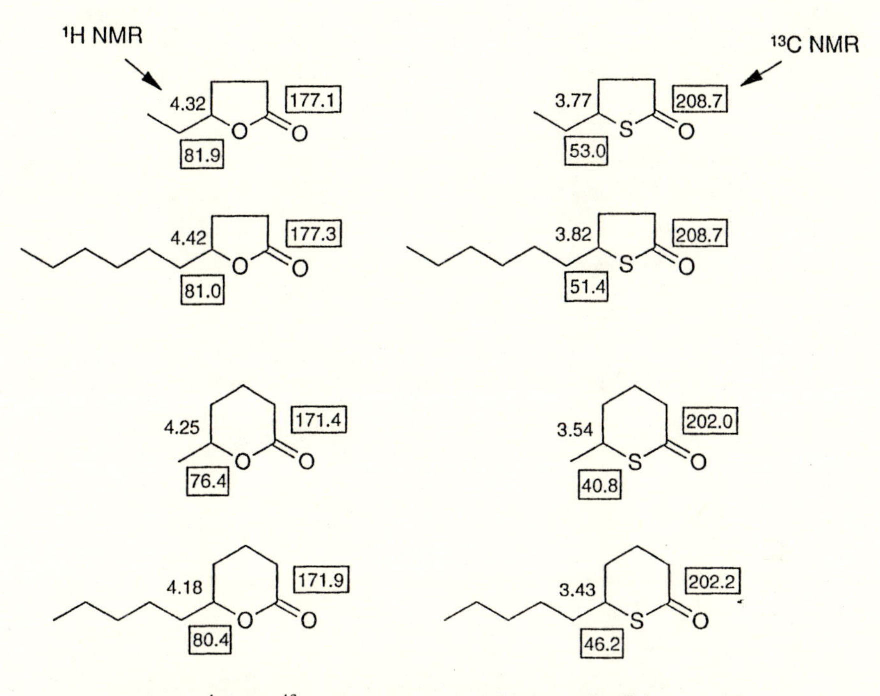

Figure 4. ^{1}H and ^{13}C NMR spectroscopic data of γ-/δ-hexa- and decathiolactones

The enantiomers of the γ- and δ-thiolactones could be separated on heptakis-(2,3-di-*O*-methyl-6-*O*-TBDMS)-β-cyclodextrin (Figure 5). The separation factors of the thiolactone enantiomers are lower than those obtained for the oxygen containing lactones on this modified cyclodextrin stationary phase (*19*). The δ-thiolactones show a minimum enantioseparation for the C_8 homologue. The assignment of the order of elution by means of optically pure enantiomers is the subject of ongoing studies.

Sensory Evaluation

Odor thresholds in air were determined by gas chromatography-olfactometry according to the method reported by Schieberle (*16*). As demonstrated in Figure 6, the odor thresholds (ng/l) in air vary with ring size and chain length. Within the series of synthesized compounds both, γ- and δ-thiolactones, exhibit the lowest odor thresholds for the C_8 and C_{10} homologues.

Odor thresholds in water were determined in Teflon squeeze bottles according to the procedure described by Guadagni and Buttery (*15*). Table II shows that the substitution of oxygen by sulfur has different effects depending on the ring size. The odor threshold (ppb) in water for γ-decathiolactone is increased compared to the oxygen containing counterpart; the threshold of the corresponding δ-lactone, however, is lowered drastically.

Table II. Impact of Oxygen and Sulfur on Odor Thresholds of C_{10}-Lactones in Water

lactone	odor threshold (H_2O), [ppb]
γ-decalactone [a]	11
γ-thiodecalactone	47
δ-decalactone [a]	100
δ-thiodecalactone	6

[a] Engel et al. (*7*)

These differences in the effects caused by the introduction of sulfur depending on the size of the lactone ring also become obvious in the sensory descriptions obtained for solutions (0.004%) of the thiolactones in water (Table III). The γ-thiolactones are characterized by mushroom-like odors (C_8) or fatty, rancid notes (C_{10}). The δ-thiolactones, on the other hand, still exhibit the pleasant lactone aroma, however, combined with a pronounced tropical fruit character.

In accordance with the typical behavior of sulfur containing volatiles, the odor impressions of thiolactones in water at higher concentrations are rather unpleasant. Only after dilution are desirable notes perceived (Table IV).

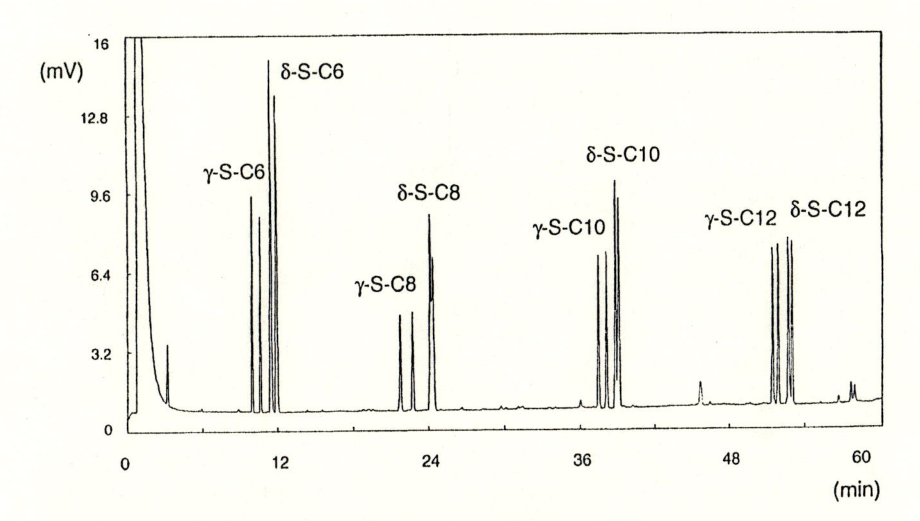

Figure 5. Capillary GC separation of the enantiomers of γ- and δ-thiolactones on heptakis-(2,3-di-*O*-methyl-6-*O*-TBDMS)-β-cyclodextrin

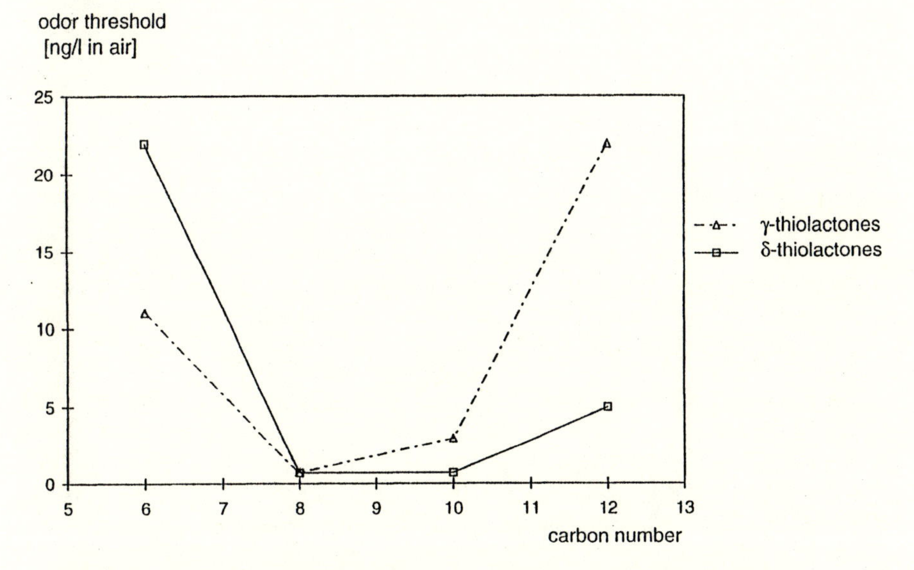

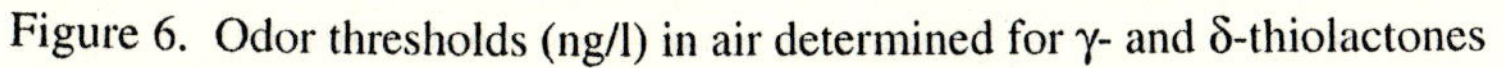

Figure 6. Odor thresholds (ng/l) in air determined for γ- and δ-thiolactones

Table III. Sensory Properties of Thiolactones

thiolactone	odor quality[a]
γ-S-C_6	sweet, sulfury, burnt
δ-S-C_6	slightly fruity, petroleum, sulfury
γ-S-C_8	mushroom homogenate, coconut, sweet, sulfury
δ-S-C_8	coconut, green, tropical fruit
γ-S-C_{10}	fruity, fatty, rancid
δ-S-C_{10}	tropical fruit, fresh
γ-S-C_{12}	slightly fruity, soapy
δ-S-C_{12}	sweet, soapy, apricot

[a] determined in 0.004% solutions in water

Table IV. Odor Profiles of Thiolactones in Water

concentration [ppm]	γ-thiodecalactone	δ-thiodecalactone
400	leek, cabbage	sweet, rancid
100	sulfury, leek	rancid, soapy
40	sweet, fatty, rancid	tropical fruit, spicy
4	cucumber, fatty	tropical fruit, sweet
1	green, cucumber, melon	sweet, fruity

Conclusion

γ- and δ-Thiolactones constitute an interesting class of sulfur containing flavor compounds with promising sensory characteristics. Especially the tropical fruit notes observed for δ-thiolactones are very attractive. For a final evaluation of the sensory properties of the thiolactones, aroma profiles at various concentrations and the characterization of the resolved enantiomers will be required. The next step will be the search for these newly described sensorially potent compounds in natural systems. Tropical fruits, many of them known for both lactone and sulfur metabolism, might be potential sources for these flavor components.

Acknowledgments

The authors thank R.G. Buttery (Western Regional Research Center, U.S. Department of Agriculture) and P. Schieberle and T. Hofmann (Institut für Lebensmittelchemie, Technische Universität München) for determining odor thresholds.

Literature Cited

1. Boelens, H.M. *Perfumer & Flavorist* **1993**, *18*, 29-39
2. *Sulfur Compounds in Foods*; Mussinan, C.J.; Keelan, M.E., Eds.; ACS Symposium Series 564; American Chemical Society: Washington, DC, 1994
3. Flament, I. *Food Rev. Int.* **1989**, *5*, 317-414
4. Engel, K.-H.; Tressl, R. *J. Agric. Food Chem.* **1991**, *39*, 2249-2252
5. Fischer, N. *dragoco report (flavoring information service)* **1996**, *41*, 137-147
6. Maga, J.A. *CRC Crit. Rev. Food Sci. Nutr.* **1976**, 1-56
7. Engel, K.-H.; Flath, R.A.; Buttery, R.G.; Mon, T.R.; Ramming, D.W.; Teranishi, R. *J. Agric. Food Chem.* **1988**, *36*, 549-553
8. Takeoka, G.R.; Flath, R.A.; Mon, R.T.; Teranishi, R.; Güntert, M. *J. Agric. Food Chem.* **1990**, *38*, 471-477
9. Horvat, R.J.; Chapman Jr. , G.W.; Robertson, J.A.; Meredith, F.I.; Scorza, R.; Callahan, A.M.; Morgens, P. *J. Agric. Food Chem.* **1990**, *38*, 234-237
10. *Riechstoffe und Geruchssinn. Die molekulare Welt der Düfte*; Ohloff, G., Ed.;Springer Verlag: Berlin, Heidelberg, 1990
11 Demole, E.; Enggist, P.; Ohloff, G. *Hel. Chim. Acta* **1982**, *65*, 1785-1794
12. Mussinan, C.J.; Katz, I. *J. Agric. Food Chem.* **1973**, *21*, 43-45
13. Shu, C.-K.; Ho, C.-T. In *Thermal Generation of Aromas;* Parliment, T.H.; McGorrin, R.J.; Ho, C.T., Eds.; ACS Symposium Series 409; American Chemical Society: Washington, DC, 1989, pp. 229-241
14. Roling, I.; Schmarr, H.-G.; Eisenreich, W.; Engel, K.-H. *J. Agric. Food Chem.* **1998**, *2*, 668-672
15. Guadagni, D.G.; Buttery, R.G. *J. Food Sci.* **1978**, *43*, 1346-1347
16. Schieberle, P. *J. Agric. Food Chem.* **1991**, *39*, 1141-1144
17. Frank R.L.; Smith, P.V. *J. Am. Chem. Soc.* **1946**, *68*, 2103-2104
18. Kharasch, N.; Langford R.B. *J. Org. Chem.* **1963**, *28*, 1901-1903
19. Maas, B.; Dietrich, A.; Mosandl, A. *J. High Resol. Chromatogr.* **1994**, *17*, 109-115

Chapter 14

Analysis of Thermal Degradation Products of Allyl Isothiocyanate and Phenethyl Isothiocyanate

Chung-Wen Chen, Robert T. Rosen, and Chi-Tang Ho

Department of Food Science and Center for Advanced Food Technology, Cook College, Rutgers, The State University of New Jersey, 65 Dudley Road, New Brunswick, NJ 08901–8520

Allyl isothiocyanate (AITC) or phenethyl isothiocyanate (PEITC) in an aqueous solution were heated and refluxed at 100 C for 1 hr. The reaction mixtures were simultaneously distilled and extracted into methylene chloride using a Likens-Nickerson (L-N) apparatus and then analyzed using gas chromatography (GC) and gas chromatography-mass spectrometry (GC-MS). The mixtures in the aqueous phase were analyzed by high performance liquid chromatography (HPLC) and liquid chromatography-mass spectrometry (LC-MS) equipped with an atmospheric pressure chemical ionization (APCI) interface. Nine thermal degradation volatile products including diallyl sulfide, diallyl disulfide, diallyl trisulfide, diallyl tetrasulfide, allyl thiocyanate, 3H-1,2-dithiolene, 2-vinyl-4H-1,3-dithiin, 4H-1,2,3-trithiin, and 5-methyl-1,2,3,4-tetrathiane were identified from AITC; while no volatile degradation products from PEITC were found. N,N′-diallylthiourea and N,N′-diphenethylthiourea, which were the major degradation products in the aqueous phase from the thermal reaction of AITC and PEITC, respectively, were identified by LC-MS (APCI+), direct probe EI-MS and H^1-NMR. A possible mechanism for the formation of these products is proposed.

In mustard and other cruciferous vegetables such as cabbage and cauliflower, allyl isothiocyanate (AITC) and phenethyl isothiocyanate (PEITC) are generated from their precursors, the glucosinolates, namely sinigrin and gluconasturtiin, respectively (*1,2*). These glucosinolates break down and release isothiocyanates such as AITC and PEITC by the action of myrosinase (thioglucoside glucohydrolase) when the plant tissue is disrupted (*1,3-5*). This action is presented in Figure 1.

AITC is a major pungent flavor component in mustard and Wasabi (*2,6*) and has been shown to be unstable. It easily decomposes to other compounds having garlic-like odor in the presence of water at both room temperature and 37 °C (*6*). AITC is also reported to be sensitive to temperature and pH. High temperature (37 °C) and alkaline conditions accelerate the decomposition of AITC (*7*).

It has been reported that PEITC is a potent inhibitor of nitrosamine-induced esophageal cancer (*8*) and N-nitrosobenzylmethylamine (NBMA)-induced esophageal tumors in rats (*9*). PEITC also inhibits the formation of lung tumors induced by 4-(methylnitrosamino)-1-(3-pyridyl)-1-butanone (NNK) in A/J mice (*10*). PEITC inhibits 7,12-dimethylbenz[a]anthracene (DMBA)-induced mammary tumors and benzo[a]pyrene (BP)-induced lung and forestomach tumors in mice (*11*).

Because no information exists about the chemical reactivity of AITC and PEITC during cooking conditions, the purpose of this study was to investigate the thermal reaction of AITC and PEITC in boiling water and to identify the thermal degradation products from them. In this study, the GC-MS was used to analyze the volatile decomposition products. LC-MS-APCI(+), direct probe EI-MS, and H^1-NMR were used to analyze the nonvolatile decomposition products.

Experimental Procedures

Thermal Reaction of Allyl isothiocyanate (AITC) and Phenethyl isothiocyanate (PEITC). One gram AITC or PEITC (Aldrich Chemical Co., Milwaukee, WI) in 200 mL of distilled water was heated and refluxed at 100 °C for one hour. The reaction mixtures were then simultaneously distilled and extracted into methylene chloride using a Likens-Nickerson (L-N) apparatus. The methylene chloride solution was dried over anhydrous sodium sulfate and concentrated to 5 mL using a stream of nitrogen prior to analysis by gas chromatography (GC) and gas chromatography-mass spectrometry (GC-MS).

The aqueous phase was concentrated to 20 mL using a vacuum rotary vaporator and analyzed by high performance liquid chromatography (HPLC) and liquid chromatography-mass spectrometry (LC-MS). The compounds of interest were collected by HPLC and identified using direct probe electron impact MS and H^1-NMR.

GC and GC-MS Analysis. A Varian Model 3400 Gas Chromatograph equipped with a flame ionization detector and a fused silica capillary column (DB-Wax, J&W 30 m x 0.25 mm i.d.) was used. The GC oven was temperature programmed from 50 to 220 °C at a rate of 4 °C/min. The carrier gas (He) flow rate was 2.0 mL/min. A split ratio of 25:1 was used. The temperatures of the injection port and detector were 240 and 250 °C, respectively. The GC-MS analysis was performed on an HP Model 5890 GC coupled with an HP 5971 mass selective detector. The same column and temperature program were used.

HPLC and LC-MS Analysis. HPLC analysis was performed using a Varian 5000 liquid chromatograph equipped with a Varian 2050 detector. A TSK-GEL ODS-

80TS reverse phase column (250 x 4.6 mm i.d., TOSOH Co., Japan) and a Whatman Partisphere amino phase column (250 x 4.6 mm i.d.) were used for the analysis of thermal degradation products from AITC and PEITC, respectively. For analysis of AITC degradation products in the aqueous phase, elution was carried out at a flow rate of 1.0 mL/min using a solvent system of acetonitrile/water (40:60%) isocratically over a 10 min period. The percentage of acetonitrile was then linearly increased to 100% over 2 min and kept there for another 15 min. A solvent system consisting of hexane/1-propanol (95:5%) pumped at a flow rate of 1.5 mL/min was used for the analysis of thermal degradation products from PEITC in the aqueous phase. The eluate was monitored at 254 nm. MS analysis was performed on a VG Platform II system (Micromass Co., MA) equipped with a Digital DECPc XL560 computer for data analysis. Positive ion mass spectra were obtained using the heated nebulizer-atmospheric pressure chemical ionization (APCI) interface. The ion source temperature was set at 150 °C and the sample cone voltage was 10V.

Direct Probe EI-MS Analysis. Direct probe EI-MS was conducted on a Finnigan MAT 8230 high resolution mass spectrometer. The temperature of the probe was increased from 35 to 350 °C at 5 °C/sec.

H^1-NMR Analysis. The H^1-NMR (360 MHz) spectrum was obtained in pyridine-*d*5 solution on a modified Nicolet NT 360 instrument. Chemical shifts are reported as parts per million relative to tetramethylsilane (TMS) and coupling constants are in Hertz.

Syntheses of N,N′-Diallylthiourea and N,N′-diphenethylthiourea. For the synthesis of N,N′-diallylthiourea, AITC (0.1 mole) was diluted with 30 mL of toluene and allylamine (0.1 mole) was added dropwise. For the synthesis of N,N′-diphenethylthiourea, the AITC and allylamine were replaced by PEITC (0.05 mole) and phenethyl amine (0.05 mole), respectively. The reaction solution was then stirred at 60 °C for 2 hr. The reaction mixture was concentrated and kept at -20 °C overnight. The resulting solid was washed with hexane three times. The purity of the synthesized product was 99% based on HPLC analysis. This method was a modified version of that reported by Pascual and Rindlisbacher (*12*).

Results and Discussion

Volatile Compounds Generated from Thermal Degradation of AITC and PEITC. AITC has a pungent flavor note. After thermal reaction in aqueous solution for 1 hr, the odor is totally changed to a strong garlic-like note. The degraded solutions of AITC and PEITC were simultaneously distilled and extracted and then analyzed using GC and GC-MS. Figure 2 shows the GC chromatogram of volatile compounds generated from the thermal degradation of AITC. Figure 3 lists these compounds which were tentatively identified by comparing their mass spectral data with references or by interpretation of the EI mass spectra (compounds 3 and 8). In total, there were 9 thermal degradation volatile products (compound 2 was the

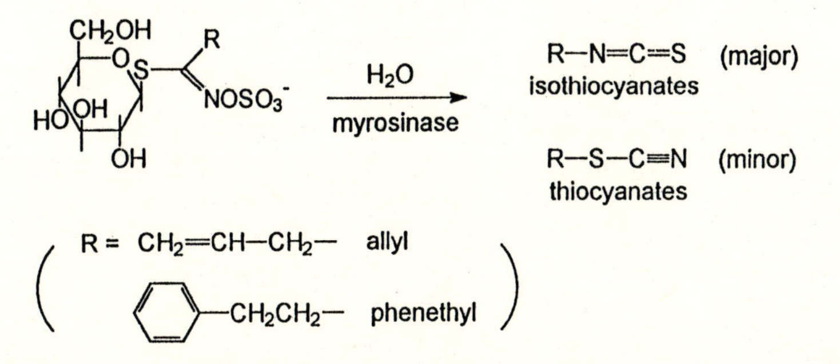

Figure 1. Scheme for the generation of isothiocyanates from glucosinolates.

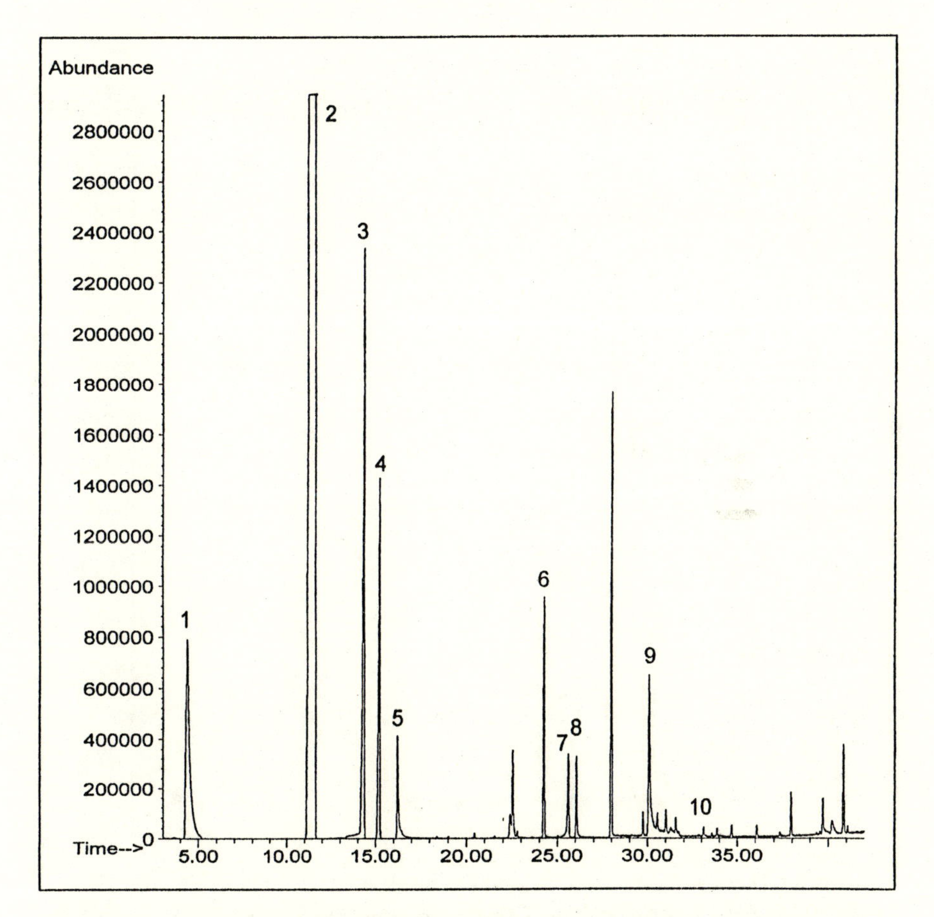

Figure 2. GC chromatogram of the volatile compounds generated from the thermal reaction of allyl isothiocyanate.

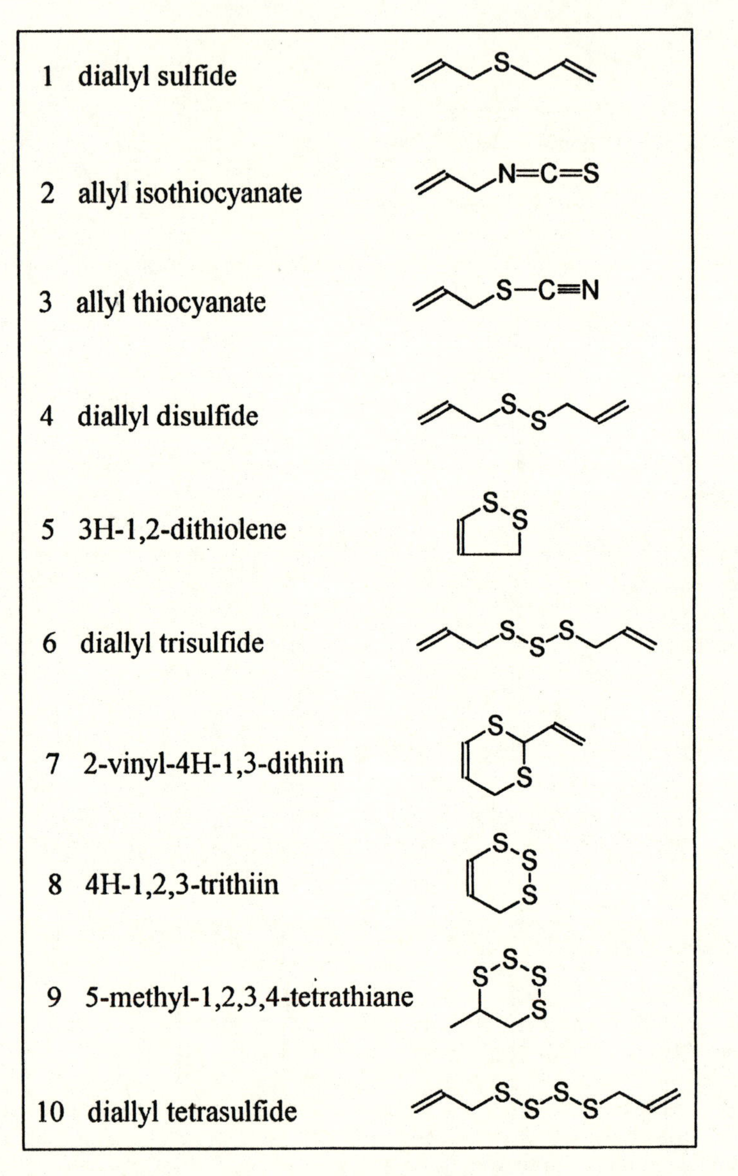

Figure 3. Thermal decomposition products identified from allyl isothiocyanate.

the starting compound) identified from AITC; however, there were no thermal degradation volatile products found from PEITC (data not shown). This suggests that the pathway for the generation of volatile compounds from the thermal reaction of AITC is different from that of PEITC although they are analogous compounds.

The mechanism for the formation of these volatile compounds from AITC is proposed in Figure 4. Compound 3 is an isomerization product from AITC. Allyl mercaptan, which could be a decomposition product from compound 3 through a reducing process and release of one molecule of HCN, is a key intermediate to generate aliphatic and cyclic sulfides. The dimerization of allyl mercaptan leads to the formation of diallyl disulfide (compound 4) as proposed by Kawakishi and Namiki (*6*). It has been shown that heating pure diallyl disulfide leads to the generation of significant amounts of diallyl sulfide, diallyl trisulfide and diallyl tetrasulfide (*13*). Diallyl sulfide (compound 1) and diallyl trisulfide (compound 6) can be generated from diallyl disulfide by a disproportionation or an allyldithio radical addition reaction (*13*). Diallyl tetrasulfide probably produced from the interaction of two allyldithio radicals. 2-Vinyl-4H-1,3-dithiin (Compound 7) is a thermal degradation product from garlic oil and can be generated from two molecules of thioacrolein (CH_2=CH-CH=S) (*13*). It is possible that thioacrolein results from the oxidation of allyl mercaptan. Compounds 9 and 5 were reported as thermal degradation products from alliin and deoxyalliin, respectively, the flavor precursors of garlic (*14*). Yu et al (*14*) proposed that compound 9 could form from the interaction of one molecule of allyl mecaptan and three molecules of hydrogen sulfide; while compound 5 could form from the interaction of one molecule of allyl mecaptan and one molecule of hydrogen sulfide followed by an oxidation reaction. Compound 8 has not been reported as a thermally degraded product of garlic oil. However, its analogous compound, 1,2,3-trithiane, has been found in an aqueous solution of the thermally degraded alliin, and its formation mechanism is proposed as involving the interaction of one molecule of allyl mecaptan and two molecules of hydrogen sulfide (*14*). 1,2,3-Trithiane can probably be further oxidized to generate compound 8.

Nonvolatile Compounds from Thermal Degradation of AITC and PEITC. After thermal reaction, the nonvolatile compounds in the aqueous phase of AITC or PEITC were analyzed by HPLC. Figures 5 and 6 show the HPLC chromatograms of the thermal degradation nonvolatile products from AITC and PEITC, respectively. There is one major compound (compound A) present in the thermal degradation of AITC (Figure 5); while two major compounds (compounds B and C) were found from the thermal degradation of PEITC (Figure 6). The compounds of interest were collected by HPLC and further identified by LC-MS, H^1-NMR and direct probe EI-MS.

In order to determine the molecular weight of the compounds of interest, an LC-MS equipped with (APCI+) was used. The quasimolecular ion (MH^+) of the compound (M) can be obtained by applying a lower cone voltage (10 V). This MH^+ is formed via a proton (H^+) transfer from a protonated solvent molecule, which is obtained when the polar solvent as a mobile phase is protonated at atmospheric pressure.

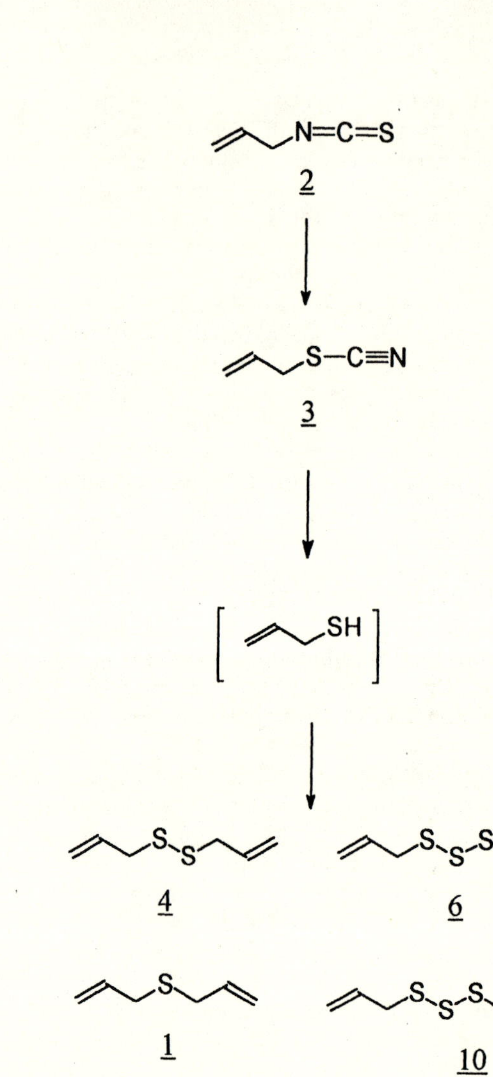

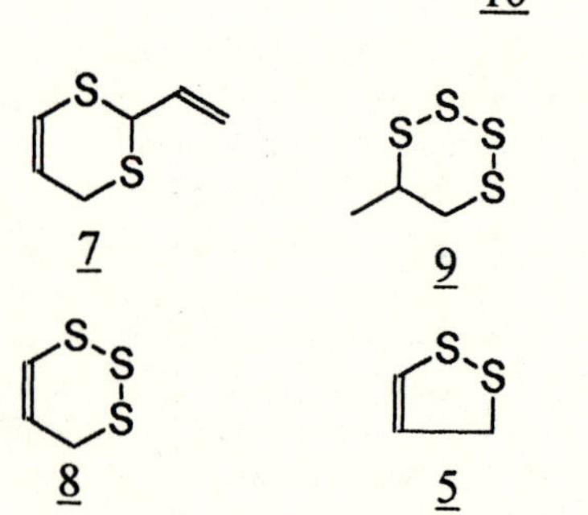

Figure 4. The possible pathways for the formation of volatile compounds from allyl isothiocyanate.

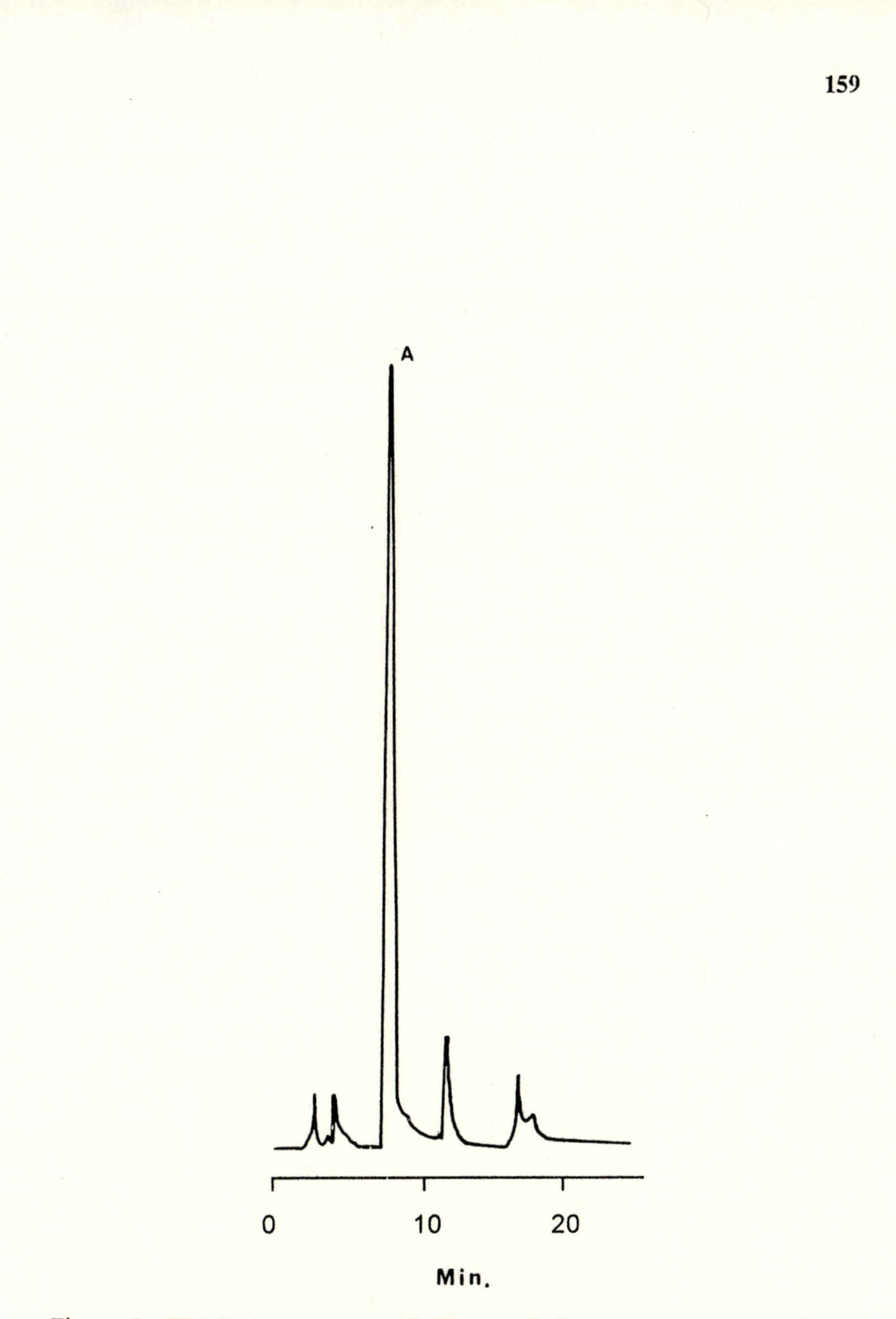

Figure 5 HPLC chromatogram of the nonvolatile products generated from thermal reaction of allyl isothiocyanate.

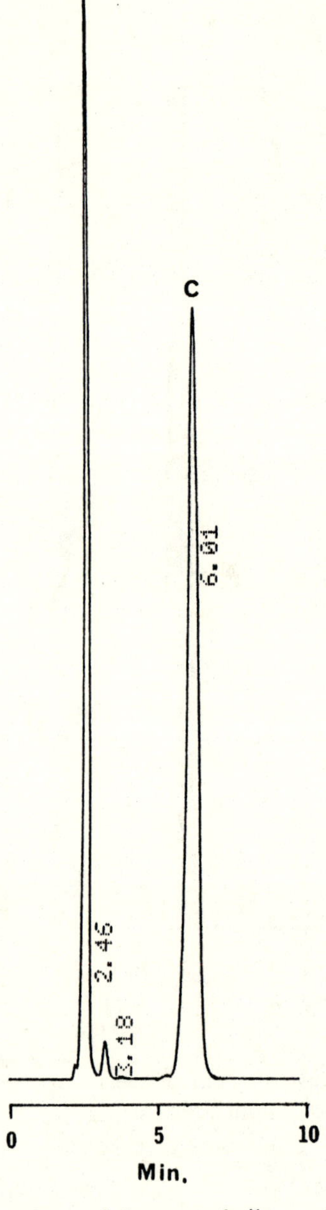

Figure 6. HPLC chromatogram of the nonvolatile products generated from thermal reaction of phenethyl isothiocyanate.

The molecular weight of compound A was 156 according to the APCI(+) mass spectral data shown in Figure 7. The strong fragment at *m/z* 41 suggests the presence of an allyl group in this structure (Figure 7). The fragment at *m/z* 56 could be CH_2=CH-CH=NH_2^+, and the fragment at *m/z* 141 could be obtained after the double bond on the allyl group is shifted to the adjacent carbon followed by loss of a methyl group. The H^1-NMR data of compound A is shown in Table 1. Combining the mass spectral and H^1-NMR data, this compound was identified as N,N′-diallylthiourea. This structure was confirmed by the mass spectral and H^1-NMR data of the synthesized compound.

Compound B was identified as the starting material, phenethyl isothiocyanate, by comparing the LC-MS spectral data and the GC and HPLC retention times to those of the standard (data not shown).

Figure 8 shows the APCI(+) and EI mass spectra of compound C. The molecular weight of compound C was 284 according to the APCI(+) mass spectral data. The EI- mass spectrum shows the molecular ion at *m/z* 284. The strong fragments at *m/z* 105, 91, and 77 suggest the presence of $C_6H_5CH_2CH_2^+$, $C_6H_5CH_2^+$, and $C_6H_5^+$, respectively. The fragment ion at *m/z* 120 is proposed as $C_6H_5CH_2CHNH_2$. Combing the EI mass spectral data and H^1-NMR data (Table 1), compound C was identified as N,N′-diphenethylthiourea. This structure was also confirmed by the mass spectral and H^1-NMR data of synthesized compound.

N,N′-diallylthiourea was reported to be one of the major degradation products from AITC when AITC was incubated at 37 C for a few days (*6*). However, there was insufficient data in the report of Kawakishi and Namiki (*6*) to support this identification. Our study confirmed that N,N′-diallylthiourea was the major nonvolatile compound in the thermal degradation of AITC.

Proposed mechanisms for the formation of N,N′-diallylthiourea and N,N′-diphenethylthiourea are shown in Figure 9. AITC (or PEITC) is hydrolyzed to allylamine (or phenethylamine), which then reacts with AITC (or PEITC) to generate N,N′-diallylthiourea (or N,N′-diphenethylthiourea). This pathway was further confirmed by reacting allylamine (or phenethylamine) and AITC (or PEITC) to produce N,N′-diallylthiourea (or N,N′-diphenethylthiourea).

Conclusion

Our study shows that AITC and PEITC can be thermally decomposed to the nonvolatile products, N,N′-diallylthiourea and N,N′-diphenethylthiourea, respectively. AITC, which possesses a pungent mustard odor, is an unstable compound that decomposes to compounds with a garlic-like flavor note. It was found that most of the volatile decomposition products from AITC were sulfur-containing aliphatic or cyclic compounds. However, there are no volatile degradation compounds found from the thermal reaction of PEITC. These results indicate that the structure of the side chain on the isothiocyanate effects the pathway for the formation of thermal reaction products.

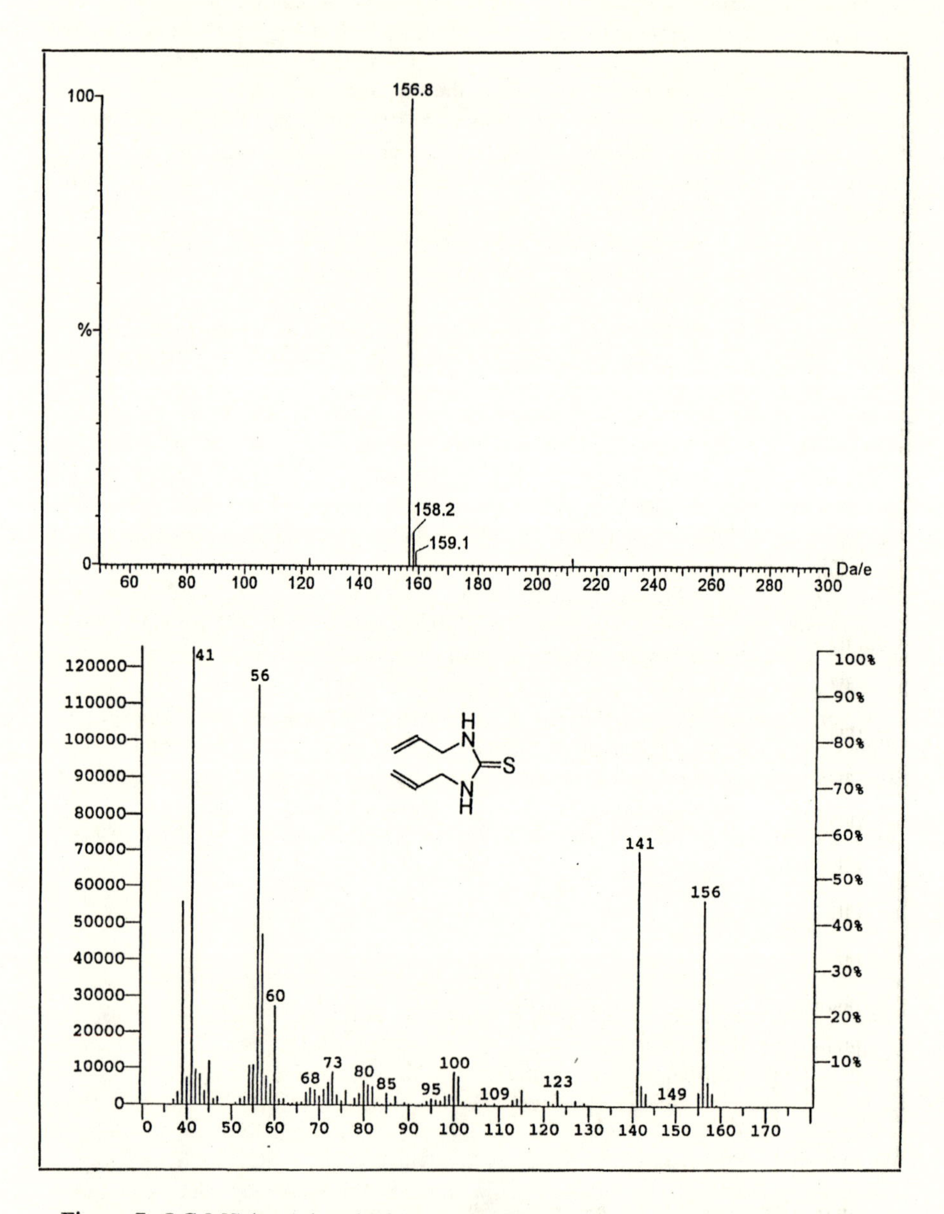

Figure 7. LC-MS (upper) and Direct probe EI-MS spectra of compound A from the thermal degradation of allyl isothiocyanate.

Table I. Mass Spectra of Diethyl Acetals

Acetal	Characteristic Mass Spectral Ions and Abundances	[1]M	[2]M-1	[3]RI
trans-2-hexenal diethyl acetal	57(100), 127(100), 85(67.2), 41(42.5), 43(32.3), 99(31.3), 129(28.4), 69(26.2)	172 (2.9)	171 (1.7)	1031
isovaleraldehyde diethyl acetal	103(100), 75(79.0), 115(66.3), 47(62.5), 69(56.4), 71(48.5), 43(38.3), 45(25.5)	160 (0)	159 (0.5)	986
citral (neral/geranial) diethyl acetal	69(100), 87(99.2), 41(52.7), 181(34.6), 103(27.9), 83(21.5), 59(20.7), 75(19.6)	226 (0.5)	225 (0.6)	1514
heptanal diethyl acetal	103(100), 143(74.8), 75(63.5), 47(58.3), 55(51.7), 97(50.6), 43(30.6), 57(28.8)	188 (0)	187 (0.8)	1186
octanal diethyl acetal	103(100), 47(41.7), 157(35.5), 69(31.9), 57(31.8), 75(29.6), 55(18.9), 41 (16.5)	202 (0)	201 (0.5)	1307
nonanal diethyl acetal	103(100), 75(31.0), 47(26.5), 171(26.0), 69(24.9), 57(15.3), 55(12.9), 41(12.5)	216 (0)	215 (0.3)	1319
dodecanal diethyl acetal	103(100), 43(57.1), 57(54.6), 41(57.1) 55(41.5), 75(28.5), 47(26.6), 69(23.1)	258 (0)	257 (0)	n.d.
cinnamaldehdye diethyl acetal	161(100), 133(72.7), 105(48.3), 55(46.8), 131(42.7), 103(39.7), 77(39.3), 115(31.6)	206 (17.3)	205 (3.3)	1518
phenyl-acetaldehyde diethyl acetal	103(100), 47(62.2), 75(46.6), 91(28.2), 121 (21.5), 149(17.6), 104(6.7), 65(6.3)	194 (0)	193 (0.1)	1307

[1]M = Molecular ion (% abundance)
[2]M-1 = Molecular ion minus H (% abundance)
[3]RI = GC retention index on non-polar, polydimethylsiloxane capillary column relative to n-paraffin series

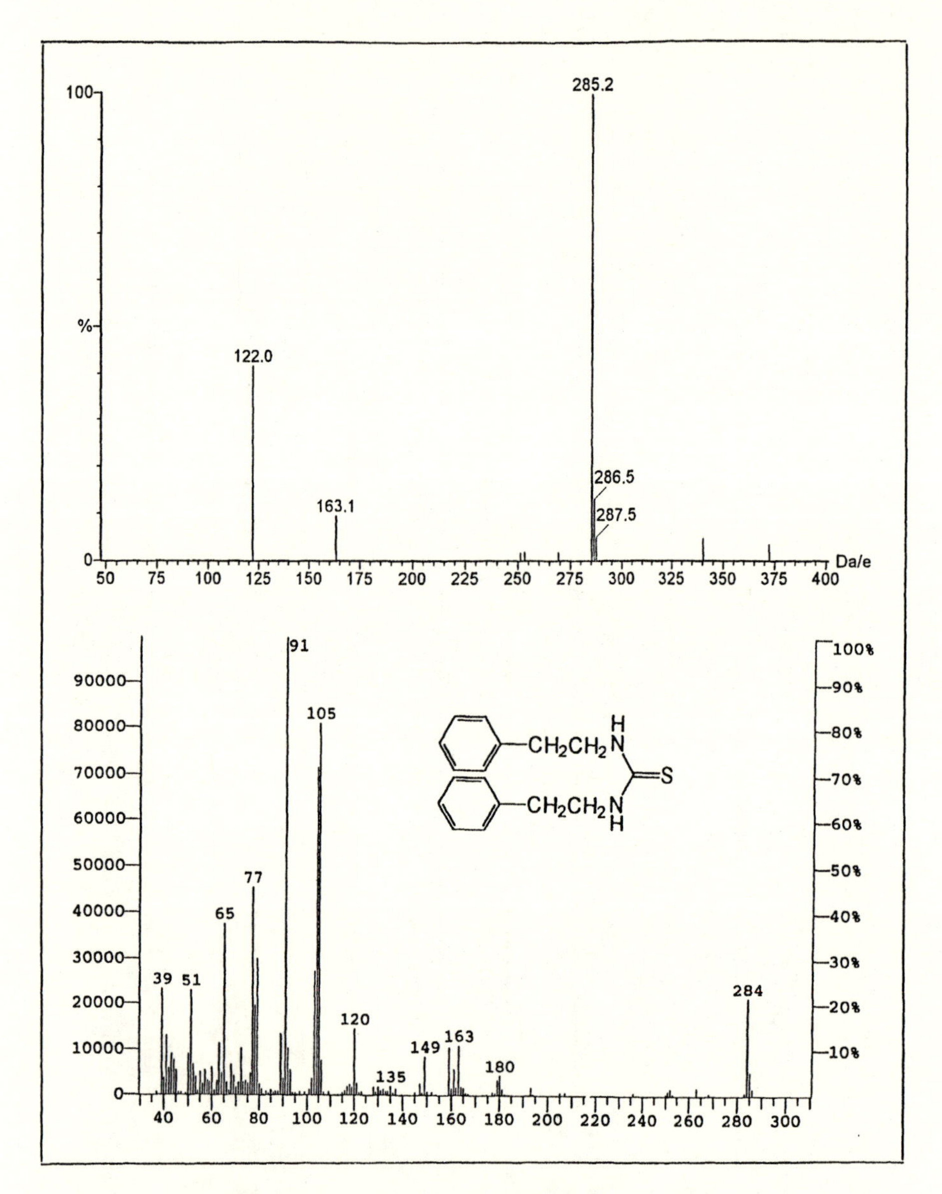

Figure 8. LC-MS (upper) and Direct probe EI-MS spectra of compound C from the thermal degradation of phenethyl isothiocyanate.

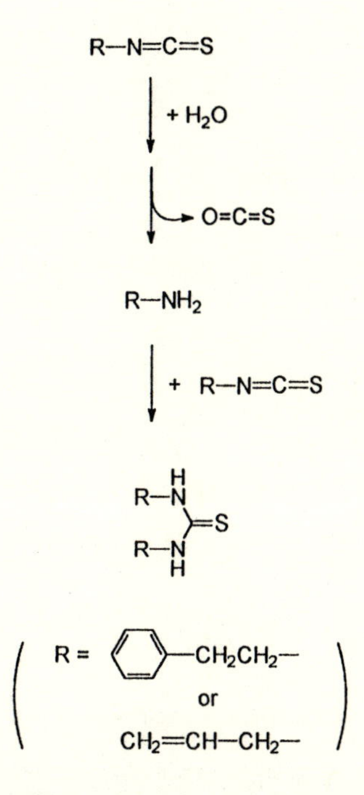

Figure 9. The possible pathways for the formation of N,N′-diallylthiourea or N,N′-diphenethylthiourea from allyl isothiocyanate or phenethyl isothiocyanate.

Literature Cited

1. VanEtten, C. H.; Daxenbichler, M. E.; Williams, P. H.; Kwolek, W. F. *J. Agric. Food Chem.* **1976**, *24*, 452-455.
2. Masuda, H.; Harada, Y.; Tanaka, K.; Nakajima, M.; Tateba, H. In *Biotechnology for Improved Foods and Flavors;* Takeoka, G. R.; Teranishi, R.; Williams, P. J.; Kobayashi, A., Eds.; ACS Symp. Ser. 637; American Chemical Society: Washington, DC, 1996, pp. 67-78.
3. Chin, H.-W.; Lindsay, R. C. *J. Food Sci.* **1993**, *58*, 835-839, 841.
4. Shahidi, F.; Gabon. J.-E. *J. Food Sci.* **1990**, *55*, 793-795.
5. Zrybko, C. L.; Rosen, R. T. In *Spices: Flavor Chemistry and Antioxidant Properties*; Risch, S. J.; Ho, C.-T., Eds.; ACS Symp. Ser. 660, American Chemical Society: Washington, DC, **1997**, pp 125-137.
6. Kawakishi, S.; Namiki, M. *Agric. Biol. Chem.* **1969**, *33*, 452-459.
7. Ina, K.; Nobukuni, M.; Sano, A.; Kishima, I. *Nippon Shokuhin Kogyo Gakkaishi.* **1981**, *12*, 627-631.
8. Stoner, G. D.; Galati, A. J.; Schmidt, C. J.; Morse, M. A. In *Food Phytochemicals for Cancer Prevention I. Fruits and Vegetables.* Hung, M.-T.; Osawa, T.; Ho, C.-T.; Rosen, R. T., Eds.; ACS Symp. Ser. 546, American Chemical Society, Washington, DC, 1994, pp. 173-180.
9. Stoner, G. D.; Morrissey, D. T.; Heur, Y.-U.; Daniel, E. M.; Galati, A. J.; Wagner, S. A. *Cancer Res.* **1991**, *51*, 2063-2068.
10. Chung, F.-L.; Morse, M. M.; Eklind, K. I.; Xu, Y. In *Nitrosamines and Related N-Nitroso Compounds. Chemistry and Biochemistry.* Loeppky, R. N.; Michejda, C. J., Eds.; ACS Symposium Series 553, American Chemical Society: Washington, DC, **1994**, pp. 232-247.
11. Wattenberg, L. W. J. *Natl. Cancer Inst.* **1977**, *58*, 295-298.
12. Pascual, A.; Rindlisbacher, A. *Pestic. Sci.* **1994**, *42*, 253-263.
13. Block, E.; Iyer, R.; Grisoni, S.; Saha, C.; Belman, S.; Lossing, F. P. *J. Amer. Chem. Soc.* **1988**, *110*, 7813-7827.
14. Yu, T.-H.; Wu, C.-M.; Rosen, R. T.; Hartman, T. G.; Ho, C.-T. *J. Agric. Food Chem.* **1994**, *42*, 146-153.

Chapter 15

Biosynthesis of Optically Active δ-Decalactone and δ-Jasmin Lactone by 13-Lipoxygenase Pathways in the Yeast *Sporobolomyces odorus*

R. Tressl, L.-A. Garbe, T. Haffner, and H. Lange

Institut für Biotechnologie, Technical University of Berlin, Seestrasse 13, 13353 Berlin, Germany

In a series of [^{2}H]- and [^{13}C]- labeling experiments the biochemical formation of linoleic acid derived (*R*)-δ-decalactone and of linolenic acid derived (*R*)-δ-jasmin lactone in cultures of *S. odorus* was investigated. Application of isotopically labeled (13*S*,9Z,11E)-13-hydroxy-[9,10,12,13-$^2H_4$]-octadeca-9,11-dienoic acid and (Z)-13-oxo-[9,10-$^2H_2$]-9-octadecenoic acid as OH-functionalized precursors of (*R*)-δ-decalactone allowed insight into the stereochemical course of the biosynthesis. Using an established chiral GC/MS method a complete inversion of the initial configuration could be characterized. (13*S*,9Z,11E,15Z)-13-Hydroxy-[9,10,12,13,15,16-$^2H_6$]-octadeca-9,11,15-trienoic acid as precursor was transformed into (*R*)- and (*S*)-jasmin lactone. In this experiment 88 % of the metabolized labeled precursor were transformed under retention of the original configuration. Administration of ethyl [1-^{13}C]-5-hydroxydecanoic acid elucidated a catabolic pathway of δ-decalactone to ethyl [1-^{13}C]-glutarate and pentanol-1 by Baeyer-Villiger oxidation.

δ-Decalactone (**3**) and δ-jasmin lactone (**6**) are known as important aroma compounds in fruits, fermented foods and essential oil. The chirospecific analysis by means of enantiomer separation by chiral GC/MS is an effective tool to prove the authenticity of natural flavorings in food. Therefore, the knowledge of the biosynthesis of lactone enantiomers and their catabolic metabolism is of importance. **3** as well as **6** are biosynthesized as enantiomers or in distinct enantiomeric distributions (*1-4*) (Table I and Table II). (*R*)-**3** is predominantly formed in peach, apricot, nectarine whereas the (*S*)-enantiomer is formed in raspberries. Milk, butter, cheddar cheese and coconut contain **3** in distinct enantiomeric compositions.

A natural source of (*R*)-**3** is massoia bark oil from which it can be obtained by enzymatic reduction of (*R*)-2-decen-5-olide (*5*). **6** is biosynthesized as (*R*)- and (*S*)-

Table I: Configuration and optical purity of naturally occurring δ - decalactone

Source	Configuration	Enantiomeric purity % e.e.
Peach	(R)	96
Apricot	(R)	90
Nectarine	(R)	90
Raspberries	**(S)**	**94**
Milk, Butter	(R)	60
Cheddar cheese	(R)	43.6
Coconut	(R)	48
S. odorus	(R)	>99
Massoia bark oil	(R)	100

Table II: Configuration and optical purity of naturally occurring δ - jasmin lactone

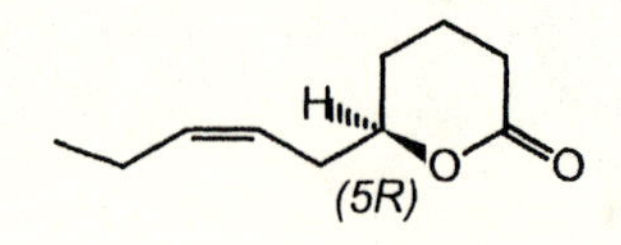

Source	Configuration	Enantiomeric purity % e.e.
Jasmin absolute	(R)	99
Tuberose absolute	(S)	99
Peach	(R)	34
Apricot	(R)	90
Sp. odorus	(R)	46

enantiomer in flowers and in distinct enantiomeric distributions in fruits, fermented products and in the yeast *S. odorus*. The biosynthesis of (*R*)-**3** in the lactone-producing yeast *S. odorus* was characterized as a (*R*)-12-hydroxylation of oleic acid, followed by β-oxidation. Parallel to this pathway a Δ^{12}-desaturase transformed [9,10-$^2H_2$]-oleic acid into [9,10-$^2H_2$]-linoleic acid ([$^2H_2$]-**1**) which was further metabolized to [3,4-$^2H_2$]-γ-dodecenlactone and (*R*)-[^{2}H]-**3**, respectivly (*6*). The biosynthesis of (*R*)-**3** ((*R*) >99 %) in *S. odorus* is initiated by a lipoxygenation at C(13) of linoleic acid (**1**). It could be shown that (13*S*,9Z,11E)-13-hydroxy-[9,10,12,13-$^2H_4$]-octadeca-9,11-dienoic acid ([$^2H_4$]-**2**) was converted to (*R*)-[2,4-$^2H_2$]-**3** under total inversion of the initial configuration (Scheme 1) (*7*). In contrast, [9,10,12,13,15,16-$^2H_6$]-α-linolenic acid ([$^2H_6$]-**4**) and the corresponding (13*S*,9Z,11E,15Z)-13-hydroxy-[9,10,12,13,15,16-$^2H_6$]-octadeca-9,11, 15-trienoic acid ([$^2H_6$]-**5**) applied to cultures of *S. odorus* were transformed into (*R*)-**6** under 88 % retention and only 12 % inversion of the stereogenic center C(5) of the lactone. For both lactones a 13-lipoxygenase / reductase pathway was characterized as initial step of the biosynthetic pathways (*8,9*). Recently, a lipoxygenase / reductase pathway was characterized in plants, degrading polyunsaturated fatty acids in cucumber and barley glyoxysomes (*10*). The multifunctional protein (**MFP**) of plant peroxisomes possesses enoyl-CoA-hydratase, 3-hydroxyacyl-CoA epimerase and L-3-hydroxyacyl-CoA dehydrogenase activities. (*11*) The β-oxidation of **2** is an important catabolic pathway of storage lipids. Investigation of the **MFP** of yeast showed a bifunctional protein with enoyl-CoA hydratase-2 and D-3- hydroxyacyl-CoA dehydrogenase activities (*12,13*). These results clearly demonstrate that the β-oxidation of fatty acids in *S. cerevisiae* follows a previously unknown stereochemical course, namely it occurs via a D-3-hydroxy-acyl-CoA intermediate. Therefore the 13-lipoxygenase / reductase pathway in *S. odorus* leading to (*R*)-**3** and (*R*)-**6** was further elucidated by a series of labeling experiments.

Biosynthesis of (*R*)-δ-Jasmin Lactone in *S. odorus*

The identification of a 13-lipoxygenase / reductase pathway in *S. odorus* was demonstrated by administration of [$^2H_6$]-**4** and [$^2H_6$]-**5** to growing cultures of *S. odorus*. The isotopomeric distribution of **6** was investigated by GC/MS. Both substrates were transformed into **6** with nearly identical labeling patterns (Table III). The enantiomeric distribution of **6** isotopomers was further characterized by chiral GC/MS (Figure 1). The configuration and the GC elution order of **6** were estimated after catalytic hydrogenation of the compound to **3** and comparison with enantiomerically pure, authentic lactone. The label which was bound to the ring moiety of both lactones, was quantified by a GC/MS method applied to γ-lactones (*6*). (*R*)-[2,3-$^2H_2$]-decano-5-lactone was synthesized from massoia lactone by PtO_2 -catalyzed reduction with $^2H_2$ and analyzed by ^{1}H-NMR and MS, respectively. (*S*)-[$^2H_6$]-**5** is transformed into (*R*)-[$^2H_5$]-**6** (CIP-rules!) with retention and into (*R*)-[$^2H_4$]-**6** with 89 % retention of the original stereocenter (Tabelle IV). The loss of one deuterium in [$^2H_4$]-(*R*)-**6** is explainable by the C=C bond isomerization steps (1,3-

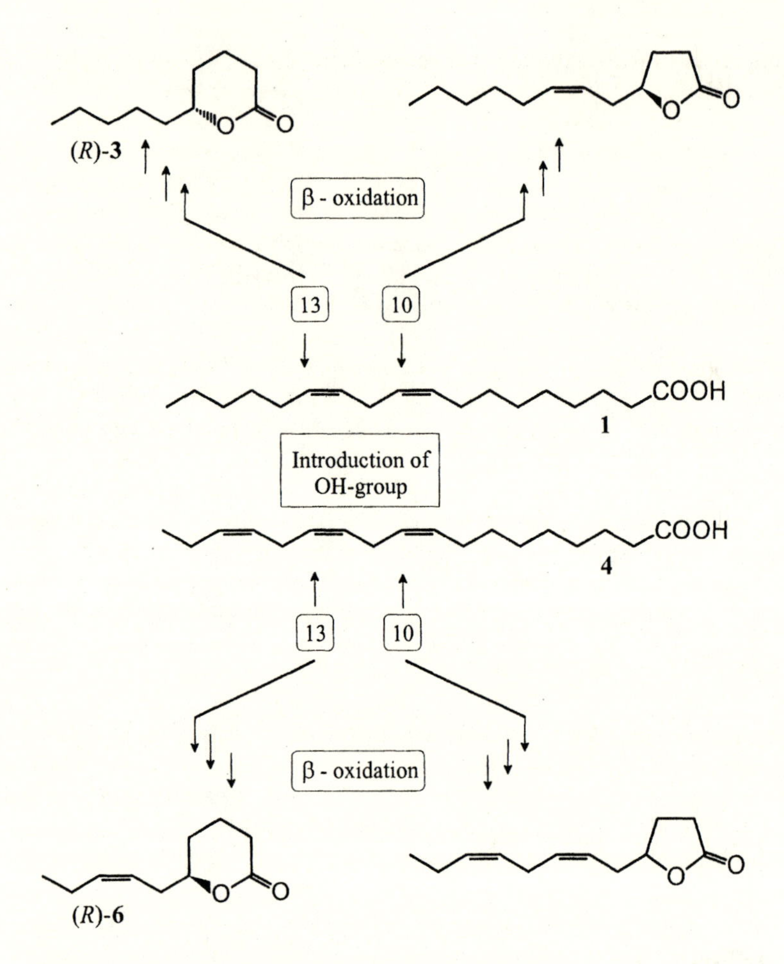

Scheme 1. Overview of lactones formed by regioselective oxygenation and metabolism of linoleic- and linolenic acid.

Table III: Isotopomeric distribution of δ-jasmin lactone (6) after addition of [$^2H_6$]-4 and [$^2H_6$]-5 to cultures of *S. odorus*

isotopomer	representing mass fragment m/z	substrate [$^2H_6$]-**4** (%)	substrate [$^2H_6$]-**5** (%)
unlabeled **6**	99	28.1	32.8
[$^2H_3$]- **6**	100	9.9	8.1
[$^2H_4$]- **6**	101	25.7	23.8
[$^2H_5$]- **6**	102	36.3	35.3

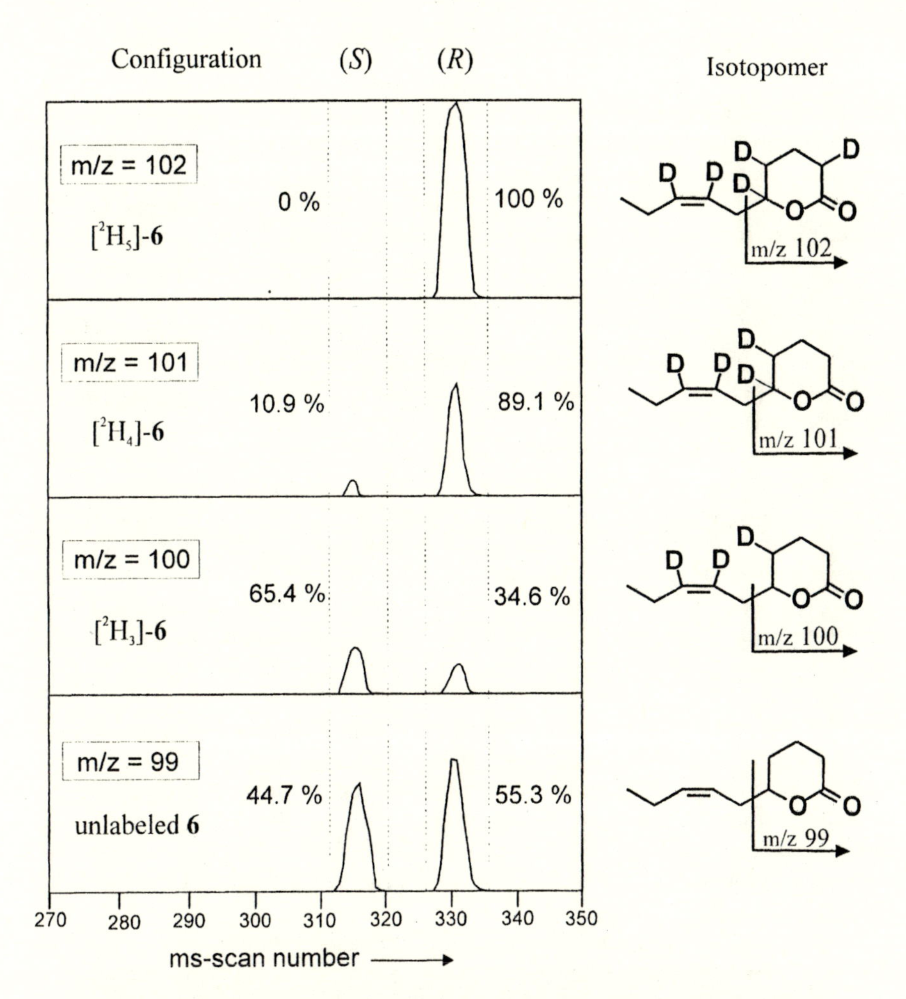

Figure 1. Enantiomeric distribution of the biosynthesized isotopomers of the δ-jasmin lactone **(6)** after administration of [$^2H_6$]-**4** in *S. odorus*.

Table IV: Enantiomeric disribution of δ-jasmin lactone (6) after addition of 5 to cultures of *S. odorus*

isotopomer	configuration (*S*) [%]	(*R*) [%]
[$^2H_5$]- **5**	-	52.5
[$^2H_4$]- **5**	3.9	31.6
[$^2H_3$]- **5**	7.8	4.2
Total Part	**11.7**	**88.3**

H(^{2}H) and 1,5-H(^{2}H) shifts) during isomerase catalysis of the β-oxidation. The formation of the (*S*)-**6** isotopomers involves an oxidation / reduction step with a total loss of the ^{2}H-atom at the stereogenic C(5) of the resulting lactone. The results clearly demonstrate a lipoxygenase / reductase pathway in *S. odorus* leading to (*R*)-**6** with 88% retention of the initial stereogenic center. Only 10 % of the administrated labeled precursor were transformed into the lactone, whereas one major part (of the labeled precursor) was metabolized by an unknown pathway.

Biosynthesis of (*R*)-δ-Decalactone in *S. odorus*

A 13-lipoxygenase / reductase pathway converting [$^2H_4$]-**1** and [$^2H_4$]-**2** into optically pure (*R*)-[2,4-$^2H_2$]-**3** had been characterized (*7*). The results clearly demonstrated that the initial (*S*)-configurated stereocenter at C(13) has to undergo a complete stereochemical inversion before the (*R*)-configurated lactone is formed. Incubation experiments for the elucidation of the biosynthesis of (*R*)-**3** identified 13-oxo-9,11-octadecadienoic acid (**13-KOD**) as an intermediate. The decisive enzyme which controls the enantioselectivity of this pathway was established to be an enonereductase, catalyzing the reduction of **13-KOD** into 13-oxo-9-octadecenoic acid (**13-KOE**). Two enonereductases were recently characterized in *Saccharomyces cerevisiae* (*14*). Synthesis of [9,10-$^2H_2$]-**13- KOE** and addition of the labeled precursor to growing cultures of *S. odorus* established the reduction of 5-oxo-decanoic acid to (*R*)-5-hydroxydecanoic acid as stereoselective step (Table V and Table VI) (Tressl, R.; Garbe, L.-A.; Haffner, T.; Lange, H., *J. Agric. Food Chem.* in prep.). [9,10-$^2H_2$]-**13-KOE** is transformed into [5,6-$^2H_2$]-9-oxo-(*Z*)-5-tetradecenoic acid, [2-^{2}H]-5-oxodecenoic acid by β-oxidation and reduced to (*R*)- [2-^{2}H]-5-hydroxydecanoic acid which undergoes a cyclization to (*R*)-[2-^{2}H]-**3**. These results clearly demonstrated a β-oxidation of **13-KOE** to 5-oxodecanoic acid and a subsequent reduction to (*R*)-5-hydroxydecanoic acid which undergoes a cyclization into (*R*)-δ-decalactone. Detailed results and the spectroscopic data will be published separately. The stereochemical course of the metabolism of [$^2H_6$]-**4** to labeled **6** (Scheme 2) via [$^2H_6$]-**5** and the corresponding transformation of [$^2H_4$]-**1** into labeled **3** (Scheme 3) clearly demonstrated that the presence of a ω-3 double bond changed the stereochemistry of this pathway almost completely. 13-Hydroxylated **1** undergoes 100 % inversion of the stereocenter, leading to (*R*)-5-hydroxydecanoic acid possessing the (*R*)-configuration which is a prerequisite for the successful ß-oxidation by the multifunctional protein of yeast peroxisomes. Only 12 % of the initial stereocenter of 13-hydroxylated **4** were inverted, whereas the predominant part was transformed under retention into (*R*)-**6**.

Degradation of (*R,S*)-5-Hydroxydecanoic Acid in *S. odorus*

3 and **6** are further metabolized by *S. odorus.* A degradative pathway of 2-deceno-5-lactone (massoia lactone) by Fusarium solani was recently identified. During this catabolic pathway (*R*)-**3** was transformed into (*R*)-4-hydroxy-2-nonanone by

Table V: Incubation experiments for the elucidation of the biosynthesis of δ-decalactone (3) in the yeast *S. odorus*

Precursor	Product	Enantio-meric purity [% ee]	Conver-sion [%]
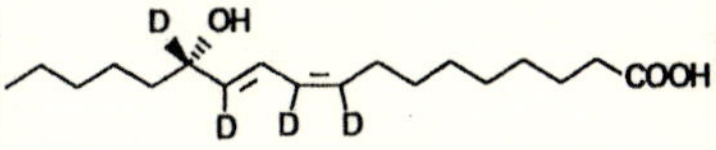 D OH / D D D / COOH [$^2H_4$]-(*S*)-**2**	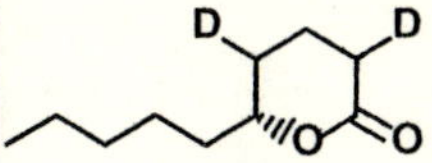 D D / O O [$^2H_2$]-(*R*)-**3**	(R) > 98	15.0
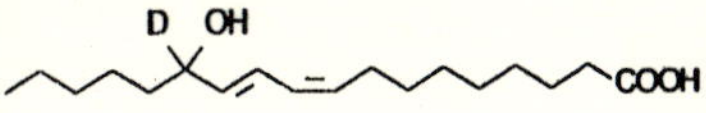 D OH / COOH [^{2}H]-(*R,S*)-**2**	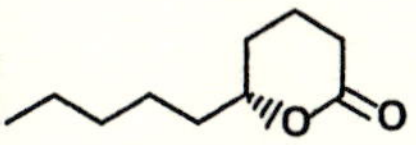 O O (*R*)-**3**	(R) > 98	14.0
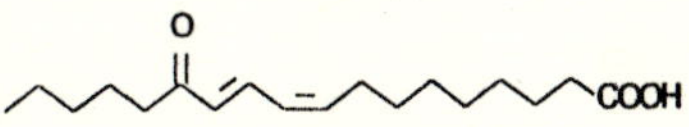 O / COOH 13-KOD	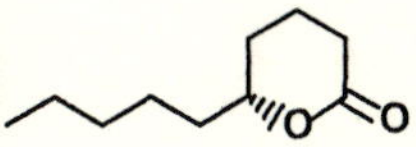 O O (*R*)-**3**	(R) > 98	10.8

Table VI: Labeled metabolites formed by *S. odorus* after administration of 250 ppm [9,10-$^2H_2$]-13-oxooleic acid

Analysed products after administration of 13-KOE (methylester)	Yield after 72 h [ppm]
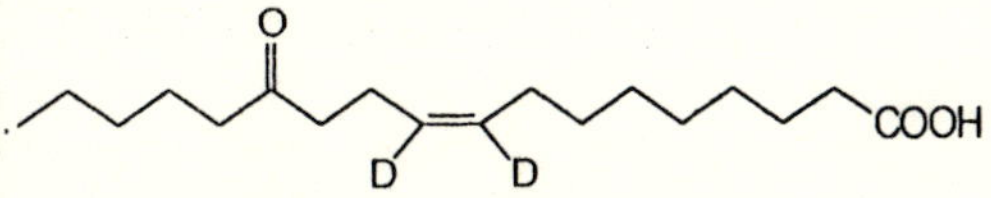 O / COOH / D D	4
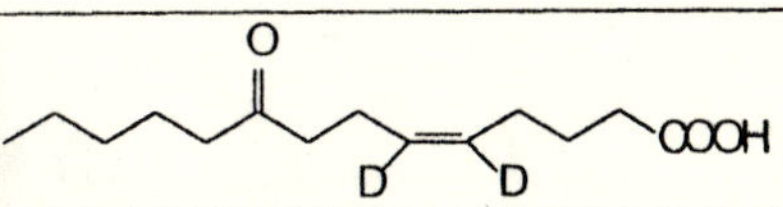 O / COOH / D D	2.5
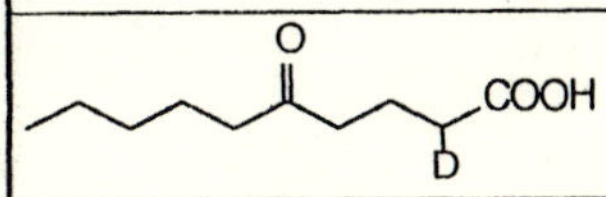 O / COOH / D	0.5
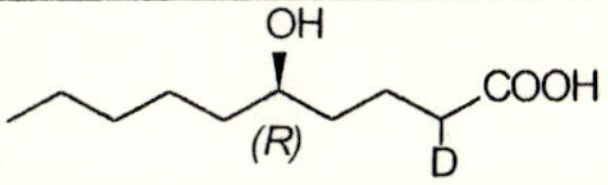 OH / COOH / (R) / D	11.5
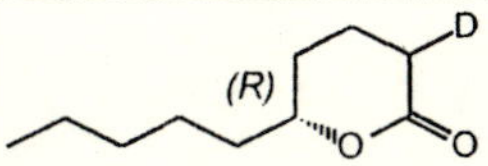 D / (R) / O O	16.5

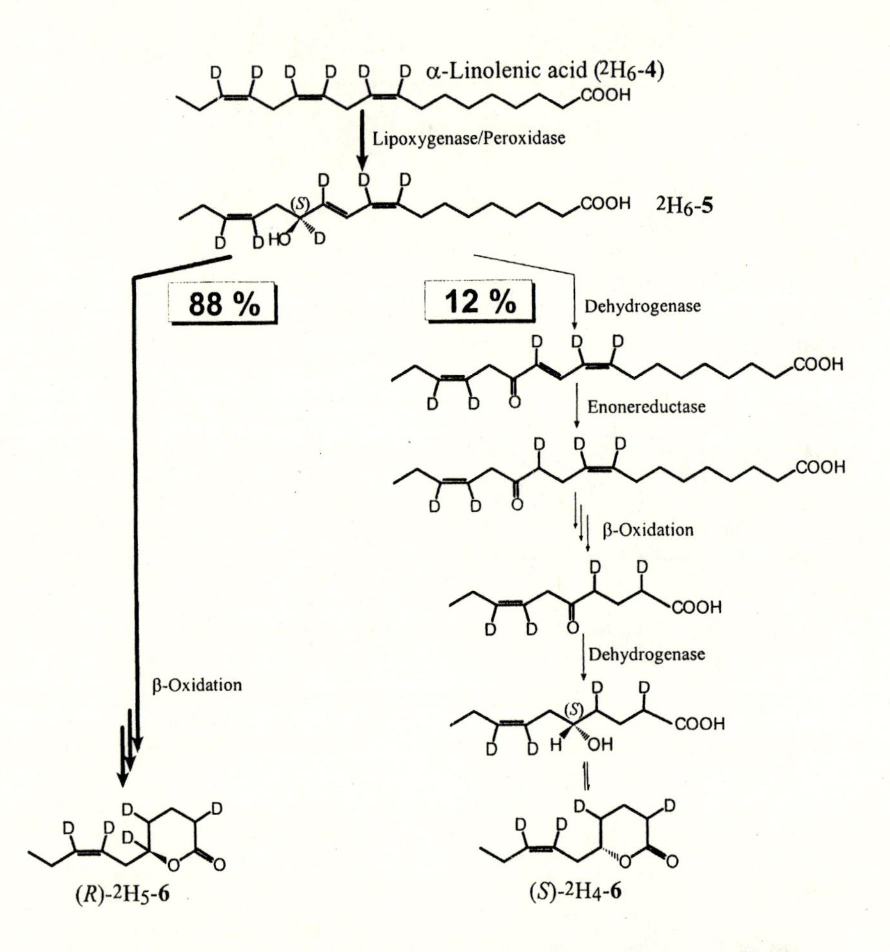

Scheme 2. Stereochemical course of the metabolism of [$^2H_6$]-**4** to δ-jasmin lactone (**6**) via (*S*)-[$^2H_6$]- 13-hydroxy-9,11,15-octadecatrienoic acid ([$^2H_6$]-**5**) in *S. odorus*.

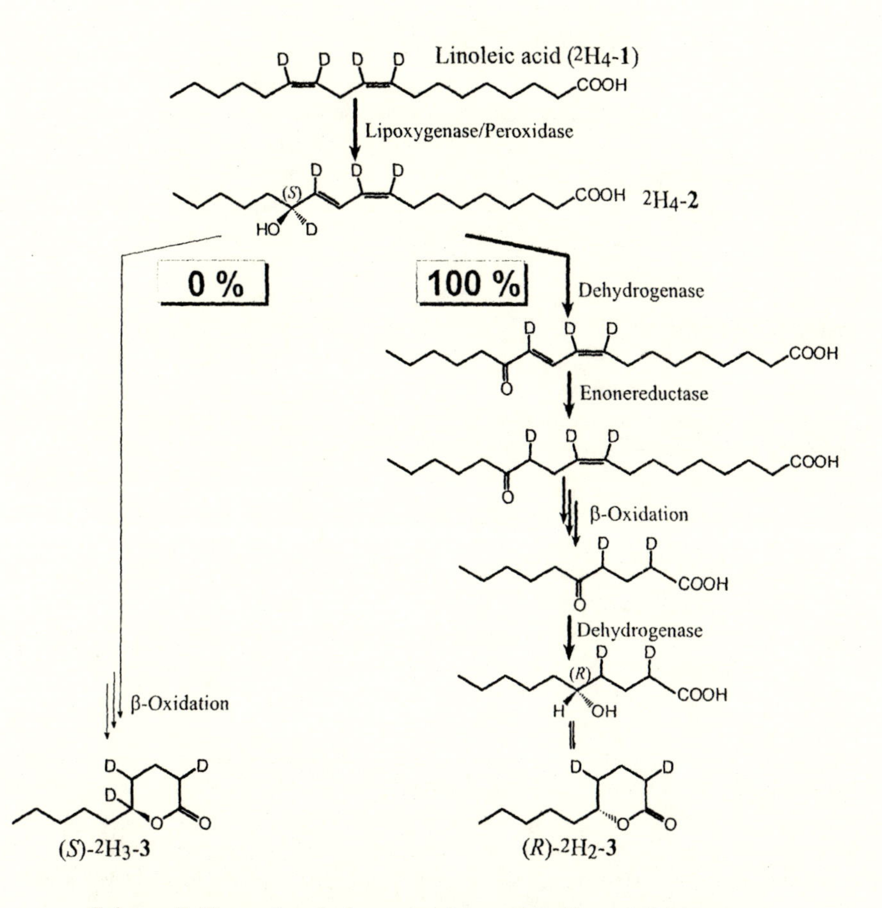

Scheme 3. Stereochemical course of the metabolism of [$^2H_4$]-**1** to δ-decalactone (*S/R* - **3**) via [9,10,12,13-$^2H_4$]-(*S*)-13-hydroxy-9,11-octadecadienoic acid ([$^2H_4$]-**2**) in *S. odorus*.

decarboxylation. In addition, 2,4-nonadienone, 3-nonen-2-one and hexanal were characterized. After administration of (*R*,*S*)-2-deceno-5-lactone, **3** was analyzed with an enantiomeric access of (*S*)-**3** (36,6 % ee.) showing, that (*R*)-**3** was slightly faster metabolized than (*S*)-**3** (*15*). In culture broths of *S. odorus* 4-hydroxy-2-nonanone, 2,4-nonandiol and 2,4-nonadione are not detectable during degradation of **3**. Labeling experiments elucidated endogenous 13 -lipoxygenase / reductase pathways for **1** and **4** in *S. odorus*. According to the stereochemical course of the peroxysomal β-oxidation in yeast only (*R*)-5-hydroxydecanoic acid should be metabolized by *S. odorus*. (*R*)-dec-7-eno-5-lactone did not accumulate during incubation experiments. Therefore, (*S*)-5-hydroxydecanoic acid must be metabolized by a hitherto unknown pathway. Syntheses of [1-^{13}C]-(*R*,*S*)-5-hydroxydecanoic acid and of the corresponding methyl- and ethyl esters were carried out. The [^{13}C]-content of the purified compounds was determineted by NMR- and MS-spectroscopy (>97 %). Incubation experiments with growing cultures of *S. odorus* demonstrated that (*R*,*S*)-5-hydroxydecanoic acid and esters are completely metabolized by *S. odorus*. Racemic ethyl 5-hydroxydecanoate was hydrolyzed by growing cultures of *S. odorus* and transformed into (*R*,*S*)-[1-^{13}C]-**3** (Figure 2 and Figure 3). The free fatty acid and the lactone were metabolized in a correlated way, indicating, that **3** is formed as an intermediate of the 13-lipoxygenase / reductase pathway in the yeast. Investigation of the enantiomeric composition of [1-^{13}C]-**3** indicated a chiral discrimination of this catabolic pathway. The enantiomeric (*S*) / (*R*)-ratio of **3** changed from 49/51 after 3 h of fermentation to 44/56 (100 h) and 33/67 (200 h). These results demonstrated that the (*S*)-enantiomer, which can not be degraded by **MFP** of yeast, is slightly better metabolized as (*R*)-5-hydroxydecanoic acid. The identification of [^{13}C]-labeled 5-oxodecanoic acid, pentyl glutarate and glutarate as well as the corresponding ethyl esters indicated a degradation via a Baeyer-Villiger oxidation in *S. odorus*. The mass spectra of unlabeled and [1-^{13}C]-labeled methyl ethyl glutarate and of methyl 5-oxodecanoate proved this unexpected catabolic pathway (Figure 4). As far as we know it has been shown for the first time that 13-hydroxylated **1** and **4** are degraded by this pathway. In contrast to the 13-lipoxygenase / reductase pathway in plant, the stereochemical course in yeast must be changed for peroxisomal β- oxidation. There are obviously two different pathways operative leading to δ-lactones as intermediates (Scheme 4) (Garbe, L.-A.; Tressl, R., *Lipids*, in prep.). The biosynthesis of (*R*)-**3** corresponds to the β-oxidation of (*S*)-13-hydroxylated **1** via enonereductase and (*R*)-5-oxodecanoic acid dehydrogenase leading to (*R*)-3-hydroxyoctanoic acid, whereas (*R*)-**6** may be biosynthesized as intermediate which is further metabolized via Baeyer-Villiger oxidation. Detailed results and spectroscopic data of the Baeyer-Villiger oxidation of 5-hydroxydecanoic acids will be the subject of another paper.

Literature Cited

(1) Albrecht, W.; Heidlas, J.; Schwarz, M.; Tressl, R.; In *Flavor Precursors-Thermal and Enzymatic Conversion*; Teranishi, R.; Takeoka, G. R.; Güntert, M.,

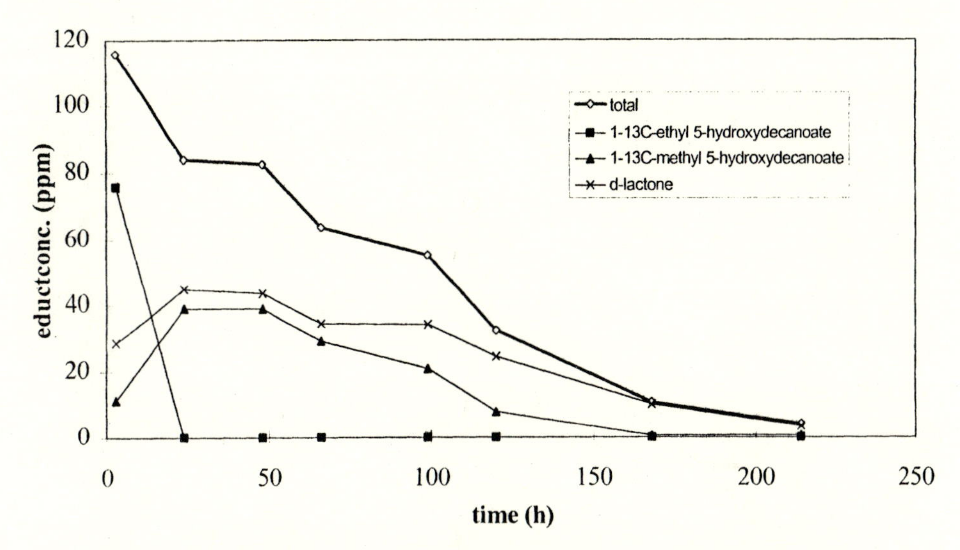

Figure 2. Biodegradation of ethyl [1-^{13}C] -5-hydroxydecanoate in *S. odorus*.

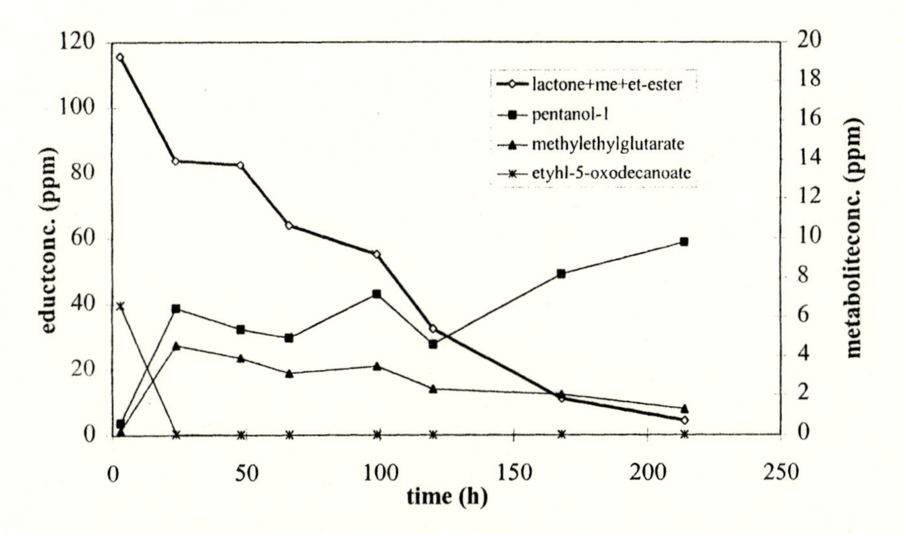

Figure 3. Formation of [1-^{13}C]- ethyl glutarate and pentanol-1 during degradation in *S. odorus*.

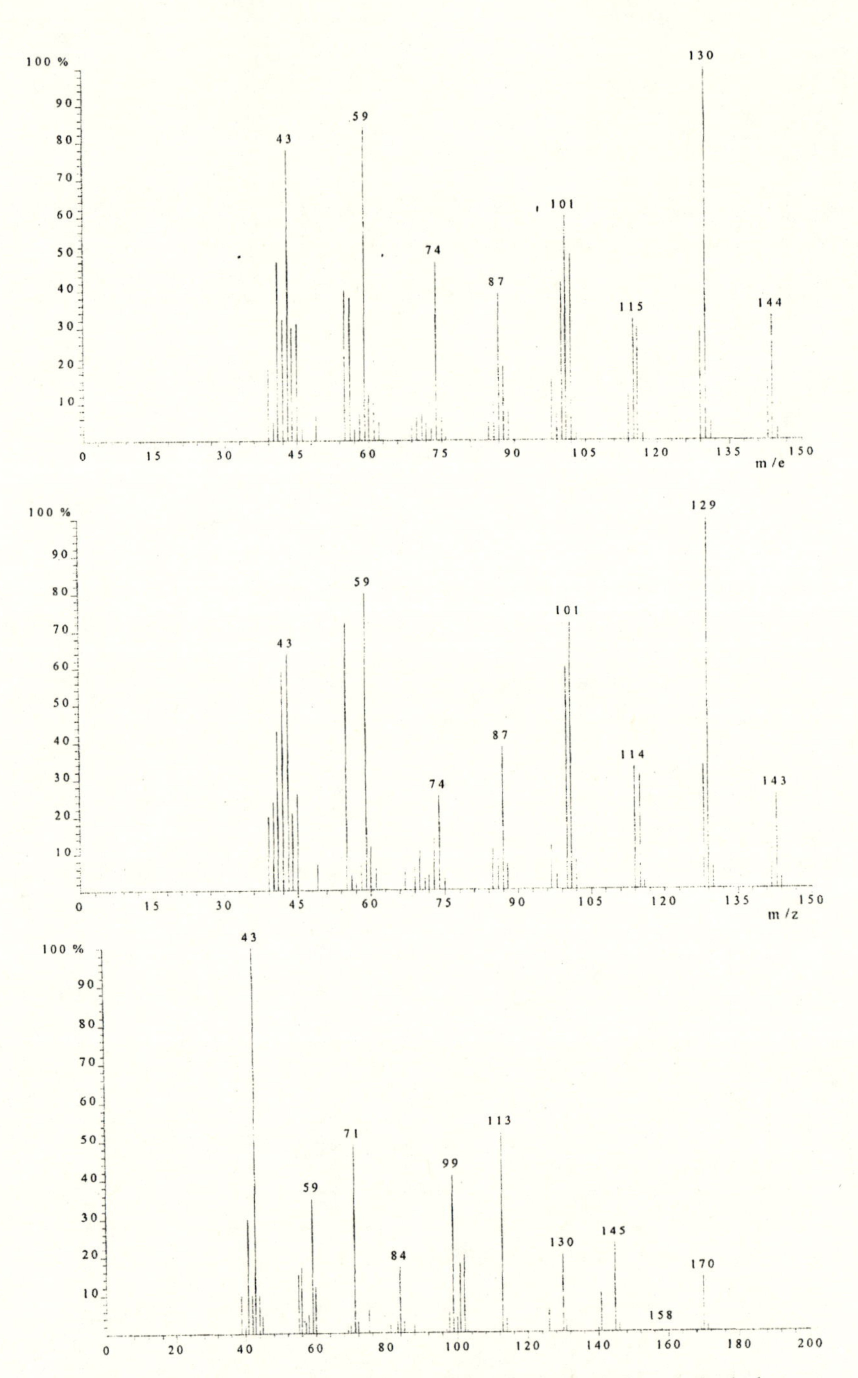

Figure 4. Mass spectra of [1-^{13}C]- labeled and unlabeled methyl ethyl glutarate and of methyl [1-^{13}C]-5-oxodecanoate.

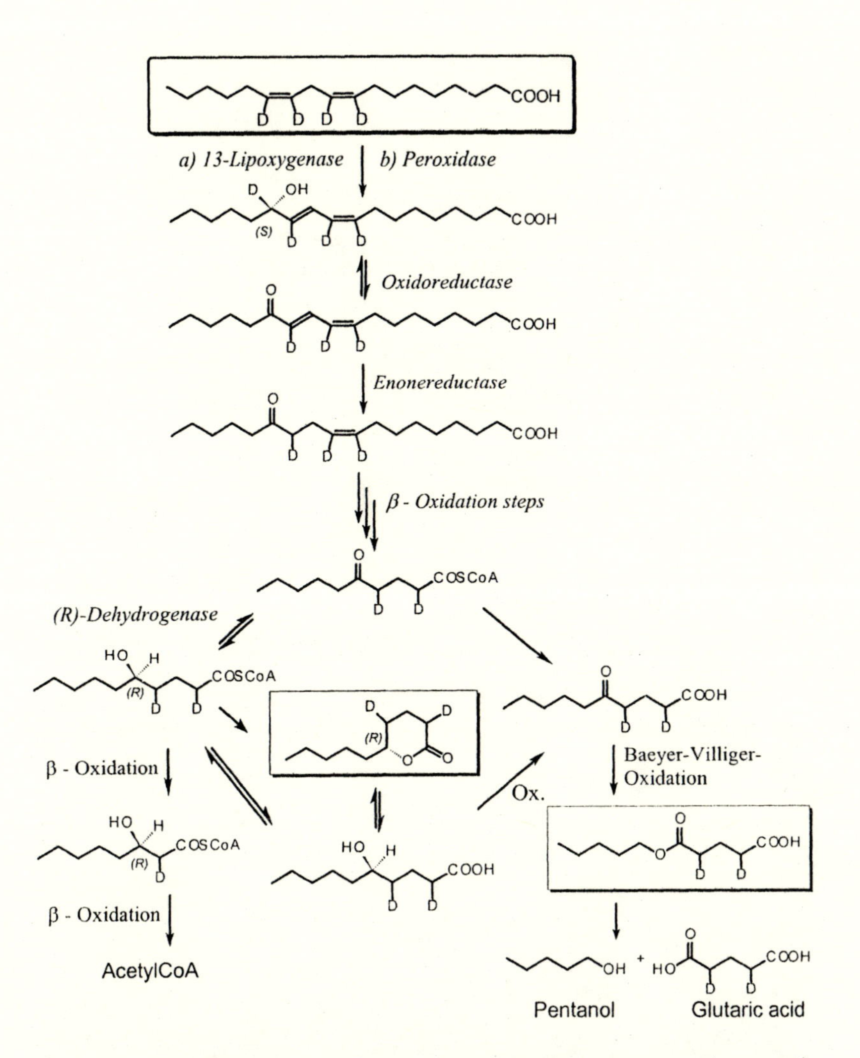

Scheme 4. Formation of δ-decalactone (**3**) via 13-lipoxygenase pathway and degradation of **3** via β-oxidation or Baeyer-Villiger oxidation in *S. odorus*.

Eds.; ACS Symposium Series 490; American Chemical Society: Washington, DC, 1992, 46-58

(2) Palm, U.; Askari, C.; Hener, U.; Jakob, E.; Mandler, C.; Geßner, M.; Mosandl, A.; König, W. A.; Evers, P. und Krebber, R., *Z. Lebensm. Unters. Forsch.* **1991,** *192,* 209-213

(3) Werkhoff, P.; Brennecke, S.; Bretschneider, W.; Güntert, M.; Hopp, R.; Surburg, H., *Z. Lebensm. Unters. Forsch.* **1993**, *196*, 307-328

(4) Lehmann, D.; Dietrich, A.; Schmidt, S.; Dietrich, H.; Mosandl A., *Z. Lebensm. Unters. Forsch.* **1993**, *196*, 207-213

(5) van der Schaft, P. H.; ter Burg, N.; van den Bosch, S.; Cohen, A. M., *Appl. Microbiol. Biotechnol.*, **1992**, *36*, 712-716

(6) Haffner, T.; Tressl, R., *J. Agric. Food Chem.* **1996**, *44*, 1218-1223

(7) Albrecht, W.; Schwarz, M.; Heidlas, J.; Tressl, R., *J. Org. Chem.* **1992,** *57*, 1954-1956

(8) Haffner, T.; Nordsieck, A.; Tressl, R., *Helv. Chim. Acta* **1996,** *79*, 2088-2099

(9) Tressl, R.; Haffner, T.; Lange, H.; Nordsieck, A., In *Flavour Science*; Taylor, A. J.; Mottram, D. S., Eds.; Recent Developments, The Royal Society of Chemistry: Cambridge, UK, 1996, 141-150

(10) Holtunan, W. L.; Vredenbregt-Heistek, J. C.; Schmitt, N. F.; Feussner, I., *Eur. J. Biochem.* **1997**, *248*, 452-458

(11) Behrends, W.; Thieringer, R.; England, K.; Kunau, W.-H.; Kindl, H., *Arch. Biochem. Biophys.* **1988**, *263*, No. 1, 170-177

(12) Hiltunen, J. K.; Wenzel, B.; Beyer, A.; Erdmann, R.; Fossa, A.; Kunau, W.-H.; *J. Biol. Chem.* **1992**, *267*, 6646

(13) Fillippula, S.; Sormunen, R. T.; Hertig, A.; Kunau, W.-H.; Hiltunen, K., *J. Biol. Chem.* **1995**, *270*, 27453

(14) Wanner, P., *Ph. D. Thesis* TU-Berlin **1996**

(15) Nago, H.; Matsumoto, M.; Nakai, S., Biosci. Biotech. Biochem. **1993**, *57* (12), 2111-2115

Chapter 16

Application of Countercurrent Chromatography to the Analysis of Aroma Precursors in Rose Flowers

Peter Winterhalter[1,3], Holger Knapp[1], Markus Straubinger[1], Selenia Fornari[1], and Naoharu Watanabe[2]

[1]Institut für Pharmazie und Lebensmittelchemie, Universität Erlangen-Nürnberg, Schuhstrasse 19, D-91052 Erlangen, Germany
[2]Department of Applied Biological Chemistry, Faculty of Agriculture, Shizuoka University, 836 Ohya, Shizuoka, Japan

After a description of the basic principle of countercurrent chromatography (CCC) as well as the instrumentation used, this chapter gives an overview on the application of CCC in flavor analysis. The focus is on the application of multilayer coil countercurrent chromatography (MLCCC) to the analysis of reactive flavor precursors from *Rosa damascena* flowers. As a result, the isolation of the ß-damascenone generating precursor 3-hydroxy-7,8-didehydro-ß-ionol 9-O-ß-D-glucopyranoside as well as a whole series of dioxygenated monoterpenes will be reported. The latter structures included a genuine progenitor of isomeric rose oxides, i.e. (*S*)-3,7-dimethyl-5-octene-1,7-diol, which was for the first time successfully isolated from rose flowers.

A major problem in the investigation of flavor compounds is the possibility of artifact formation during the isolation and separation process. 'Secondary' volatile formation through enzymatic and nonenzymatic reactions is diverse, thus, knowledge about enzymatic activities (e.g., enzymatic conversion of glucosinolates or alkenyl cystein S-oxides) in plant tissues is one important prerequisite to ensure the isolation of an artifact-free aroma extract (*1*). Moreover, suitable isolation and separation techniques are needed so as to reduce the risk of chemical or thermal artifact formation (*2*). Comparative studies on different standard procedures used for the isolation of volatile constituents from flowers and fruits have been published (*3,4*). These studies have shown that the composition of the aroma concentrates differed both quantitatively and, in many cases, also qualitatively depending on the isolation technique applied. Many of the discrepancies observed could be explained by assuming a degradation of labile precursor compounds during sample work-up. For example, in rose oil, remarkable differences have been observed in the composition of various aroma concentrates (*3*). The aim in this study was the isolation of the genuine progenitors of two key aroma compounds of rose essential oil, ß-damascenone and isomeric rose oxides (*5*), by application of the all-liquid chromatographic technique of multilayer coil countercurrent chromatography (*6*).

[3]Current address: Institut für Lebensmittelchemie, Technische Universität Braunschweig, Schleinitzstrasse 20, D-38106 Braunschweig, Germany

EXPERIMENTAL PROCEDURES

Plant Material. Rose flowers (*Rosa damascena* Mill.) were harvested at their full bloom stage in the Shizuoka prefecture, Japan.

Extraction of Rose Flowers and Preparation of a XAD-2 Extract. Rose flowers (10 kg) were homogenized in ice-cooled 80 % aqueous MeOH. After filtration, the organic solvent was concentrated *in vacuo*. Freeze-drying of the aqueous residue yielded 206 g of a crude extract. This extract was subjected to XAD-2 column chromatography (*7*) in portions of 26 g. The column was rinsed with water. Subsequent elution of the retained material with MeOH and concentration under reduced pressure yielded approximately 40 g of an aroma precursor concentrate.

MLCCC Separation. For the initial fractionation of the residue, multilayer coil countercurrent chromatography (MLCCC) was used (Multilayer Coil Separator-Extractor, P.C. Inc.; Potomac, USA; equipped with a 85 m x 2.6 mm i.d. PTFE tubing; solvent system: $CHCl_3/MeOH/H_2O$ 7:13:8; injection volume: 2.5 g of extract dissolved in 5 mL of aqueous phase, for details cf. ref. *8*). To facilitate the screening of aroma precursors, sequential MLCCC fractions were pooled into five groups, i.e. combined MLCCC fractions I-V.

Isolation of Compounds **4, 8, 11** and **23.** A screening of frs. I-V was carried out by acid hydrolysis (SDE, pH 2.5, 1 h) as well as enzymatic hydrolysis (Rohapect D5L, Röhm, Darmstadt, Germany). The generated compounds from these hydrolyses were then determined by GC-MS. Fraction V was further separated by MLCCC ($BuOH/MeOH/H_2O$ 10:1:10), and the major compounds were finally purified by flash chromatography and normal phase HPLC using hexane/MTBE gradients.

Reference Compounds. Compounds **2-13, 15, 17-22** were prepared according to refs. *9-15*. Diol **16** was donated and compounds **23** and **26** were commercially obtained.

Capillary Gas Chromatography-Mass Spectrometry (GC-MS). GC-MS was performed with a Hewlett-Packard GCD system equipped with a PTV-injector (KAS-system, Gerstel, Mülheim, Germany). The system was equipped with a J&W fused silica DB-5 capillary column (30 m x 0.25 mm i.d., film thickness 0.25 µm). The temperature program was from 60°C (2 min isothermal) to 300°C at 5°C/min. The flow rate of the carrier gas was 1.0 mL/min of He. Other conditions were as follows: temperature of ion source, 180°C; electron energy, 70eV. The linear retention index (R_i) is based on a series of n-hydrocarbons.

Nuclear Magnetic Resonance (NMR). ^1H and ^{13}C NMR spectral data were recorded on Fourier transform Bruker AM 360 and AC 250 spectrometers with TMS as internal reference standard.

RESULTS AND DISCUSSION

An isolation and separation technique that has seen a renaissance in natural product analysis - including flavor as well as flavor precursor analysis - is the all liquid chromatographic technique of countercurrent chromatography. In the following, the principles and features of modern countercurrent chromatography are briefly discussed. Detailed descriptions of CCC-techniques can be found in refs. *16-20*.

Principles of Countercurrent Chromatography. Countercurrent chromatography is a separation method originally defined as liquid-liquid partition chromatography without a solid support. The name was derived from two classical partition methods, i.e. countercurrent distribution (CCD) and liquid chromatography (LC). CCC utilizes two immiscible solvent phases, one as a stationary phase and the other as a mobile phase. The separation is dependent on the partition coefficients (K) of the solutes, i.e., the ratio of the solute concentration between the stationary (C_s) and mobile phase (C_m):

$$K = C_s / C_m$$

A successful separation necessitates a careful search for suitable two-phase systems which provide an ideal range of partition coefficients for the solutes of interest. Strategies for the systematic search of suitable two-phase solvent systems for CCC have been published (*21,22*). The resolution (R_s) of a solute mixture is according to the fundamental equation of chromatography, i.e. $R_s = 1/4\ (\alpha\text{-}1)\ N^{1/2}\ [k' / (1 + k')]$, obtained from the selectivity factor (α), the number of theoretical plates (N) and the solute retention expressed as the capacity factor ($k' = K\ \ V_s / V_m$), where V_s and V_m are the volume of the stationary and the mobile phase, respectively:

$$R_s = 1/4\ (\alpha\text{-}1)\ N^{1/2}\ \{K / [(V_m / V_s) + K]\}$$

From this equation the influence of the selectivity factor α, the efficiency N as well as the partition coefficient K is obvious; the higher the values of α, N and K, the higher the peak resolution R_s. It is important to note, that the ratio of the volumes of the stationary and mobile phase is an additional parameter that directly influences R_s. In CCC large volumes of stationary phase are used. Therefore, the term V_m/V_s is much smaller in CCC (0.1 - 1) compared to, e.g., preparative HPLC (~ 15). Or in other words, controlling V_s provides a key for adjusting the solute capacity factor k'. Hence, CCC provides in many cases good peak resolution although the efficiency - expressed as theoretical plate number N - may well be in the low range of a thousand plates (*17,23*).

Advantages of Countercurrent Chromatography. As a major advantage of CCC one can consider its versatility. Since there is a tremendous variety of solvent systems available, the selectivity factor α and, thus, the peak resolution R_s is easily influenced. Furthermore, due to a large volume of stationary phase, CCC enables large sample loads (gram-range) without overloading problems. Finally, as an all-liquid chromatographic technique, CCC has a good sample recovery rate, since problems arising from a solid stationary phase (e.g., irreversible adsorption or artifact formation) are eliminated or at least minimized.

Instrumentation. Modern CCC started during the 1970s with the introduction of the so-called *hydrostatic* CCC-techniques of droplet countercurrent chromatography (DCCC) and rotation locular countercurrent chromatography (RLCC) by Ito and co-workers (*24* and refs. cited). DCCC has later seen an improvement by applying a strong centrifugal force which reduced separation time considerably (*25*). Details on this latter technique which is called centrifugal countercurrent chromatography (CDCCC) as well as overviews on the hydrostatic techniques have been presented earlier (*19,20,26*).

The major breakthrough of CCC came with the discovery of the so-called *hydrodynamic* phenomenom and the subsequent development of the coil planet centrifuge (*6*), a technique which was marketed as multilayer coil countercurrent chromatography (MLCCC) or synonymously as high speed countercurrent chromatography (HSCCC). Since this technique has been applied in the present study, its principle will be briefly outlined. In MLCCC the column used consists of a PTFE tubing which is wrapped around a holder in several layers (multilayer coil). The radius (r) of the coiled column depends on the number of layers. In essence, the coil undergoes a synchronous planetary motion while the column holder revolves around the central axis of the instrument. The revolution radius is ***R***, and the ratio of ***r/R*** is defined as ß, which may vary from 0.25 to 0.75. The apparatus is designed with an anti-twist mechanism that ensures a continuous solvent flow without requiring a rotating seal. During rotation, an Archimedean screw effect is achieved causing the migration of the stationary phase towards one end of the column. The mobile phase is now pumped in the opposite direction. During operation the interfacial friction force of the mobile phase and the Archimedean screw force are counteracting and create a hydrodynamic equilibrium system depending on the rotation speed, flow rate and viscosity of the solvents. The mixing of the two phases is automatically achieved. When the ratio ***r/R*** exceeds 0.5, the trajectory forms a loop in which the force field is much lower compared with the opposite part of the loop. This difference leads to the following behaviour of a two phase system in the coil: when the force field is strong, the two phases are separated (settling step); when the force field is weak (in the loop), mixing of the two phases occurs (*17*). Hence, during rotation, mixing and settling steps are automatically produced that ensure an efficient partitioning of the solutes.

Application of Countercurrent Chromatography in Flavor Analysis - An Overview. The usefulness of MLCCC for flavor analysis was first demonstrated by Fischer and co-workers (*27*) in their study on celery flavor. A major task in separating flavor compounds by CCC is to find solvent systems that can directly be used for the subsequent GC analysis. On the basis of the two-phase system hexane/acetonitrile (1:1), the authors developed a series of non-aqueous solvent mixtures that have been successfully applied to the analysis of phthalides in celery oil. Peak purity of the compounds which were isolated in amounts of 1-30 mg was in a range of 60 to 85 %. As a consequence of the large amount of sample being separated in a single run, the localization of aroma active compounds was reported to be easily performed by checking the smell of each separated fraction. In this way, a trace component with a strong celery-like odor could be enriched from a barely detectable level in the original oil which enabled its structure to be elucidated as 3-butyl-5,6-dihydro-4*H*-isobenzofuran-1-one (*28*).

Other examples for successful separations of aroma compounds using CCC have been reported by our group (*29,30*). Separation of diastereomeric theaspirones (300 mg) was achieved by a single-run MLCCC using pentane/methyl *tert.* butyl ether/methanol/water (10:1:10:1) as solvent system (elution mode: tail to head). The solvent sytem hexane/ethyl acetate/methanol/water (70:30:14:10) was used for the purification of (*E*)-3-oxo-retro-α-ionol from a mixture of isomers. Most recently, isomeric theaspiranes have been successfully separated on a semi-preparative scale using hexane/methyl *tert.* butyl ether/methanol/water (10:1:10:1) or alternatively hexane/methyl *tert.* butyl ether/acetonitrile (10:1:10) as solvent system.

Application of Countercurrent Chromatography to the Analysis of Flavor Precursors in Rose Flowers. Contrary to the only limited application of CCC to volatile constituents, the technique has proven to be an indispensable tool for the analysis of labile aroma precursors. This topic has been the subject of review (*31*) and numerous applications can be found in the literature (*32,33* and refs. cited). The absence of a solid stationary phase and the gentle operation conditions of CCC predestinate this technique for the analysis of labile natural products. Due to this reason, we decided to apply MLCCC to the investigation of aroma precursors in rose flowers. Although rose petals were the first substrate in which terpenoid aroma precursors were detected (*34*), little information was available about the presence of additional aroma precursors in roses. Especially for the key aroma compounds of rose essential oil, i.e. the intensely odorous ß-damascenone as well as isomeric rose oxides, the genuine progenitors still remain to be elucidated. The protocol for the work-up of rose flowers is outlined in Fig. 1.

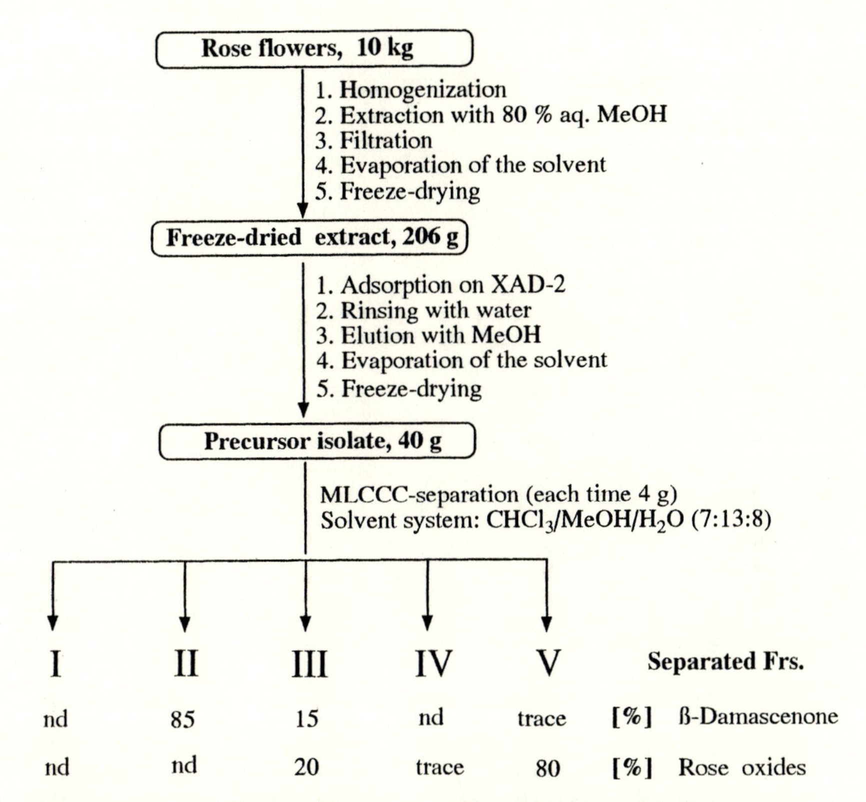

Figure 1. Protocol for the isolation of labile progenitors of ß-damascenone and isomeric rose oxides and relative proportions of ß-damascenone and rose oxides formed under SDE (pH 2.5) conditions (nd = not detected).

Sequential fractions from the MLCCC separation were pooled in five groups. Aliquots of these grouped fractions I-V were then subjected to SDE (pH 2.5) as well as glycosidase treatment. Upon SDE, the major portion of ß-damascenone came out of the medium polar fractions II and III, whereas 80 % of the total rose oxide formation was observed from the least polar fraction V. From preceding studies on aroma progenitors of rose flowers (Straubinger, M. *et al.*, *J. Agric. Food Chem.*, in press), it was known that the medium polar CCC-fractions, i.e. frs. II-IV, mainly contain glycosidically bound compounds. The glycosides that have been successfully isolated with the aid of MLCCC included *inter alia* the ß-D-glucopyranoside of 3-hydroxy-7,8-didehydro-ß-ionol **1**, which has earlier been established as an important progenitor of ß-damascenone (*35,36*).

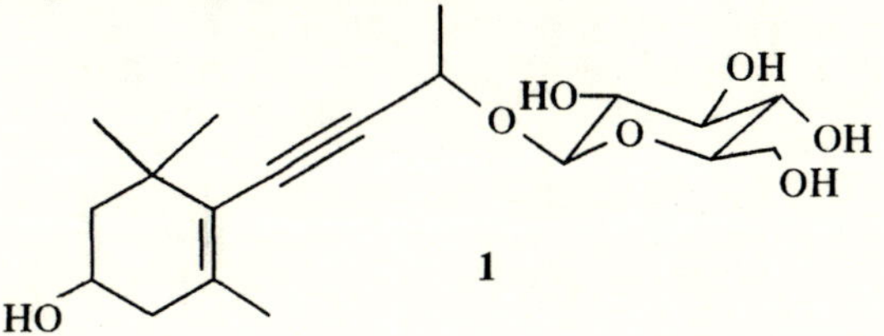

In the present study, we were mainly interested in the composition of the least polar MLCCC-fraction V, which was expected to contain a major progenitor of isomeric rose oxides. Extraction of fraction V with diethyl ether and subsequent GC-MS analysis revealed that fraction V contains non-glycosidic constituents. In an effort to further separate the diethyl ether extract, an additional CCC-fractionation with BuOH/MeOH/H_2O (10:1:10) as solvent system has been carried out. Due to the complexity of the Et_2O isolate (approx. 50 different compounds), however, further work-up by flash chromatography and purification by HPLC was necessary. In this way, four oxygenated monoterpenes could be isolated whose structures were elucidated by NMR spectroscopy (cf. Table I). Importantly, the isolated compounds included 3,7-dimethyl-5-octene-1,7-diol **4**, the known synthetic precursor of isomeric rose oxides **24a/b**. The configuration at the stereogenic center C-3 of rose oxide precursor **4**, which has for the first time been isolated from a natural source, was determined by acid-catalyzed conversion of diol **4** into isomeric rose oxides. Since chiral MDGC analysis (DB-Wax / 6-*tert.* butyl-dimethyl-silyl-2,3-dimethyl-ß-cyclodextrin) revealed (2*S*,4*R*) and (2*R*,4*R*) configuration for cyclic ethers **24a/b**, the configuration of precursor diol **4** was assigned as *S*.

In addition to diols **4**, **8**, **11** and **23**, sixteen structurally related compounds were identified by GC-MS analysis and comparison of the mass spectral data with those of authentic references. Only in the case of diol **14**, which showed essentially the same fragmentation pattern as observed for the (*E*)-configured isomer **16**, does the identification have to be considered tentative. The structures of all positively identified diols are shown in Fig. 2. Their mass spectral data and GC retention indices are gathered in Table II.

Table I. NMR Spectral Data ($CDCl_3$) for Isolated Monoterpene Diols

4	^{1}**H-NMR** (360 MHz, ppm, *J* in Hz): δ 0.89 (3H, d, *J* = 6.5, Me-C3); 1.30 (6H, s, 2Me-C7); 1.35 (1H, m, H_aC2); 1.59 (1H, m, H_bC2); 1.67 (1H, m, HC3), 1.91 (1H, m, H_aC4); 2.01 (1H, m, H_bC4); 2.60 (1H, br s, -OH); 2.87 (1H, br s, -OH); 3.60 (1H, ddd, *J* = 12.0, 7.0, 2.0, H_aC1); 3.67 (1H, ddd, *J* = 12.0, 7.0, 3.0, H_bC1); 5.56 (1H, ddd, *J* = 15.0, 5.5, 1.0, HC5); 5.61 (1H, d, *J* = 15.0, HC6). ^{13}**C-NMR** (90.6 MHz): δ 19.5 (Me-C3), 29.6 (2Me-C7), 29.7 (C3), 38.9 (C2), 39.5 (C4), 60.5 (C1), 70.5 (C7), 125.1 (C5), 139.4 (C6).
8	^{1}**H-NMR** (250 MHz, ppm, *J* in Hz): δ 1.30 (3H, s, Me-C6); 1.59 (2H, m, H_2C5); 1.65 (3H, d, *J* = 1.0, Me-C2); 2.08 (2H, m, H_2C4); 3.98 (2H, s, H_2C1); 5.07 (1H, dd, *J* = 11.0, 1.5, H_aC8); 5.22 (1H, dd, *J* = 17.5, 1.5, H_bC8); 5.41 (1H, tq; *J* = 7.0, 1.0, HC3); 5.91 (1H, dd, *J* = 17.5, 11.0, HC7). ^{13}**C-NMR** (60 MHz): δ 13.6 (Me-C2), 22.3 (C4), 27.8 (Me-C6), 41.7 (C5), 68.7 (C1), 73.3 (C6), 111.8 (C8), 125.8 (C3), 135.9 (C2), 144.8 (C7).
11	^{1}**H-NMR** (360 MHz, ppm, *J* in Hz): δ 1.26 (3H, s, Me-C3); 1.56 ((2H, m, H_2C4); 1.63 (3H, d, *J* = 1.0, Me-C7); 1.67 (1H, ddd, *J* = 14.5, 6.5, 4.5, H_aC2); 1.69 (3H, d, *J* = 1.0, Me-C7); 1.81 (1H, ddd, *J* = 14.5, 7.5, 4.5, H_bC2); 2.06 (2H, m, H_2C5); 3.86 (1H, ddd, *J* = 11.0, 6.5, 4.5, H_aC1); 3.92 (1H, ddd, *J* = 11.0, 7.5, 4.5, H_bC1); 5.15 (1H, tqq, *J* = 7.0, 1.0, 1.0, HC6). ^{13}**C-NMR** (90.6 MHz): δ 17.7 (Me-C7), 22.7 (C5), 25.7 (Me-C7), 26.7 (Me-C3), 41.7 (C2), 42.4 (C4), 59.9 (C1), 74.0 (C3), 124.1 (C6), 132.1 (C7).
23	^{1}**H-NMR** (250 MHz, ppm, *J* in Hz): δ 1.64 (3H, d, *J* = 1.0, Me-C2); 1.66 (3H, d, *J* = 1.0, Me-C6); 2.02 - 2.23 (4H, m, H_2C4, H_2C5), 2.88 (2H, br s, -OH); 3.95 (2H, s, H_2C1); 4.11 (2H, d, *J* = 7.0, H_2C8); 5.37 (1H, tq, *J* = 7.0, 1.0, HC3); 5.38 (1H, tq, *J* = 7.0, 1.0, HC7). ^{13}**C-NMR** (60 MHz): δ 13.6 (Me-C2), 16.0 (Me-C6), 25.4 (C4), 38.9 (C5), 58.9 (C8), 68.3 (C1), 123.8 (C7), 125.0 (C3), 134.9 (C2), 138.3 (C6).

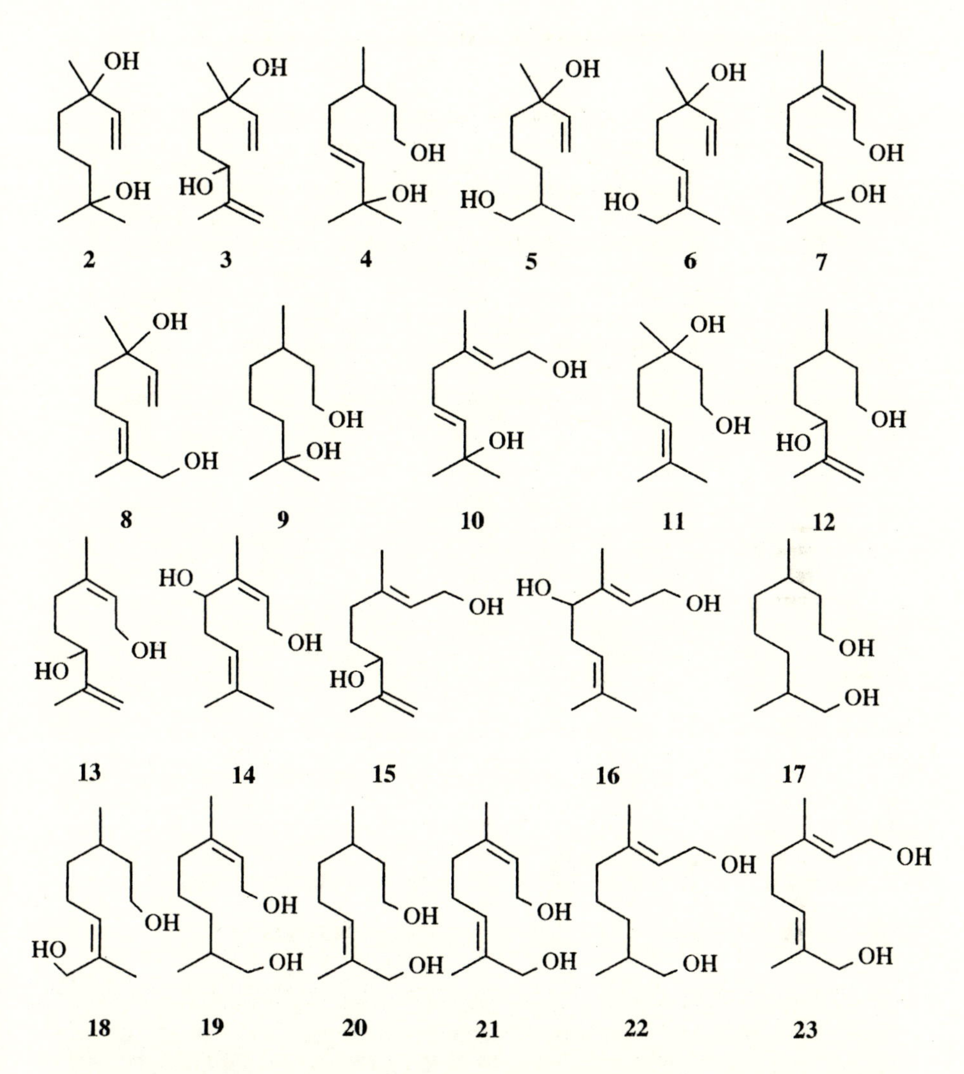

Figure 2. Structures of monoterpene diols **2-23** identified in a rose flower extract: 2,6-dimethyl-7-octene-2,6-diol **2**, 2,6-dimethyl-1,7-octadiene-3,6-diol **3**, 3,7-dimethyl-5-octene-1,7-diol **4**, 2,6-dimethyl-7-octene-1,6-diol **5**, (*Z*)-2,6-dimethyl-2,7-octadiene-1,6-diol **6**, (2*Z*,5*E*)-3,7-dimethyl-2,5-octadiene-1,7-diol **7**, (*E*)-2,6-dimethyl-2,7-octadiene-1,6-diol **8**, 3,7-dimethyloctane-1,7-diol **9**, (2*E*,5*E*)-3,7-dimethyl-2,5-octadiene-1,7-diol **10**, 3,7-dimethyl-6-octene-1,3-diol **11**, 3,7-dimethyl-7-octene-1,6-diol **12**, (2*Z*)-3,7-dimethyl-2,7-octadiene-1,6-diol **13**, (2*Z*)-3,7-dimethyl-2,6-octadiene-1,4-diol (tent.) **14**, (2*E*)-3,7-dimethyl-2,7-octadiene-1,6-diol **15**, (2*E*)-3,7-dimethyl-2,6-octadiene-1,4-diol **16**, 2,6-dimethyloctane-1,8-diol **17**, (*Z*)-2,6-dimethyl-2-octene-1,8-diol **18**, (*Z*)-3,7-dimethyl-2-octene-1,8-diol **19**, (*E*)-2,6-dimethyl-2-octene-1,8-diol **20**, (2*E*,6*Z*)-2,6-dimethyl-2,6-octadiene-1,8-diol **21**, (*E*)-3,7-dimethyl-2-octene-1,8-diol **22**, and (2*E*,6*E*)-2,6-dimethyl-2,6-octadiene-1,8-diol **23**.

Table II. Mass Spectral Data (70 eV) of Monoterpene Diols 2-23

Compound	R_i*	m/z (%)
2	*1223*	154 (M^+- H_2O, < 1), 139 (7), 136 (1), 126 (3), 121 (11), 109 (5), 97 (3), 96 (5), 84 (9), 83 (8), 81 (29), 71 (100), 69 (22), 68 (44), 59 (32), 56 (31), 55 (16), 43 (57).
3	*1277*	155 (M^+- CH_3, < 1), 137 (9), 125 (3), 123 (4), 110 (7), 109 (8), 96 (7), 95 (6), 84 (10), 83 (8), 82 (43), 81 (15), 71 (83), 69 (16), 68 (32), 67 (100), 55 (37), 43 (70).
4	*1331*	154 (M^+- H_2O, 1), 139 (25), 121 (10), 112 (2), 111 (5), 109 (8), 99 (3), 97 (4), 95 (8), 93 (3), 85 (15), 81 (11), 71 (18), 69 (20), 67 (11), 59 (13), 57 (7), 55 (17), 43 (100).
5	*1335*	157 (M^+- CH_3, 0.5), 139 (1), 121 (2), 111 (2), 109 (2), 97 (3), 96 (5), 83 (4), 81 (5), 72 (6), 71 (100), 69 (9), 56 (8), 55 (16), 43 (29).
6	*1340*	152 (M^+- H_2O, < 1), 137 (7), 134 (2), 123 (4), 121 (4), 119 (12), 110 (9), 109 (7), 105 (3), 96 (10), 93 (10), 91 (5), 84 (19), 82 (18), 81 (14), 79 (13), 71 (82), 69 (13), 68 (24), 67 (56), 57 (10), 55 (39), 53 (12), 43 (100).
7	*1345*	152 (M^+- H_2O, 1), 137 (6), 134 (2), 123 (2), 121 (2), 119 (5), 109 (22), 107 (4), 96 (7), 95 (13), 93 (12), 91 (8), 85 (15), 84 (12), 83 (33), 81 (16), 79 (13), 71 (10), 69 (15), 68 (27), 67 (21), 59 (17), 55 (13), 53 (8), 43 (100).
8	*1361*	152 (M^+- H_2O, < 1), 137 (8), 123 (4), 121 (6), 119 (9), 110 (9), 109 (7), 96 (11), 93 (14), 84 (14), 82 (20), 81 (16), 79 (17), 71 (75), 69 (11), 68 (24), 67 (58), 57 (10), 55 (36), 53 (12), 43 (100).
9	1365	159 (M^+- CH_3, 1), 141 (1), 123 (7), 109 (1), 98 (7), 95 (2), 89 (1), 85 (2), 83 (8), 81 (4), 71 (7), 70 (13), 69 (8), 59 (100), 55 (16), 43 (22).
10	*1371*	170 (M^+, < 1), 155 (1), 152 (0.5), 137 (7), 123 (2), 119 (4), 109 (15), 107 (3), 95 (12), 93 (14), 85 (12), 82 (10), 81 (11), 79 (11), 77 (6), 71 (13), 69 (11), 68 (11), 67 (18), 59 (16), 55 (12), 53 (7), 43 (100).
11	*1391*	154 (M^+- H_2O, 17), 139 (4), 123 (5), 121 (41), 111 (10), 110 (11), 109 (65), 97 (4), 95 (15), 93 (11), 89 (17), 84 (8), 83 (7), 81 (15), 71 (37), 69 (60), 68 (12), 67 (23), 56 (7), 55 (19), 53 (8), 43 (100).
12	*1398*	154 (M^+- H_2O, 1), 139 (3), 121 (2), 111 (3), 99 (7), 97 (4), 86 (16), 84 (10), 83 (10), 81 (5), 72 (19), 71 (100), 69 (30), 67 (10), 58 (12), 57 (12), 56 (30), 55 (33), 43 (49).
13	*1404*	155 (M^+- CH_3, < 1), 137 (7), 134 (6), 123 (6), 121 (8), 119 (13), 109 (12), 107 (5), 105 (6), 97 (21), 95 (11), 93 (15), 91 (10), 84 (100), 83 (33), 82 (26), 81 (35), 79 (20), 72 (12), 71 (49), 69 (48), 68 (49), 67 (96), 57 (14), 56 (11), 55 (39), 53 (20), 43 (67).
14	*1412*	152 (M^+- H_2O, < 1), 121 (2), 101 (27), 84 (11), 83 (100), 81 (7), 71 (18), 70 (69), 69 (28), 67 (11), 57 (14), 55 (73), 53 (12), 43 (49).
15	*1432*	155 (M^+- CH_3, < 1), 137 (7), 134 (7), 123 (6), 121 (10), 119 (14), 109 (12), 107 (6), 105 (7), 97 (25), 95 (11), 93 (19), 91 (11), 84 (100), 83 (26), 82 (20), 81 (37), 79 (21), 72 (16), 71 (56), 69 (59), 68 (43), 67 (85), 57 (18), 56 (12), 55 (39), 53 (21), 43 (74).

16	*1443*	152 (M^+- H_2O, < 1), 121 (1), 109 (1), 101 (18), 84 (7), 83 (51), 71 (13), 70 (90), 69 (18), 59 (8), 57 (6), 55 (100), 53 (10), 43 (30).
17	*1466*	138 (M^+- $2H_2O$, < 1), 123 (5), 111 (13), 110 (7), 109 (13), 99 (8), 97 (27), 96 (19), 95 (16), 84 (9), 83 (35), 82 (25), 81 (45), 71 (31), 70 (45), 69 (88), 68 (26), 67 (25), 57 (32), 56 (40), 55 (100), 45 (8), 43 (40).
18	*1474*	154 (M^+- H_2O, 1), 139 (10), 136 (2), 123 (5), 121 (26), 111 (13), 109 (21), 107 (11), 97 (16), 95 (18), 94 (9), 93 (21), 85 (16), 84 (35), 83 (22), 82 (15), 81 (42), 79 (17), 71 (45), 69 (44), 68 (45), 67 (47), 57 (26), 56 (27), 55 (69), 53 (16), 45 (11), 43 (100).
19	*1479*	154 (M^+- H_2O, < 1), 139 (3), 136 (1), 123 (5), 121 (9), 111 (5), 107 (11), 97 (23), 96 (34), 95 (16), 94 (11), 93 (18), 85 (15), 84 (35), 83 (26), 82 (13), 81 (38), 79 (22), 71 (100), 70 (19), 69 (48), 68 (37), 67 (31), 57 (22), 56 (28), 55 (58), 53 (16), 43 (41).
20	*1494*	154 (M^+- H_2O, 2), 139 (11), 136 (3), 123 (6), 121 (28), 111 (15), 110 (10), 109 (19), 107 (12), 97 (16), 95 (21), 94 (9), 93 (20), 85 (17), 84 (28), 83 (20), 82 (14), 81 (50), 79 (15), 71 (46), 69 (46), 68 (41), 67 (43), 57 (26), 56 (27), 55 (73), 53 (14), 45 (7), 43 (100).
21	*1502*	152 (M^+- H_2O, 1), 137 (5), 134 (3), 123 (6), 121 (14), 119 (10), 109 (12), 107 (9), 97 (11), 96 (10), 95 (14), 94 (11), 93 (18), 91 (8), 84 (44), 83 (20), 82 (11), 81 (25), 79 (15), 71 (41), 69 (31), 68 (58), 67 (46), 57 (20), 56 (15), 55 (42), 53 (18), 43 (100).
22	*1515*	154 (M^+- H_2O, < 1), 139 (2), 136 (1), 123 (5), 121 (9), 111 (3), 107 (11), 97 (18), 96 (33), 95 (13), 94 (10), 93 (16), 85 (13), 84 (32), 83 (21), 82 (9), 81 (32), 79 (19), 71 (100), 70 (13), 69 (40), 68 (36), 67 (27), 57 (18), 56 (24), 55 (45), 53 (14), 43 (35).
23	*1523*	152 (M^+- H_2O, < 1), 137 (6), 134 (3), 123 (5), 121 (12), 119 (7), 109 (9), 107 (5), 97 (5), 96 (7), 95 (11), 94 (14), 93 (11), 91 (5), 84 (39), 83 (12), 82 (5), 81 (17), 79 (9), 71 (17), 69 (15), 68 (72), 67 (44), 57 (18), 55 (21), 53 (13), 43 (100).

*Linear retention index on a J&W DB-5 capillary column (30 m x 0.25 mm i.d., film thickness 0.25 μm).

Precursor Function of Monoterpene Diols. Polyhydroxylated terpenes (polyols) together with terpene glycoconjugates (*33*) are involved in the generation of volatile compounds. The specific role of polyols in the acid-catalyzed production of volatiles in fruits and wine has been the subject of several reviews (*37-39*). As for rose polyols, only the identification of diols **11** and **12** has so far been reported (*40-41*). The remaining compounds **2-10** and **13-23** are, to the best of our knowledge, reported here for the first time in rose flowers. It is furthermore noteworthy that the structures of five additional diols remain to be elucidated. In view of the reactivity of terpene diols and the well-recognized aroma properties of the volatiles formed, their contribution to rose odor has obviously been underestimated.

Concerning rose oxide formation, it can be concluded that a major portion of isomeric ethers **24a/b** is chemically formed from the labile progenitor **4**. In a similar reaction, racemic nerol oxides **25a/b** are likely to be formed from the dehydroderivative **7** (cf. Fig. 3). These results confirm earlier published observations of Surburg and co-workers (*3*) who reported two remarkable differences between the composition of steam distillates and vacuum headspace concentrates of rose flowers, i.e. the absence of (i) rose oxides and of (ii) ß-damascenone in the latter concentrates. With regard to the biochemical formation of allyldiols **4** and **7** it is still unclear whether the introduction of the second alcohol function into the parent alcohol (-)-citronellol and nerol is an enzyme-mediated reaction or due to photooxidation processes (*41*).

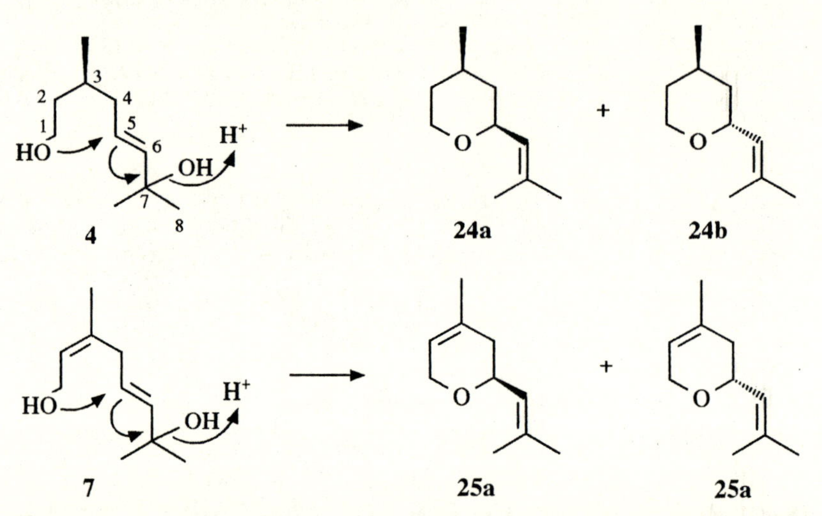

Figure 3. Formation of rose oxides **24a/b** and nerol oxides **25a/b** through acid-catalyzed cyclization of precursor diols **4** and **7**, respectively (*42,43*).

During the fractionation of the monoterpene diol containing MLCCC fraction V, additional oxygenated compounds have been recognized (cf. Fig. 4), i.e. two monoterpenoid acids **26** and **27** as well as a series of carotenoid degradation products **28-32**. Most recently, monoterpene acid **27** has been elucidated as progenitor of the intensely odorous wine lactone (*44*), and the free acetylenic diol **28** can be considered as being responsible for the small amount of ß-damascenone generated from fraction V (cf. Fig. 1).

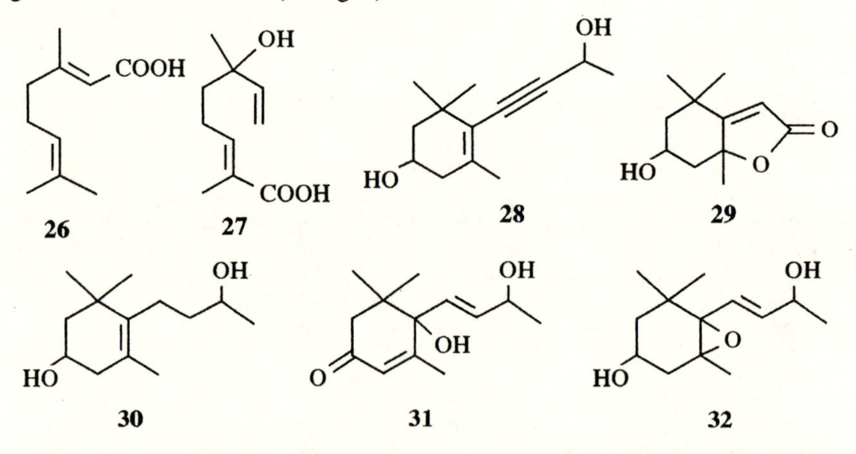

Figure 4. Additional oxygenated terpenoids identified in MLCCC fraction V.

CONCLUSIONS

It has been demonstrated that MLCCC facilitates the isolation of aroma-relevant precursor compounds from rose flowers. For both glycosidic and non-glycosidic precursors, MLCCC readily enabled an enrichment of components which are responsible for the formation of key aroma compounds of rose odor. Whereas ß-damascenone was found to be mainly derived from a glycosidic progenitor, isomeric rose oxides are to a large extent generated from non-conjugated (*S*)-3,7-dimethyl-5-octene-1,7-diol **4**. The isolation of a second rose oxide precursor in the medium polar MLCCC fraction III is the subject of ongoing studies.

ACKNOWLEDGMENTS

The skillful assistance of M. Messerer and N. Oka is gratefully acknowledged. We are indebted to Prof. P. Schreier, University of Würzburg, for a reference of (2*E*)-3,7-Dimethyl-2,6-octadiene-1,4-diol and to Dr. Schmarr, Technical University of München-Weihenstephan, for the enantio-MDGC analysis of rose oxides. The Deutsche Forschungsgemeinschaft, Bonn, is thanked for funding the research.

LITERATURE CITED

1. Schreier, P. *Lebensmittelchem. Gerichtl. Chem.* **1987**, *41*, 25-34.
2. Teranishi, R.; Kint, S. In *Flavor Science - Sensible Principles and Techniques*; Acree, T.E.; Teranishi, R., Eds.; American Chemical Society: Washington, DC, 1993, pp. 137-167.
3. Surburg, H.; Güntert, M.; Harder, H. In *Bioactive Volatile Compounds from Plants*; Teranishi, R.; Buttery, R.G.; Sugisawa, H., ACS Symp. Ser. 525; American Chemical Society: Washington, DC, 1993, pp. 168-186.
4. Takeoka, G.R.; Flath, R.A.; Buttery, R.G.; Winterhalter, P.; Güntert, M.; Ramming, D.W.; Teranishi, R. In *Flavor Precursors - Thermal and Enzymatic Conversions*; Teranishi, R.; Takeoka, G.R.; Güntert, M., Eds.; ACS Symp. Ser. 490; American Chemical Society: Washington, DC, 1992, pp. 116-138.
5. Ohloff, G.; Demole, E. *J. Chromatogr.* **1987**, *406*, 181-183.
6. Ito, Y. *CRC Crit. Rev. Anal. Chem.* **1986**, *17*, 65-143.
7. Günata, Y.Z.; Bayonove, C.L.; Baumes, R.L.; Cordonnier, R.E. *J. Chromatogr.* **1985**, *331*, 83-90.
8. Roscher, R.; Winterhalter, P. *J. Agric. Food Chem.* **1993**, *41*, 1452-1457.
9. Williams, P.J.; Strauss, C.R.; Wilson, B. *Phytochemistry* **1980**, *19*, 1137-1139.
10. Matsuura, T.; Butsugan, Y. *J. Chem. Soc. Japan* **1968**, *89*, 513-516.
11. Behr, D.; Wahlberg, I.; Toshiaka, N.; Enzell, C.R. *Acta Chem. Scand. B* **1978**, *32*, 228-229.
12. Gramatica, P.; Giardina, G.; Speranza, G.; Manitto, P. *Chem. Lett.* **1985**, 1395-1398.
13. Gramatica, P.; Manitto, P.; Poli, L. *J. Org. Chem.* **1985**, *50*, 4625-4628.
14. Brunerie, P.; Benda, I.; Bock, G.; Schreier, P. *Appl. Microbiol. Biotechnol.* **1987**, *27*, 6-10.
15. Bock, G.; Benda, I.; Schreier, P. *Appl. Microbiol. Biotechnol.* **1988**, *27*, 351-357.
16. Mandava, N.B.; Ito, Y. *Countercurrent Chromatography: Theory and Practice*; Marcel Dekker: New York, NY, 1988.
17. Conway, W.D. *Countercurrent Chromatography - Apparatus, Theory & Applications*; VCH: Weinheim, Germany, 1990.
18. *Modern Countercurrent Chromatography*; Conway, W.D.; Petroski, R.J., Eds.; ACS Symp. Ser. 593; American Chemical Society: Washington, DC, 1995.

19. *Centrifugal Partition Chromatography*; Foucault, A.P., Ed.; Chromatographic Science Ser. 68, Marcel Dekker: New York, NY, 1994.
20. Marston, A.; Slacanin, I.; Hostettmann K. *Phytochem. Anal.* **1990**, *1*, 3-17.
21. Oka, F.; Oka, H.; Ito, Y. *J. Chromatogr.* **1991**, *538*, 99-108.
22. Abbott, T.P.; Kleiman, R. *J. Chromatogr.* **1991**, *538*, 109-118.
23. Conway, W.D. *J. Chromatogr.* **1991**, *538*, 27-35.
24. Ito, Y. *J. Chromatogr.* **1991**, *538*, 3-25.
25. Murayama, W.; Kobayashi, T.; Kosuge, Y.; Yano, H.; Nunogaki, Y.; Nunogaki, K. *J. Chromatogr.* **1982**, *239*, 643-649.
26. Hostettmann, K.; Hostettmann, M.; Marston, A. *Preparative Chromatography Techniques - Applications in Natural Product Isolation.* Springer Verlag: Berlin, 1986.
27. Fischer, N.; Weinreich, B.; Nitz, S.; Drawert, F. *J. Chromatogr.* **1991**, *538*, 193-202.
28. Nitz, S.; Spraul, M.H.; Drawert, F. *J. Agric. Food Chem.* **1992**, *40*, 1038-1040.
29. Herion, P.; Full, G.; Winterhalter, P.; Schreier, P.; Bicchi, C. *Phytochem. Anal.* **1993**, *4*, 235-239.
30. Schmidt, G.; Full, G.; Winterhalter, P.; Schreier, P. *J. Agric. Food Chem.* **1995**, *43*, 185-188.
31. Winterhalter, P. In *Progress in Flavour Precursor Studies - Analysis, Generation, Biotechnology*; Schreier, P.; Winterhalter, P., Eds.; Allured Publ.: Carol Stream, IL, 1993; pp. 31-44.
32. Krammer, G.E.; Buttery, R.G.; Takeoka, G.R. In *Fruit Flavors - Biogenesis, Characterization and Authentication*; Rouseff, R.L.; Leahy, M.M., Eds; ACS Symp. Ser. 596; American Chemical Society: Washington, DC, 1995, pp. 164-181.
33. Winterhalter, P.; Skouroumounis, G. *Adv. Biochem. Engin./Biotech.* **1997**, *55*, 73-105.
34. Francis, M.J.O.; Allcock, C. *Phytochemistry* **1969**, *8*, 1339-1347.
35. Skouroumounis, G.K.; Massy-Westropp, R.A.; Sefton, M.A.; Williams, P.J. In *Progress in Flavour Precursor Studies - Analysis, Generation, Biotechnology*; Schreier, P.; Winterhalter, P., Eds.; Allured Publ.: Carol Stream, IL, 1993, pp. 275-279.
36. Skouroumounis, G.K.; Massy-Westropp, R.A.; Sefton, M.A.; Williams, P.J. *J. Agric. Food Chem.* **1995**, *43*, 974-980.
37. Rapp, A.; Mandery, H.; Güntert, M. In *Flavour Research of Alcoholic Beverages*, Poceedings of the Alko Symposium, Helsinki 1984; Nykänen, L.; Lehtonen, P., Eds.; Foundation for Biotechnical and Industrial Fermentation Research: Helsinki, Finland, 1984, pp. 255-274.
38. Williams, P.J.; Strauss, C.R.; Wilson, B.; Dimitriadis, E. In *Topics in Flavour Research*; Berger, R.G.; Nitz, S.; Schreier, P., Eds.; H. Eichhorn: Marzling, Germany, 1985, pp. 335-352.
39. Strauss, C.R.; Wilson, B.; Gooley, P.R.; Williams, P.J. In *Biogeneration of Aromas*; Parliment, T.H.; Croteau, R., Eds.; ACS Symp. Ser. 317; American Chemical Society: Washington, DC, 1986, pp. 222-242.
40. Ohloff, G.; Flament, I.; Pickenhagen, W. *Food Rev. Int.* **1985**, *1*, 99-148.
41. Ohloff, G.; Giersch, W.; Schulte-Elte, K.H.; Enggist, P.; Demole, E. *Helv. Chim. Acta* **1980**, *63*, 1582-1588.
42. Ohloff, G.; Schulte-Elte, K.-H.; Willhalm, B. *Helv. Chim. Acta* **1964**, *47*, 602-626.
43. Ohloff, G.; Lienhard, B. *Helv. Chim. Acta* **1965**, *48*, 182-189.
44. Winterhalter, P.; Messerer, M.; Bonnländer, B. *Vitis* **1997**, *36*, 55-56.

Chapter 17

Mass Spectrometry of the Acetal Derivatives of Selected Generally Recognized as Safe Listed Aldehydes with Ethanol, 1,2-Propylene Glycol and Glycerol

Keith Woelfel[1] and Thomas G. Hartman[2]

[1]M&M Mars, High Street, Hackettstown, NJ 07840
[2]Center for Advanced Food Technology and Department of Food Science, Cook College, Rutgers, The State University of New Jersey, New Brunswick, NJ 08901–8520

The FEMA-GRAS list offers flavor chemists a repertoire of nearly 2000 chemicals for use in compounding natural and synthetic flavors for the U.S. marketplace. Aldehydes constitute an important class of these potential flavorants and are widely utilized to impart specific nuances. Alcohols such as ethanol, 1,2-propylene glycol and glycerol are commonly employed as solvents in compounded flavor systems due to their low odor and miscibility in a wide range of aqueous and organic matrices. However, alcohols and aldehydes react rapidly under anhydrous conditions to form acetal derivatives which often possess different sensory properties. This well known reaction is reversible and its equilibrium is influenced by time, temperature, pH and moisture content. Mass spectra of acetals are currently under represented in commercial databases and few literature references are available. Our investigation involved a systematic mass spectrometric study of the acetal derivatives of selected GRAS aldehydes reacted with ethanol, 1,2-propylene glycol and glycerol. Aldehydes from different chemical classes representing saturated and unsaturated aliphatics, aromatics, heterocyclics, terpenoids and others were included for characterization. The corresponding acetals were synthesized, analyzed by GC-MS in electron ionization mode and their retention indices on a non-polar (polydimethylsiloxane) capillary column were determined. A database of mass spectra was produced which includes many previously unreported species. In total, over 60 individual mass spectra were recorded. The characteristic mass spectral fragmentation pathways for each class of acetal are described.

Naturally occurring or synthetically produced, flavoring substances are numerous and diverse in foods and beverages which we consume. However, intentional addition of flavoring ingredients to a food or beverage is limited to less than 2000 federally regulated compounds or materials. Commonly referenced as Generally Recognized As Safe (GRAS) substances, this list is regulated by the Food and Drug Administration (FDA) (1,2). The FDA is in a continual state of monitoring additions or deletions to the GRAS list of flavoring substances based upon concerns for public safety and recommendations by the Flavor and Extract Manufacturers Association of the United States (FEMA). FEMA, an industry trade association, commissions an expert panel of scientists, which includes toxicologists, pharmacologists, biochemists, nutritionists, statisticians, analytical chemists and other independent experts to determine the relative safety of flavoring substances. Based on the recommendations of the expert panel FEMA coordinates the GRAS affirmation process of FDA approval resulting in the FEMA GRAS list of flavor compounds (3).

Aldehydes are an important class of compounds for providing certain flavors in foods and beverages. Currently, there are 126 aldehydic substances on the FEMA GRAS list (1). In many instances aldehydes are important keynotes in producing characterizing flavors. For example, vanillin is an aromatic aldehyde which is the most important contributor to vanilla flavor, the worlds most popular. Ethyl vanillin, although not found naturally, also has an intense vanilla-like flavor. Methional, a Strecker aldehyde produced in the Maillard reaction via the thermal interaction of methionine with reducing sugars, is characterizing for potato flavor. Cinnamic aldehyde largely forms the basis for the flavor of cinnamon. Safranal, the major aldehyde found in saffron, is responsible for the spice's characteristic flavor impact. Benzaldehyde can be characterizing for both cherry and bitter almond depending on its concentration. The lipid oxidation derived compound, 2,4-decadienal, is known to impart a powerful deep fried fat flavor at low concentration. Cis-3-hexenal, also known as leaf aldehyde, is a compound recognized for its fresh green leafy odor and is widely used in compounding a myriad of fruit flavors. The aldehydes neral and geranial (citral) produced in the terpene synthesis pathway impart characteristic citrus notes and are present at high concentration in essential oils of several citrus varieties (4-9). Examples of aldehydes as characterizing species such as these abound in the flavor literature.

Because of the relative insolubility of many flavor raw materials in the aqueous environments typical of most foods and beverage systems, flavor compounding often requires the use of alcoholic cosolvents. Ethanol is extensively employed for this purpose. This universal solvent is miscible in water as well as most organic phases, it provides "lift" for flavorings, imparts little odor and safeguards against microbial growth. Another popular solvent is 1,2-propanediol, commonly referred to as propylene glycol (PG). Imparting a mild, sweet mouth warming sensation, it exhibits low toxicity and is used extensively as a base for formulating compounded flavors (5,10). To a somewhat lesser extent glycerol is employed as a solvent, cosolvent or ingredient in flavorings. Sixty percent as sweet as sucrose, this triol is often employed in beverages to build viscosity and add mild sweetness. It possesses the added ability to act as a plasticizer and humectant and is used occasionally in the production of process flavors (11).

It is well known to flavor chemists that the presence of carbonyl compounds in combination with alcohols leads to the consequential formation of acetals in flavor formulations. Acetals form via a well known simple reaction which is documented in the most basic organic chemistry textbooks. Temperature, time, pH and moisture content influence the rate of reaction. Small, unbranched aldehydes react fastest due to the lack of steric hindrance (12). Acetal formation proceeds by a two step mechanism, the first step of which can be both acid or base catalyzed (13,14). However, the second step, involving the conversion of a hemiacetal intermediate to an acetal, is strictly an acid catalyzed reaction. In the acid catalyzed reaction, protonation of the carbonyl group occurs and a carbocation intermediate forms. Nucleophilic attack by an alcohol group produces a hemiacetal intermediate. Following the loss of a single mole of water, a second carbocation intermediate is produced. Additional nucleophilic attack by a second alcohol group results in the formation of an acetal (14). The base catalyzed reaction involves the reaction of the carbonyl group of the aldehyde with an alkoxide ion of the alcohol leading to formation of a carboanion intermediate which reacts with a second alcohol group to form a hemiacetal. Diols and triols are capable of reacting intramolecularly to form cyclic acetals and these reactions generally proceed rapidly due to their low activation energies (13).

Acetal formation is a reversible reaction, the equilibrium of which will be influenced by the concentration of reaction products under a given set of conditions. Since water is a product of acetal formation, its increasing concentration shifts the equilibrium toward reversal of the reaction. To obtain a high yield synthesis of acetals, sequesterants or azeotropic distillations are often commercially employed to remove water from the reaction as it forms, and reactions are carried out at elevated temperatures to increase their rate (13).

Reaction of an aldehyde with ethanol produces a diethyl acetal which is an R-1-substituted 1,1-diethoxy derivative. The simplified reaction scheme is illustrated in figure 1. The reaction of an aldehyde with PG produces a cyclic acetal which is an R-2-substituted, 4-methyl- 1,3-dioxolane. This reaction produces equal quantities of two geometrical isomers called the *syn* and *anti* conformations according to the positions of the R-group and methyl substitution relative to the planar dioxolane ring (14). These isomers are normally resolved by high resolution capillary gas chromatography (GC) yielding characteristic tightly spaced, doublet peaks of equal intensity, the mass spectra of which are indistinguishable from each other (15-17). The generic reaction pathway for the formation of cyclic PG acetals is shown in figure 2.

Glycerol is capable of forming two different types of cyclic acetals when reacted with an aldehyde. Since this compound is a triol, cyclic 1,2- or 1,3- addition products result. The 1,2- glyceryl acetals are R-2-substituted-4-methanol-1,3-dioxolanes and the 1,3- glyceryl acetals are R-2-substituted-5-hydroxy-1,3-dioxanes (12,13,14). Once again, syn and anti geometrical isomers of both the dioxolane and dioxane structures are produced in equal quantities. A capillary GC chromatogram of a typical glyceryl acetal therefore contains at least four peaks pertaining to the dioxolane and dioxane ring forms and their corresponding geometrical isomers. The generic pathway to the formation of glyceryl acetals is shown in figure 3. The glyceryl 1,2- and 1,3-cyclic acetals can be differentiated by their characteristic mass spectra.

The conversion of an aldehyde to an acetal profoundly changes its vapor pressure, solubility and aroma characteristics (5,10,11) and generally attenuates or qualitatively alters its flavor impact. One example is isovaleraldehyde which contains a strong, unpleasant stench at high concentration. However, formation of a propylene glycol acetal imparts a pleasant, chocolate-like aroma (5). A second example is benzaldehyde, often a critical substance in almond or artificial cherry flavorings. Upon formation of an acetal, this substance becomes virtually flavorless (16,17). The presence of propylene glycol in imitation vanilla flavors where it is used as a cosolvent or for control of water activity leads to formation of vanillin and/or ethyl vanillin cyclic PG acetals which causes flavor attenuation. However, the loss of flavor to acetal formation is inconsequential in many systems since they may rapidly regenerate the original aldehydes upon hydration in a high moisture food or beverage end use application. The chemistry and sensory properties of acetals have been described as important to some perfumers and flavorists and as mere curiosities or unimportant to the development efforts of others.

Although limited, a few investigations regarding formation and mass spectral identification of acetals in food and flavor systems have been published. Heydanek and Min (17) investigated carbonyl-propylene glycol interactions in flavor systems to identify the presence of acetals and determine the rate of formation. Mass spectrometry (MS) was employed to assist in identification of 11 acetals. Electron ionization (EI) mass spectra of 11 acetals, including those derived from reaction of PG with benzaldehyde, cinnamic aldehyde, isovaleraldehyde, octanal, citral, vanillin and others were reported. Several mass spectrometric studies of selected acetals have been published (18). However, these investigations did not involve aldehydic flavoring substances.

Shu and Lawrence (16), also investigated several carbonyl containing flavor substances for interactions in PG based flavor systems. Rate of acetal formation and implications in quality control of flavors were discussed. Heliotropin (piperonal), ethyl vanillin, vanillin, levulinic acid and 2,4-dimethyl-4-hydroxy-3(2H)-furanone (DMHF) were investigated. Due to the lack of reference EI spectra, the authors relied on NMR and IR instrumentation for positive identification.

The work of Shu et al. and other investigators demonstrates the need to reduce the informational gap present in reference mass spectral libraries with respect to flavor derived acetals. Although mass spectral databases, such as National Institute of Standards and Technology (NIST) and Wiley contain over 80,000-200,00 entries (19,20), the spectra of many acetals are still absent. The lack of reference spectra impedes the rapid identification of these substances in flavor formulations or finished products. The purpose of this research was to narrow the informational gap by

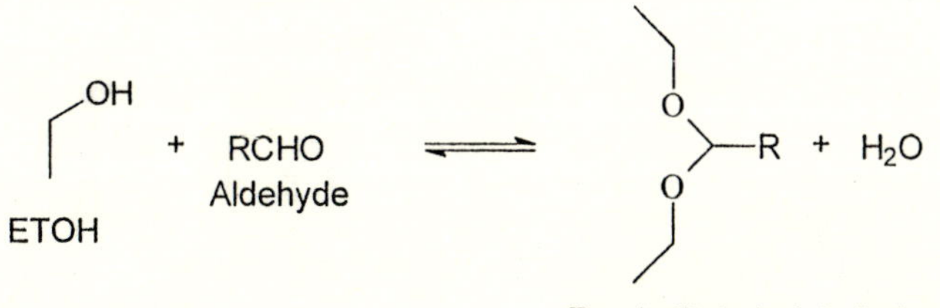

Figure 1. Diethyl acetal formation leading to the production of an R-substituted-1,1-diethoxy derivative.

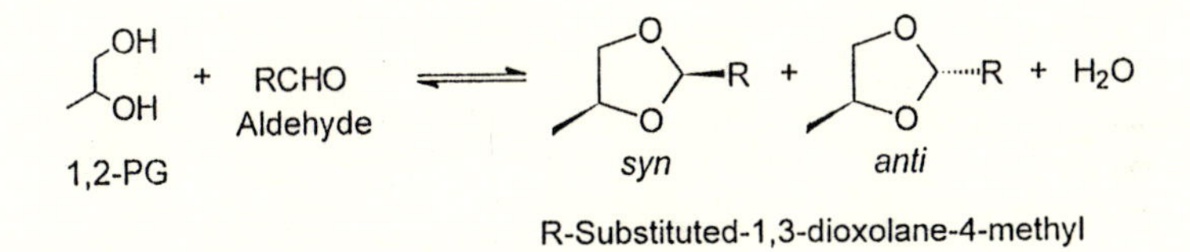

Figure 2. 1,2-propylene glycol cyclic acetal formation leading to the production of an R-substituted-4-methyl-1,3-dioxolane derivative with a racemic mixture of *syn* and *anti* geometrical isomers.

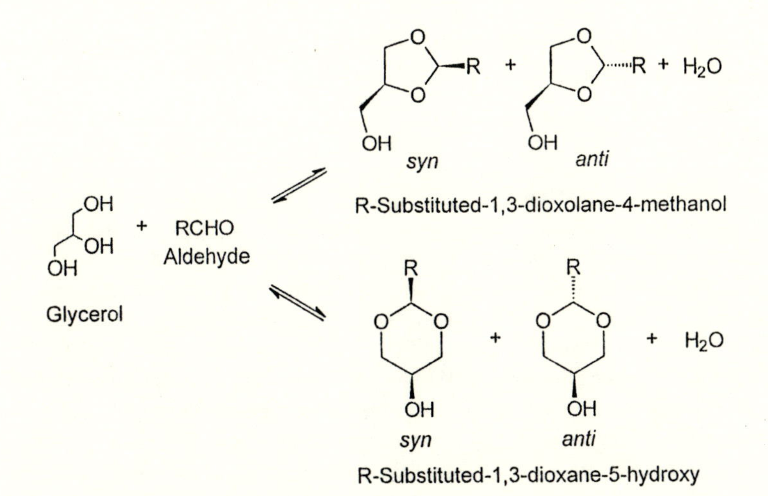

Figure 3. Glyceryl acetal formation showing the production of 1:2 (R-substituted-4-methanol-1,3-dioxolane) and 1:3 (R-substituted-5-hydroxy-1,3-dioxane) addition products with *syn* and *anti* geometrical isomer conformations.

conducting a systematic mass spectrometric study of the acetal derivatives of some common GRAS-listed aldehydes. Twenty five GRAS aldehydes selected from various chemical classes were reacted with ethanol, 1,2-propylene glycol and glycerol to form the corresponding acetals. Subsequent to acetal synthesis and extraction, GC and MS instrumentation was utilized to obtain EI spectra. A condensed subset of this data including the eight most abundant ions, including M+ and M-1 (if present) for each acetal, were recorded in tabular form. Aromatic, aliphatic, unsaturated and terpenoid aldehydes were included in this investigation. In addition, the characteristic mass spectral fragmentation pathways for each acetal class are described. The formation of base peak ions, presence or absence of molecular ions, and other characteristic fragments is discussed. EI fragmentation of organic molecules is addressed in detail by Mc Lafferty(15) and Budzikiewicz (21). However, specific discussions pertaining to acetals are limited. Generally, linear and cyclic acetals are treated as ether linkages. This research was intended to develop interpretation guidelines for acetals, supplementing the observations of previous investigators with some novel spectra described in this manuscript. Retention indices on a non-polar, bonded and cross-linked polydimethylsiloxane fused silica capillary column were also calculated.

Materials

Benzaldehyde, phenylacetaldehyde, vanillin, ethyl vanillin, heliotropine (piperonal), cinnamic aldehyde, anisaldehyde, hexanal, heptanal, octanal, nonanal, decanal, undecanal, dodecanal, t-2-hexenal, 4-decenal, 10-undecenal, 2,4-decadienal, methional, isovaleraldehyde, furfural, citral (neral & geranial), perillal, safranal, 1,2-propylene glycol and glycerol were all obtained from Aldrich Chemical Co., Milwaukee, WI. Methylene chloride (Optima grade) and water (HPLC grade) were purchased from Fisher Scientific, Springfield, NJ. Ethanol (USP, 200 proof) was obtained from Pharmco, Brookfield, CT.

Methods

Acetal Synthesis and Isolation. Diethyl, PG and Glyceryl acetals were prepared by reacting each aldehyde with ethanol, 1,2-propylene glycol and glycerol individually. In all cases 0.9 ml of alcohol and 0.1ml of aldehyde were placed in 1.8 ml borosilicate glass vials sealed with Teflon-lined screw caps. The vials were heated to 70°C overnight in a heating block. Following incubation, the vial contents were pipetted into 20 ml borosilicate glass test tubes sealed with Teflon-lined screw closures along with 15 ml of distilled water and 5 ml of methylene chloride. The samples were vigorously extracted and then centrifuged at 2000 rpm for 10-15 minutes to promote complete phase separation. The lower methylene chloride layers containing acetals and any unreacted aldehydes were aspirated from the tubes using a Pasteur pipet and were transferred to 5 ml vials. The vials were refrigerated prior to GC and GC-MS analysis. The aqueous phases, containing primarily water and unreacted alcohols, were discarded.

GC-FID and GC-MS Analysis. GC-MS analyses were performed using a Varian 3400 GC directly interfaced to a Finnigan MAT 8230 high resolution, double focusing magnetic sector mass spectrometer. Data were acquired and processed using a Finnigan MAT SS300 data system. Injections (1.0 μl) of acetal solutions in methylene chloride were made on a 60 meter DB-1 (non-polar, bonded & cross linked polydimethylsiloxane phase) J&W capillary column with a 0.32 mm inside diameter and a 0.25 μm film thickness. Injections were made in split mode using a split ratio of 100:1; the carrier gas was helium; and a column flow rate of 1.0 ml/minute was employed. The injector and GC-MS transfer line temperatures were 250 and 280°C respectively. The column was temperature programmed from 50°C (hold 3 min.) to 280°C at a rate of 10°C per minute with a 10-15 minute hold at the upper limit. The mass spectrometer was operated in electron ionization mode (70 eV) scanning masses 35-450 at a rate of 1.0 second per decade with a 0.8 interscan time.

The acetals were also analyzed by GC with flame ionization detection (FID) using the same chromatographic conditions as described above to establish retention time indices. In this instance a Varian 4290 integrator was used to record the data. Retention indices of the acetals relative to a mixture of C-6 through C-36 n-paraffin standards were calculated using the equation described by Majlat (21).

Results and Discussion

Diethyl Acetals. The EI mass spectra of the diethyl acetals produced in this investigation are summarized in condensed form in table I. The table lists the eight peaks of highest intensity in each spectrum along with their percent abundance. Included in the table are the molecular weights, percent abundance of the molecular ions if present and the M-1 peaks which are diagnostic fragment ions for many diethyl acetals. The GC retention time indices for each compound relative to n-paraffin reference standards on a non-polar, polydimethylsiloxane, bonded and cross linked fused silica capillary column are also provided.

The mass spectra of the diethyl acetals of aliphatic aldehydes usually do not contain molecular ions. However, a low intensity M-1 peak is typically present in the mass spectrum owing to rapid neutral loss of a proton from the ionized species. The absence of molecular ions in the mass spectra of these compounds is due to the inability of these species to stabilize the energy of ionization through resonance hybridization owing to a lack of conjugation and the weak bond energies associated with the ether linkages and paraffinic bonds. The base peak in the mass spectra of these compounds is always m/z 103 which results from rapid cleavage of the R-substituted alkyl chain in the number one carbon position accompanied by charge localization on the 1,1-diethoxy radical. This ion then ejects one and two units of ethylene to form characteristic fragment ions at m/z 75 and 47 which are found in virtually all of the spectra. Other diagnostic features of the fragmentation pattern include rapid neutral loss of an ethoxyl fragment from the molecular ion leading to charge retention on the M-45 fragment. Classic fragmentation of linear alkyl chains from the R-group produce the familiar m/z 43, 57, 85, 99, 113 ... pattern of ions as methylene units are sequentially cleaved from the backbone (15,21). In branched alkyl chains a higher intensity fragment ion is usually observed due to preferred cleavage at the branch point. However, the series of hydrocarbon fragments are always of lower intensity due to preferred charge localization on the oxygenated fragments. An example of a typical mass spectrum from an aliphatic aldehyde diethyl acetal is that derived from nonanal. The mass spectrum of nonanal, diethyl acetal does not exhibit a molecular ion but shows a weak (0.3%) M-1 fragment and loss of ethoxy producing the m/z 171 ion. As expected, the base peak is m/z 103 and the two successive losses of ethylene from this ion produce the characteristic oxygenated fragments at m/z 75 and 47 respectively. Lower intensity fragments from the linear alkyl chain are also present in the background of the mass spectrum.

The presence of a single double bond in the R-substituted alkyl chain of a diethyl acetal begins to impart some degree of charge stabilization, especially when this bond is conjugated to the 1,1-diethoxy group. For instance, the mass spectrum of trans-2-hexenal, diethyl acetal exhibits a molecular ion at m/z 172 (2.9%) while the spectrum of hexanal, diethyl acetal does not. The M-1 peak is also elevated in this spectrum compared to saturated analogues. The typical base peak for saturated linear aldehyde diethyl acetals at m/z 103 is conspicuously absent from the spectrum of the hexenal derivative and is replaced by m/z 127 which is the M-45 ion. Furthermore, the linear hydrocarbon backbone fragment ions are much more intense in this spectrum as compared to the hexanal diethyl acetal. This clearly illustrates the site directed fragmentation incurred by charge localization at the unsaturation site of the alkyl chain.

Increasing unsaturation in the R-substituted alkyl chains greatly increases the charge stabilization effect. For instance, in citral diethyl acetal, the two double bonds present in the terpenoid side chain stabilize the molecular ion which is observed at 226 (0.5%) and promote charge localization. As a consequence, the base peak (m/z 69) and major fragment ions in the spectrum are dominated by the alkyl chain fragments and the oxygenate fragments (m/z 103, 75 & 47) exhibit reduced abundance.

Aromatic diethyl acetals generally exhibit higher intensity molecular ions, especially if the aromatic rings are conjugated through the 1,1-diethoxy moiety affording the opportunity for resonance hybridization. This is illustrated by the mass spectrum of cinnamic aldehyde diethyl acetal in which the propenyl bridge between the 1,1-diethoxy group and the benzene ring provides for resonance stabilization. This compound exhibits a more stable molecular ion at m/z 206 with an abundance of 17%. In the case of phenylacetaldehyde diethyl acetal no unsaturated "bridge" exists and a molecular ion does not occur. Surprisingly, in this example charge localization at the oxygenate ions occurs at the expense of the aromatic ring and the tropyllium ion at m/z 91 is relatively weak in comparison.

Of the compounds listed in table I, the mass spectra of the diethyl acetal derivatives of t-2-hexenal, nonanal, dodecanal, cinnamic aldehyde and

Table I. Mass Spectra of Diethyl Acetals

Acetal	Characteristic Mass Spectral Ions and Abundances	[1]M	[2]M-1	[3]RI
trans-2-hexenal diethyl acetal	57(100), 127(100), 85(67.2), 41(42.5), 43(32.3), 99(31.3), 129(28.4), 69(26.2)	172 (2.9)	171 (1.7)	1031
isovaleraldehyde diethyl acetal	103(100), 75(79.0), 115(66.3), 47(62.5), 69(56.4), 71(48.5), 43(38.3), 45(25.5)	160 (0)	159 (0.5)	986
citral (neral/geranial) diethyl acetal	69(100), 87(99.2), 41(52.7), 181(34.6), 103(27.9), 83(21.5), 59(20.7), 75(19.6)	226 (0.5)	225 (0.6)	1514
heptanal diethyl acetal	103(100), 143(74.8), 75(63.5), 47(58.3), 55(51.7), 97(50.6), 43(30.6), 57(28.8)	188 (0)	187 (0.8)	1186
octanal diethyl acetal	103(100), 47(41.7), 157(35.5), 69(31.9), 57(31.8), 75(29.6), 55(18.9), 41 (16.5)	202 (0)	201 (0.5)	1307
nonanal diethyl acetal	103(100), 75(31.0), 47(26.5), 171(26.0), 69(24.9), 57(15.3), 55(12.9), 41(12.5)	216 (0)	215 (0.3)	1319
dodecanal diethyl acetal	103(100), 43(57.1), 57(54.6), 41(57.1) 55(41.5), 75(28.5), 47(26.6), 69(23.1)	258 (0)	257 (0)	n.d.
cinnamaldehdye diethyl acetal	161(100), 133(72.7), 105(48.3), 55(46.8), 131(42.7), 103(39.7), 77(39.3), 115(31.6)	206 (17.3)	205 (3.3)	1518
phenyl-acetaldehyde diethyl acetal	103(100), 47(62.2), 75(46.6), 91(28.2), 121 (21.5), 149(17.6), 104(6.7), 65(6.3)	194 (0)	193 (0.1)	1307

[1]M = Molecular ion (% abundance)
[2]M-1 = Molecular ion minus H (% abundance)
[3]RI = GC retention index on non-polar, polydimethylsiloxane capillary column relative to n-paraffin series

phenylacetaldehyde are not present in any commercial databases and to the best of the author's knowledge are reported for the first time in this investigation. The mass spectrum for citral (racemic neral/geranial) diethyl acetal recorded in the current investigation is thought to be superior to that in the NIST and Wiley libraries as we observe a molecular ion and none is reported in the commercial databases. It is noteworthy that many compounds investigated did not readily form diethyl acetals under the stated reaction and isolation conditions.

Propylene Glycol Cyclic Acetals. The mass spectra of PG acetals are summarized in table II. Chromatograms of PG acetals typically contained doublet peaks of equal intensity with one or two second spacing between them corresponding to the *syn* and *anti* geometrical isomers produced during synthesis. The mass spectra of the *syn* and *anti* isomers were always indistinguishable so only one of the two are presented in the tables. This has been the experience of other investigators as well (16,17).

The mass spectra of PG acetals derived from linear, saturated, aliphatic aldehydes are typified by a weak or nonexistent molecular ion, a weak M-1 peak, a base peak of 87, a significant m/z 59 fragment and characteristic paraffinic fragments of lower intensity. The signature ions from these spectra are the m/z 87 and 59 fragments which are diagnostic for the 4-methyl-1,3-dioxolane ring (15,19,20,21). The m/z 87 fragment arises from alpha cleavage of the R-substituted alkyl chain at the number two carbon of the dioxolane ring accompanied by charge retention in on the cyclic moiety. Ejection of an ethylene fragment from this ion generates the m/z 59 signal. Quite predicably, as the R-group of the aldehyde becomes unsaturated the fragmentation pattern changes. The molecular ions become stabilized and competition for charge retention between different regions of the molecule leads to site directed fragmentation. For example, the PG acetals of decanal, 4-decenal and 2,4-decadienal yield molecular ion abundances of 0, 4.0 and 12.5 % respectively. The 87 ion is the base peak for the PG acetal of decanal and remains so for 4-decenal, but in 2,4-decadienal the base peak becomes m/z 67 reflecting the charge retention at the conjugated double bonds of the alkyl chain. Aromaticity in the R-group of the aldehyde virtually guarantees the presence of a molecular ion. Accordingly, the PG acetals of benzaldehyde, phenylacetaldehyde, anisaldehyde, vanillin, ethyl vanillin and cinnamic aldehyde all contain significant molecular ions. The PG acetals of terpenoid or heterocyclic aldehydes yield mass spectra with fragmentation characteristics intermediate between the aliphatics and aromatics. In these compounds, molecular ions will almost always be present but the base peaks may oscillate between the 87 peak from the dioxolane ring and ions generated from the R-group depending upon the degree of unsaturation and conjugation. For instance, furfural and safranal PG acetals exhibit base peaks which derive from the R-group rings rather than the dioxolane but the opposite is true for citral and perillal which are less conjugated. Of the compounds listed in table II, the mass spectra of the PG acetal derivatives of heptanal, nonanal, decanal, 4-decenal, 2,4-decadienal, 10-undecenal, methional, furfural, safranal, perillal, and phenylacetaldehyde are not present in any commercial databases and to the best of the author's knowledge are reported for the first time in this investigation.

Table II. Mass Spectra of Propylene Glycol Acetals

Acetal	Characteristic Mass Spectral Ions and Abundances	[1]M	[2]M-1	[3]RI
trans-2-hexenal PG acetal	113(100), 55(90.6), 69(87.9), 127(43.8), 87(40.8), 59(30.0), 97(24.8), 83(20.1)	156 (1.5)	155 (13.6)	1136, 1144
4-decenal PG acetal	87(100), 41(72.1), 59(70.6), 113(49.5), 55(32.5), 43(16.0), 39(15.3), 67(14.6)	212 (4.0)	211 (1.2)	1528, 1538
2,4-decadienal PG acetal	67(100), 41(83.9), 54(80.4), 81(63.4), 100(62.1), 55(38.1), 139(36.4), 95(32.8)	210 (12.5)	209 (2.5)	1657
10-undecenal PG acetal	87(100), 41(71.4), 59(66.9), 55(26.7), 39(13.1), 43(12.7), 42(11.8), 67(6.6)	226 (0.1)	225 (0.3)	1647, 1658
isovaleraldehdye PG acetal	43(100), 101(85.0), 85(22.0), 41(16.8), 59(13.1), 87(9.1), 42(5.3), 102(4.8)	144 (0.4)	1473 (0)	966, 974
methional PG acetal	87(100), 59(97.7), 61(75.6), 114(58.9), 41(54.0), 56(38.3), 45(22.3), 75(22.2)	162 (41.7)	161 (3.2)	1234, 1244
furfural PG acetal	95(100), 81(55.4), 52(40.4), 80(39.3), 68(31.8), 126(31.4), 39(27.2), 41(18.2)	154 (89.4)	153 (48.7)	1133
citral PG acetal	87(100), 69(98.3), 41(79.6), 55(52.8), 141(32.1), 59(26.2), 83(23.3), 84(19.0)	210 (4.8)	209 (4.8)	1507, 1542
perillal PG acetal	87(100), 41(98.7), 139(61.7), 59(61.3), 79(53.3), 39(51.6), 81(49.6), 93(47.2)	208 (19.8)	207 (11.1)	1615
safranal PG acetal	107(100), 87(64.4), 91(45.5), 59(40.1), 105(23.0), 122(22.2), 121(21.1), 77(19.3)	208 (5.2)	207 (0)	n.d.
hexanal PG acetal	87(100), 59(20.3), 41(9.2), 71(5.1), 43(4.8), 86(4.7), 39(2.5), 42(2.2)	158 (0.3)	157 (3.6)	1116, 1126
heptanal PG acetal	87(100), 43(34.7), 41(18.1), 59(13.3), 42(11.9), 45(10.4), 39(2.5), 88(7.1)	172 (1.1)	171 (7.8)	1224, 1234

Table II. Mass Spectra of PG acetals contd.

octanal PG acetal	87(100), 41(39.6), 59(28.2), 43(13.7), 39(8.7), 42(8.0), 57(6.2), 55(5.1)	186 (0)	185 (0.4)	1316
nonanal PG acetal	87(100), 59(33.2), 41(24.1), 43(9.2), 55(7.4), 57(5.6), 71(3.5), 69(2.9)	200 (0.1)	199 (1.0)	1387, 1390
decanal PG acetal	87(100), 59(29.0), 41(21.2), 43(10.1), 55(6.9), 57(5.2), 88(4.8), 42(4.1)	214 (0)	213 (0)	1563, 1584
dodecanal PG acetal	87(100), 59(20.3), 41(16.2), 43(10.2), 55(7.1), 57(4.6), 88(4.4), 42(3.4)	242 (0)	241 (0)	n.d.
cinnamaldehyde PG acetal	104(100), 115(37.0), 131(26.6), 103(21.0), 105(16.5), 77(16.3), 107(11.3), 78(11.0)	190 (59.7)	189 (7.5)	1613, 1623
benzaldehyde PG acetal	163(100), 105(43.6), 77(56.7), 78(48.6), 91(37.3), 51(33.7), 90(33.0), 89(26.3)	164 (43.2)	163 (100)	1318
phenylacetaldehyde PG acetal	87(100), 59(43.6), 91(33.8), 41(20.7), 85(9.1), 65(8.3), 105(6.0), 92(5.5)	178 (0.3)	177 (0.7)	1380
vanillin PG acetal	151(100), 87(93.8), 124(58.6), 59(55.2), 109(46.6), 137(35.0), 41(29.7), 92(29.1)	210 (7.0)	209 (36.0)	1761
ethyl vanillin PG acetal	87(100), 43(92.0), 41(89.8), 110(86.4), 137(72.7), 165(67.6), 59(56.8), 138(48.3)	224 (45.5)	223 (50.6)	1818
anisaldehyde PG acetal	108(100), 77(64.9), 135(59.8), 51(35.6), 91(30.6), 39(29.9), 41(29.7), 92(29.1)	194 (13.7)	193 (29.3)	1588

[1]M = Molecular ion (% abundance)

[2]M-1 = Molecular ion minus H (% abundance)

[3]RI = GC retention index on non-polar, polydimethylsiloxane capillary column relative to n-paraffin series

Glyceryl Acetals. The mass spectra of the glyceryl acetals are summarized in table III. Glyceryl acetals often generate chromatograms possessing quadruplet peaks which correspond to the *syn* and *anti* geometrical isomers of the 1,2- and 1,3- cyclic addition products (13,14). The geometrical isomers of the substituted dioxane or dioxolanes yield identical mass spectra but the ring systems may occasionally be differentiated by slight changes in the fragmentation pathway. Specifically, an M-31 ion characteristic for neutral loss of -CH_2-OH from R-2-substituted-4-methanol-1,3-dioxolanes is often present in the mass spectra of the 1,2-glyceryl acetals but this feature is usually absent in spectra of the 1,3- addition products. The 1,2-glyceryl acetals were observed to elute from the GC prior to the 1,3-glyceryl acetals. When the spectra in table III were clearly assignable to the glyceryl-1,2 or 1,3 isomer then the exact compounds are specified. In some instances only one peak was evident in the chromatogram or the isomers were indistinguishable and in these cases the entries are listed as glyceryl acetals with no further designation.

Glycerol acetals are characterized by the presence of a strong m/z 103 ion fragment. This ion is capable of two isomeric structures, a five member, 4-methylol-1,3-dioxolane ring or a six member, 5-hydroxy-1,3 dioxane ring (15,19,20,21). Aliphatic glyceryl acetals typically yield mass spectra with weak or nonexisting molecular ions, a weak M-1 peak and a base peak at m/z 103. The distinguishing feature of a low intensity M-31 is present in the 1,2-glyceryl acetals of these compounds. The glyceryl acetals of hexanal, heptanal, octanal, nonanal, decanal and dodecanal all fall into this series. As was the case for diethyl and PG acetals, as the R-group of the aldehyde in glyceryl acetals becomes increasingly unsaturated and conjugated then fragmentation is altered. In these instances, the m/z 103 peak is usually still prominent but the base peak becomes associated with the R-group side chain. The glyceryl acetals of the unsaturated, terpenoid and heterocyclic aldehydes fall within this category.

The aromatic series of glyceryl acetals typically generate spectra containing strong molecular and M-1 ions (15,21). Anisaldehyde, ethyl vanillin, vanillin, benzaldehyde, cinnamic aldehyde and phenylacetaldehyde are good examples. Base peak ions vary, usually representing a decomposition of the parent compound by ejecting part or all of the dioxolane or dioxane moiety. Similar to aromatic PG acetals, the presence of substituted, stabilizing features on the aromatic ring will attenuate the abundance of m/z 103 ions. Phenylacetaldehdye, vanillin, and anisaldehyde glyceryl acetals give this ion as the base peak. With increasing propensity to lose the diether, heterocyclic ring in the form of an uncharged radical, the presence of 103 m/z ions will decrease. This is evident in cinnamaldehyde (31.9%) and benzaldehyde (10.4%) glyceryl acetals.

A single sulfur containing aldehyde, methional, was included in this investigation. Its glyceryl-1,2-acetal isomer gives a modest molecular ion (2.9%), a prominent 103 m/z (43.3%) ion and a significant M-31 (5.8%) ion which is diagnostic for the 5-methylol-1,3-dioxolane ring. This is in contrast to the glyceryl-1,3-acetal

Table III. Mass Spectra of Glyceryl Acetals

Acetal	Characteristic Mass Spectral Ions and Abundances	[1]M	[2]M-1	[3]RI
t-2-hexenal glyceryl-1,2-acetal	129(100), 57(93.6), 41(82.4), 55(74.5), 69(54.6), 143(48.7), 43(46.3), 99(46.2)	172 (1.4)	171 (10.5)	1348, 1361,
4-decenal glyceryl acetal	57(100), 41(55.5), 129(54.8), 103(50.8), 55(38.1), 69(29.0), 43(21.6), 67(18.8)	228 (1.8)	227 (0.7)	1753, 1794
10-undecenal glyceryl-1,2- acetal	103(100), 57(46.9), 41(22.7), 55(17.0), 43(7.3), 67(7.3), 45(6.3), 39(5.1)	242 (0)	241 (1.5)	1878, 1918
isovaleraldehyde glyceryl-1,2- acetal	103(100), 57(57.8), 43(52.2), 69(38.5), 41(29.5), 45(27.3), 44(27.2), 87(21.7)	160 (0.8)	159 (8.6)	1167, 1191
methional glyceryl-1,2-acetal	61(100), 57(97.6), 103(43.4), 56(30.6), 47(28.5), 45(25.7), 160(23.4), 75(19.5)	178 (2.9)	177 (2.6)	1340, 1445
methional glyceryl-1,3-acetal	57(100), 61(70.7), 48(53.0), 43(51.5), 56(49.2), 103(45.9), 45(36.2), 47(31.9)	178 (10.2)	177 (2.3)	1470, 1497
furfural glyceryl acetal	81(100), 95(71.1), 97(61.8), 116(51.2), 39(44.7), 139(39.2), 52(27.6), 43(24.2)	170 (34.7)	169 (14.9)	1295, 1330
citral glyceryl acetal	69(100), 41(93.2), 103(65.1), 57(55.0), 43(22.3), 55(35.7), 83(19.0), 84(18.4)	226 (5.1)	225 (1.8)	1696, 1747
cinnamaldehyde glyceryl-1,2-acetal	104(100), 115(63.6), 131(40.9), 103(31.9), 107(31.0), 77(26.1), 57(24.5), 105(22.5)	206 (57.7)	205 (7.7)	1858, 1883
benzaldehyde glyceryl-1,2-acetal	91(100), 105(60.8), 77(47.7), 149(30.1), 79(25.5), 78(25.3), 51(20.7), 57(18.6)	180 (40.0)	179 (86.3)	1549
benzaldehyde glyceryl-1,3-acetal	107(100), 79(83.7), 105(79.5), 77(72.1), 43(35.7), 51(32.1), 44(30.7), 57(14.8)	180 (57.4)	179 (22.8)	1619

Table III. Mass Spectra of Glyceryl Acetals Contd.

phenylacetaldehyde glyceryl-1,2-acetal	103(100), 57(50.6), 91(22.9), 47(9.5), 105(6.0), 104(6.0), 45(5.3), 65(5.1)	194 (0.2)	193 (1.0)	1605
phenylacetaldehyde glyceryl-1,3-acetal	103(100), 57(44.3), 91(32.9), 47(8.6), 65(8.2), 92(6.3), 45(6.3), 104(5.5)	194 (0)	193 (1.8)	1644
vanillin glyceryl-1,2-acetal	151(100), 152(73.7), 57(48.4), 103(45), 137(43.2), 81(35.3), 109(33.4), 65(33.0)	226 (37.2)	225 (23.3)	1985
ethyl vanillin glyceryl-1,2-acetal	137(100), 138(97.5), 166(73.6), 57(55.2), 103(50.0), 165(36.4), 81(36.0), 110(34.2)	240 (51.4)	239 (33.4)	2071
anisaldehyde glyceryl-1,2-acetal	135(100), 108(89.1), 77(77.7), 121(72.9), 57(38.5), 51(36.2), 65(32.2), 179(28.3)	210 (18.9)	209 (33.4)	1834
heliotropine glyceryl-1,2-acetal	149(100), 135(64.1), 122(58.1), 150(55.2), 121(37.9), 65(36.3), 103(34.1), 63(28.4)	224 (45.2	223 (29.5)	n.d.
heliotropine glyceryl-1,3-acetal	149(100), 150(65.7), 122(49.3), 121(41.3), 135(39.8), 103(32.0), 65(30.5), 63(26.2)	224 (41.4)	223 (27.0)	n.d.
perillal glyceryl acetal	57(100), 103(94), 79(60.5), 155(56.9), 105(52.4), 67(50.0), 81(49.6), 41(46.4)	224 (46.5)	223 (12.5)	1867
safranal glyceryl acetal	57(100), 107(98.3), 103(77.1), 91(70.1), 135(57.2), 105(47.4), 121(35.5), 77(28.7)	224 (33.0)	223 (1.3)	n.d.
hexanal glyceryl-1,2-acetal	103(100), 57(72.9), 43(19.8), 41(18.7), 55(17.2), 45(11.5), 47(11.4), 83(9.2)	174 (0)	173 (1.6)	1316
hexanal glyceryl-1,3-acetal	103(100), 57(44), 43(33.8), 83(32.7), 44(28.7), 41(26.6), 45(13.1), 101(12.8)	174 (0.7)	173 (5.7)	1361

Table III. Mass Spectra of Glyceryl Acetals Contd.

heptanal glyceryl-1,2-acetal	103(100), 97(27.1), 55(25.9), 57(23.5), 43(17.8), 41(14.7), 44(10.2), 115(9.8)	188 (1.0)	187 (9.6)	1437
octanal glyceryl-1,2-acetal	103(100), 57(46.7), 41(19.0), 43(15.2), 69(12.2), 55(10.9), 47(8.1), 45(7.6)	202 (0.3)	201 (2.5)	1541
octanal glyceryl-1,3-acetal	103(100), 57(47.0), 69(42.0), 43(26.2), 41(24.6), 55(23.1), 44(21.7), 111(11.7)	202 (0.6)	201 (4.0)	1580
nonanal glyceryl-1,2-acetal	103(100), 87(63.0), 41(17.6), 55(15.1) 69(14.0), 43(13.7), 83(7.6), 104(7.5)	216 (0.5)	215 (3.3)	1612
decanal glyceryl-1,2-acetal	103(100), 57(42.6), 41(15.1), 43(13.5), 55(9.5), 69(5.5), 45(5.3), 47(5.2)	230 (0)	229 (1.2)	1776
dodecanal glyceryl-1,2-acetal	103(100), 57(33.2), 43(19.9), 55(18.3), 41(15.5), 69(15.4), 83(11.2), 97(10.9)	258 (0.4)	257 (3.4)	n.d.
dodecanal glyceryl-1,3-acetal	103(100), 57(31.0), 43(9.5), 41(8.8), 55(8.0), 104(4.8), 69(4.5), 47(3.5)	258 (0.1)	257 (0.7)	n.d.

[1]M = Molecular ion (% abundance)

[2]M-1 = Molecular ion minus H (% abundance)

[3]RI = GC retention index on non-polar, polydimethylsiloxane capillary column relative to n- paraffin series

isomer which lacks the M-31 loss of its isomeric counterpart and contains a stronger molecular ion (10.2%). In general, the sulfur moiety in a carbon-sulfur bond is more likely to retain the charge than its oxygen counterpart, resulting in a propensity for increased abundance of a molecular ion (21).

A search of the published literature and commercial mass spectral databases revealed very few reference spectra for glyceryl acetals and virtually none derived from aldehydes of flavor value. Therefore, all of the glyceryl acetals presented in table III are considered novel spectra which are not present in any commercial databases and to the best of the author's knowledge are reported for the first time in this investigation.

The mass spectra of the acetals presented in this manuscript, although by no means comprehensive, provide insight to the products formed by the reaction of GRAS aldehydes with several common alcohols. We hope that the characteristic fragmentation pathways described for each type of acetal in this study can serve as an interpretive aid for other investigators who may encounter unknown spectra from this interesting class of compounds.

Acknowledgments

We acknowledge the Center for Advanced Food Technology (CAFT) Mass Spectrometry Lab facility for providing instrumentation and other resources to support this research. CAFT is an initiative of the New Jersey Commission of Science and Technology. This is NJAES publication No. F-10569-1-97.

Literature Cited

1. Smith, R.L.; Newberne, P.; Adams, T.A. Food Techol. **1993**, 52, 104-117.
2. Oser, B.L.; Ford, R.A. Food Technol. **1977**, 31(1):65-74.
3. Hall, R.L.; Oser, B.L. Food Techol. **1970**, 24:25.
4. Maarse, H.; Visscher, C.A. *Volatile Compounds in Food*; TNO Biotechnology and Chemistry Institute: Utrechtseweg 48, The Netherlands, 1994.
5. Arctander, S. *Perfume and Flavor Chemicals (Aroma Chemicals)*; S. Artander: Montclair, NJ, 1969.
6. Burdock, G.A.; Wagner, B.M.; Smith, R.L.; Munro, I.C.; Newberne, P.M. Food Techol. **1990**, 44(2), 78,80,84,86.
7. Bauer, K.; Garbe, D. *Common Fragrance and Flavor Materials, Preparation, Properties and Uses; VCH:* Deerfield Beach, FL, 1985.
8. Pickenhagen, W.; Ho, C.T.; Spanier, A.M. *Contribution of Low- and Non-Volatile Materials to the Flavor of Foods;* Allured Publishing: Carol Stream, Il, 1996; 37-43.
9. Ho, C.T.;Hartman, T.G. *Lipids in Food Flavors*; American Chemical Society: Washington, DC, 1994;118-125.
10. *Fenaroli's Handbook of Flavor Ingredients*; Burdock, G.A., Eds.; CRC Press: New York, NY, 1995; Vol II.

11. *The Merck Index, An Encyclopedia of Chemicals, Drugs, and Biologicals*; Budauari, S., Eds.; Merck: Rahway, NJ, 1989; 11th ed.; pp 594, 705, 1247.
12. Migrdichian, V. *Organic Synthesis, Open-Chain Saturated Compounds*; Reinhold: New York, NY, 1957; Vol. 1, 194-200.
13. March, J. *Advanced Organic Chemistry-Reactions, Mechanisms and Structure*; McGraw-Hill: New York, NY, 1977; 810-812.
14. Carey, F. A. *Organic Chemistry*; McGraw-Hill: New York, NY, 1992; 689-693.
15. Mc Lafferty, F. W.; Turro, J. *Interpretation of Mass Spectra*; Nicholas, Ed.; University Science Books: Mill Valley, CA, 1980.
16. Shu, C.-K.; Lawrence, B.M. J. Agric. Food Chem. **1995**, 43, 782.
17. Heydanek, M.; Min, D. J. of Food Science, **1976**, 41, 145-147.
18. De la Ruelle, H.; Klok, J.; Rinken, M.; Felix, M. Rapid Communications in Mass Spectrometry, **1995**, 9, 1507-1511.
19. Mc Lafferty, F. W.; Stauffer, D. B. *The Wiley/NBS Registry of Mass Spectral Data*; John Wiley and Sons: New York, NY, 1988.
20. Lias, S.G.; Stein, S.E. *NIST/EPA/MSDC Mass Spectral Database PC*; U.S. Department of Commerce: Gaithersburg, MD, 1995; Version 3.0.
21. Budzikiewicz, H.; Djerassi, C.; Williams, D. *Interpretation of Mass Spectra of Organic Compounds;* Holden-Day: San Francisco, CA, 1964.
22. Majlat, P. J. of Chrom. **1974**, 91, 89-103.

Chapter 18

Comparison of HPLC and GC-MS Analysis of Furans and Furanones in Citrus Juices

R. Rouseff[1], K. Goodner[1], H. Nordby[1], and M. Naim[2]

[1]Citrus Research and Education Center, University of Florida, Lake Alfred, FL 33850
[2]Institute of Biochemistry, Food Science and Human Nutrition, Hebrew University of Jerusalem, Rehovot, Israel

The lack of quantitative agreement between HPLC and GC-MS values for 2,5-dimethyl-4-hydroxy-3(2H)-furanone (Furaneol), furfural and 5-hydroxymethylfurfural reported in the literature is due to incomplete HPLC resolution. Furanoids of interest were not resolved from other citrus juice components using reversed phase HPLC with detection at 290 nm. Attempts to develop a rugged chromatographic separation to resolve all components were unsuccessful. A more accurate procedure was developed which employed a second detecting wavelength (335 nm) to provide maximum response from the interfering compounds and minimum response from the furanoids of interest. Reconstructed difference chromatograms (290-335 nm) negated interfering compounds thus providing the spectral selectivity to accurately quantify the furanoids of interest. GC-MS studies demonstrated that contrary to literature reports, Furaneol is thermally stable under GC conditions and can be chromatographically or spectrally resolved from other juice components.

Furans can have considerable impact on the aroma and taste of citrus juices that have been stored at elevated temperatures. They are formed to a minor extent as the result of natural biosynthetic pathways during fruit ripening but most are formed as a result of thermal processing and/or storage at elevated temperatures. Furfural and 5-hydroxymethylfurfural are produced during the Maillard reaction in acid environment from the 1,2 enolization of Amadori products formed from pentoses and hexoses respectively. Furaneol is formed in the same reaction from the 2,3 enolization of Amadori products from hexoses under less acid conditions(1). Since citrus juices are highly acidic, furfural and 5-hydroxymethyl-furfural will be the favored Maillard reaction products.

Even though produced at much lower levels than furfural or 5-HMF, Furaneol (2,5-dimethyl-4-hydroxy-3(2H)-furanone) is the most important because of its much lower aroma threshold and striking sensory properties. Furaneol's taste threshold is 0.05 ppm in orange juice (2) whereas the corresponding values for furfural and 5-HMF are 80 and

200 ppm respectively(3). Furaneol or DMHF, has been found in arctic bramble (4-6), raspberry (Rubus idaeus, L.)(5), strawberries, pineapples, mangoes (7), grapefruit (8) coffee (9), grapes and wine (10). Furaneol has also been found as a nonvolatile glycoside in: tomatoes (11), pineapple (*Ananas comosus L.* Merr.) (12) and strawberries(13).

Furaneol, furfural and 5-HMF are all products of the Maillard reaction and as such have been proposed as markers of thermal abuse or excessive elevated temperature storage. Even though the furfural content of orange juice was observed to increase in proportion to increased storage temperature and time (14), it was not until several years later that it was proposed as an index of thermal abuse (15). Mori and coworkers (16) suggested that 5-HMF could also be used as a quality index for citrus juices. They reported that the 5-HMF content of mandarin juices increased during storage, especially at higher temperatures. Furthermore, its content correlated well with sensory acceptance. Tatum and coworkers (2) reported that Furaneol was one of the three major off-flavors produced in temperature abused canned orange juices. As a result of their studies with grapefruit juice, Lee and Nagy (8) proposed that Furaneol could also be used as a quality deterioration index.

Furaneol, furfural and 5-MHF have been analyzed using a variety of analytical techniques beginning with colorimetric procedures and currently are usually determined using chromatographic procedures such as GC or HPLC. Colorimetric tests for furfural were based on relatively non specific reactions such as the reaction of aldehydes with aniline (15) or benzidine (17). Since the color forming reactions were non specific, relatively selective sample preparation techniques were required. 5-hydroxymethylfurfural (5-HMF) was usually determined using the Winkler method (18). This procedure was based on reaction of 5-HMF with p-toluidine and barbituric acid reagent to produce a red colored product which was measured spectrophotometrically at 550 nm. However, both furfural and 5-HMF react. Values obtained thus represent the sum of both compounds in citrus juices (19). No colorimetric procedures for Furaneol have been found.

Concentrations of these compounds reported in the literature appear to be method dependent. In the case of Furaneol, HPLC values are considerably higher than those obtained using GC or GC-MS. For example, Wu and coworkers reported finding 1.6 ppm free DMHF in Costa Rican pineapple and only 0.7 ppm for the same product using GC-MS. In a similar manner, Sanz and coworkers (20) employed HPLC to determine mesifuran and DMHF concentrations in strawberries. They reported concentration values considerably higher than any previously reported for the identical cultivars analyzed using GC or GC-MS. The discrepancy between these results was attributed to the use of a non-thermal method of analysis (HPLC). It was implied that the GC results were low because of decomposition of the thermally labile Furaneol in the elevated temperature environment of the gas chromatograph. The purpose of this study was to investigate sources for the apparent lack of quantitative agreement between HPLC and GC methods for furans in food products using orange and grapefruit juices as examples. Our goal was to determine if HPLC results were high because of inadequate resolution or selectivity which might have quantified co-eluted matrix compounds or if GC results were low due to thermal decomposition of Furaneol in the elevated temperatures of typical gas chromatographs. This study will be limited to the analysis of furfural, 5-hydroxymethyl furfural and Furaneol in orange and grapefruit juices.

Materials and Methods

Solvents and Standards. Solvents were of HPLC grade from Fisher Scientific (Pittsburgh, PA). Standard furfural, Furaneol, 5-HMF, and mesifurane were obtained from Aldrich Chemical (Milwaukee, WI). Stock solutions of standards at approx. 3000 ppm were prepared in methanol. Working solutions of standards were prepared as methanolic dilutions.

Sample Preparation. Ten mL of orange or grapefruit juice were centrifuged for 5 min at 3000 rpm. Three mL of supernatant juice were passed through a C-18 cartridge (Whatman, SPE ODS 500 mg, Fairfield, NJ) which had been conditioned with methanol and washed with water. The cartridge and juice were washed with 2 mL water to remove sugars and eluted with 2 mL methanol. Samples were filtered through a .45 μ filter, and stored in an amber vial until injected into the HPLC.

HPLC Instrumentation. A Perkin Elmer (Norwalk, CN) model 410 four solvent low pressure gradient pump, with a Hewlett Packard (Palo Alto, CA) model 1050 autosampler, and either a Waters (Milford, MA) model 490E multiwavelength or 990 photodiode array detector was used to analyze the juice extracts.

HPLC Chromatographic Conditions. OJ and GFJ extracts as well as standards (25 μL) were separated on a 5μ Bondesil (Varian, Sugarland, TX) C-18, 4.0 mm i.d.x 25 cm column. The solvent gradient is shown below. All concentration changes were linear. The column was equilibrated for at least 15 minutes at initial conditions before each injection.

Time (min)	Flow (mL/min)	% methanol	% acetonitrile	% pH 4.0 Buffer
0	1.00	7	0	93
16	1.00	0	20	80
20	1.00	5	20	75
40	1.00	5	35	60
50	1.00	5	85	10
60	1.00	5	85	10
62	1.00	7	0	93

GC-MS Instrumental and Chromatographic conditions. All GC-MS data were collected using a Finnigan GCQ Plus system (Finnigan Corp, San Jose, CA). A 30 m x 0.25 mm i.d. DB-5 column was used with helium (99.999%) as the column carrier gas (flow rate 31.9 cm/sec) as well as the collision/bath gas in the ion trap. All chemical ionization experiments were conducted using 99.99% pure methane at approximately 60 torr. Injector temperature was 250°C. The thermal gradient was as follows: initial temp. 35°C, hold for 3 minutes followed by an 8°C/min temperature ramp to 127°C which was in turn followed by a 40°C/min ramp to 275°C. which was held for at constant temperature 1.8 minutes. The MS transfer line was held at 275°C and the ion source was held at 170°C. The mass spectrometer had a delay of 4 minutes to avoid the solvent peak, and then scanned from m/z 40 to m/z 300 in order to achieve 2 spectra per second with each spectra being the sum of 5 microscans. Injections were 0.2 μL in the splitless mode unless otherwise stated.

Results and Discussion

Sample Preparation Techniques. Furan derivatives such as furfural, 5-HMF and Furaneol are relatively polar compounds which are difficult to extract from food matrices with organic solvents. One of the earliest separation techniques was simple distillation of liquid products. Furfural was distilled from orange juice before it was reacted with 10% aniline in acetic acid-ethanol (15). A later HPLC technique (21) which employed this sample preparation procedure found just a single peak. Although the distillation procedure was highly selective in separating furfural from the myriad other components in orange juice, it only recovered 38% of the total furfural. Distillation has also been employed to separate and analyze furfural and 5-HMF in wines using UV absorption spectrophotometry (22). The 5-HMF is determined after the furfural has been distilled off.

The unfavorable partition coefficients for these furans can be overcome using continuous liquid-liquid extraction with a slightly polar solvent such as diethyl ether. Buttery and coworkers (23) were able to isolate water-soluble volatile compounds such as Furaneol in tomato juice using this technique. Alternatively, dynamic headspace purge and trap using a solid adsorbent and either solvent elution or thermal desorption (24-26) can be used to separate these compounds from food matrices. Final separation and analysis is usually carried out using GC, often in combination with MS. Another approach has been to use C-18 solid phase adsorbent to separate these compounds from citrus juices (8, 27). Sample preparation involved three steps. The first step is to add Carriz reagent to juice and centrifuge. The supernatant is passed through a C-18 cartridge and eluted with successive washes of ethyl acetate. Finally the ethyl acetate is dried with sodium sulfate and reduced in volume under a stream of nitrogen before injection into the HPLC. In this study we have employed a simplified technique developed earlier for the determination of 4-vinyl guaiacol (28) and later modified for Furaneol (29). In this procedure the centrifuged juice is passed through a C-18 cartridge, washed with water and eluted with a small amount of methanol. It can be injected directly into the HPLC without further concentration. Furfural recoveries using this procedure averaged 97%.

Methods of Analysis. Chromatography appears to be the method of choice in determining these compounds in foods. The earlier colorimetric methods have largely been abandoned because of their lack of specificity and the time consuming sample preparation. In terms of chromatography, these compounds can be determined using either HPLC or GC. The choice is often made on the basis of what equipment is available and what other compounds might be of interest. Gas chromatography offers the advantage of being less expensive to purchase and allows for a greater number of other volatile components to be determined. HPLC is more expensive to purchase and maintain but also allows for non volatile components such as flavanone glycosides and hydroxycinnamic acids to be determined. HPLC does not have the resolving power of GC but it is a non thermal separation process which is less subject to artifact formation. In addition, the classical concerns for selectivity, resolution, speed, and sensitivity (minimum detectable amounts) for a particular food will influence the final chromatographic choice.

HPLC (Non Thermal Analysis). Reverse phase chromatography has been the mode of choice to separate furan derivatives in citrus juices (21, 30, 31). Under reverse phase conditions the elution order is typically, 5-HMF, Furaneol, furfural and finally, mesifurane (the 4-methoxy analog of Furaneol). Shown in Figure 1 is the plot of Log k' vs solvent composition for methanol-water (buffered to pH 4). (The plot is essentially identical for acetonitrile - pH 4.0 acetate buffer combinations). It should be noted that furfural and Furaneol elute at very similar retention times and are not resolved at concentrations greater than 25% methanol. Optimum separation between these two compounds is somewhere between 5-10% methanol. One procedure (8) reported the separation of furfural and Furaneol using a Zorbax ODS column with a mobile phase of 0.05M sodium acetate (pH 4.0) and 30% methanol. This separation was achieved due to the unique composition of the stationary phase. At that time Zorbax C-18 columns were not endcapped. This left polar silanol sites exposed to the mobile phase. The net result was a "mixed mode" separation that allowed small, relatively polar molecules such as furfural to be retained much more than they would under purely reversed phase C-18 chromatography. Whereas the lack of endcapping allowed for some interesting applications, it also produced columns with shorter lifetimes. Currently all reverse phase columns are endcapped and this approach is no longer available. However the same authors have employed a new column which is more polar than normal C-18 to separate Furaneol from the other components in orange juice (32). Since only Furaneol was identified in the chromatogram provided, it is not known if Furaneol and furfural can be resolved with their new column.

Separation vs Solvent Strength. The separation obtained using a typical C-18 column is shown in Figure 2 with four isocratic solvent conditions (the aqueous portion of each mobile phase was buffered to pH 4.0). It can be seen that at 30% acetonitrile, Furaneol and furfural peaks are merged into a single symmetric peak. This combination peak becomes somewhat asymmetric at 20%, is somewhat resolved at 15% and completely resolved at 10% acetonitrile. Although resolution is improved as solvent strength decreases, analysis time increases. We have chosen to employ a solvent gradient to reduce analysis time and also increase the range of compounds which can be analyzed. The gradient and resulting chromatogram for an orange juice sample stored for 15 weeks at 40 °C are shown in Figure 3. The amount of methanol in the mobile phase remains low throughout the chromatogram and is present to provide additional selectivity to help separate the many components found in orange juice. The resolution shown in Figure 2 demonstrates that a simple 10% acetonitrile solvent composition can resolve Furaneol and furfural. However it is also necessary to resolve these compounds from the many other components in citrus juices which elute in this region. It can be seen that the interfering compounds are numerous and as large or larger than the peaks of interest.

Single vs Dual Wavelength Detection. Furaneol and furfural are small peaks that are not resolved at 290 nm from the other components in OJ. Furaneol and furfural have been detected at wavelengths which range from 280 nm (21) to 292 nm (29). Because absorbance maxima can be solvent dependent, we determined the absorbance

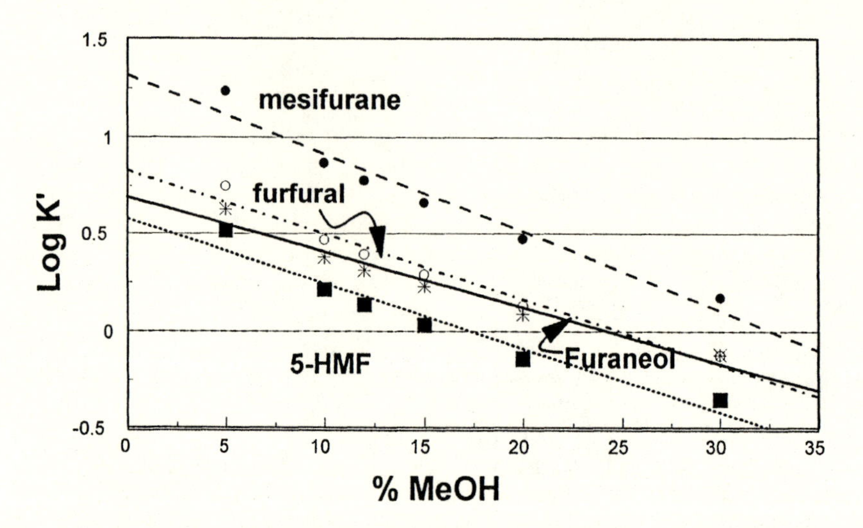

Figure 1. Log of capacity factor (k') versus methanol pH 4 acetate buffer solutions.

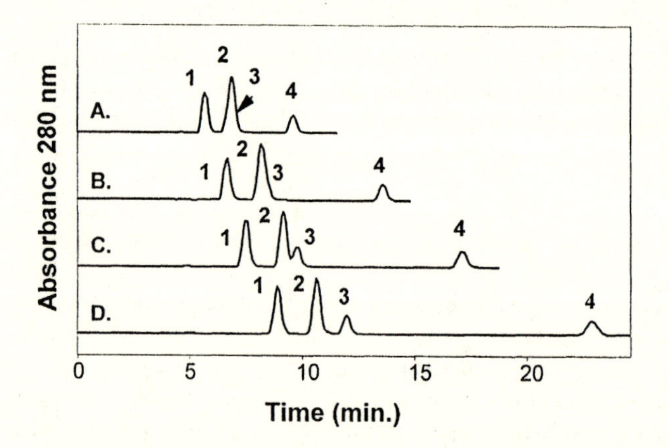

Figure 2. Four chromatograms from various combinations of acetonitrile: pH 4 aqueous buffer showing the affect of changing the solvent strength with (A) 30% CH_3CN, (B) 20% CH_3CN, (C) 15% CH_3CN, and (D) 10% CH_3CN. Flow rate in all cases was 1 mL/min where (1)-5-HMF, (2)-Furaneol, (3)-furfural, and (4)-mesi-furane.

spectra for each compound using a photodiode array detector. The advantage of this technique is that it determines the spectra for each compound of interest in the exact solvent composition it elutes in. In the above gradient we found the absorbance maxima to be 278, 287 and 285 nm for 5-HMF, Furaneol and furfural respectively. As shown in Figure 3, 290 nm wavelength is not selective for furanoids because it detects both furanoids and numerous extraneous substances. In an earlier study (29) we found that by taking the difference between two wavelengths, greatly enhanced selectivity could be obtained with little loss in sensitivity. The second wavelength (335 nm) was chosen to provide maximum response from the interfering compounds and minimum response from the furanoids of interest. The region of interest in the previous chromatogram is shown in Figure 4 showing the responses at 290 and 335 nm and the trace due to the difference between the two signals plotted as a third chromatogram. In our earlier studies we had determined the absorbance maxima for some of the surrounding non furanoid peaks and found their average maximum in the region of 335 nm. We also knew from our photodiode array spectra of standards, that the furanoids of interest had essentially no absorbance at 335 nm. It can be seen in Figure 4 that the furfural peak appears to be a well resolved peak but the 335 chromatogram indicates that there is a serious impurity in this peak and that the true peak area is considerably smaller. Furaneol could not be accurately quantified at 290 nm because it appears on the shoulder of a larger impurity and has a second impurity which elutes shortly behind it. However, it can be seen that Furaneol is completely resolved from the impurities in the "difference" chromatogram. Likewise the 5-HMF peak had an impurity that gave the appearance of a tailing peak at 290 nm. The interfering peak can be clearly distinguished at 335 nm. Thus, the "difference" peak produces a single, symmetric peak for 5-HMF which can be accurately quantified. Therefore, it appears that analytical values obtained from single wavelength studies would produce erroneously high values because integration would likely have included coeluting peaks. Whereas it was not possible to chromatographically resolve all the peaks of interest from interfering peaks from stored orange juice, the furanoids of interest could be spectrally resolved using the "difference" chromatogram.

GC-MS (Thermal Analysis). Because reports (33) identifying 4-methoxy-2,5-dimethyl-3(2H)-furanone (mesifurane) as an aroma constituent of strawberries made special note of the absence of 2,5-dimethyl-4-hydroxy-3(2H)-furanone (Furaneol), Pickenhagen and coworkers (7) employed high resolution capillary GC to determine if Furaneol was present. They found significant amounts of both compounds in strawberries, pineapple and mango. It was speculated that a possible reason that Furaneol had not been detected in earlier studies was that steam distillation was used as the extraction method, or that capillary columns other than fused silica or acid leached soft glass had been employed.

Thermal Stability. Furaneol had been reported to be a thermally unstable compound (34, 35). Thus, we were concerned that it might decompose at the elevated temperatures of most GC injector ports. To determine if decomposition might be a problem, we injected standard Furaneol solutions with injector temperatures of 150, 175, 200, 225, 250, 275 and 300°C. We found only a single peak in all cases whose height did not diminish with increasing temperature as one might expect from thermal decomposition. The single peak appeared at the proper Kovat's retention index and its resulting mass

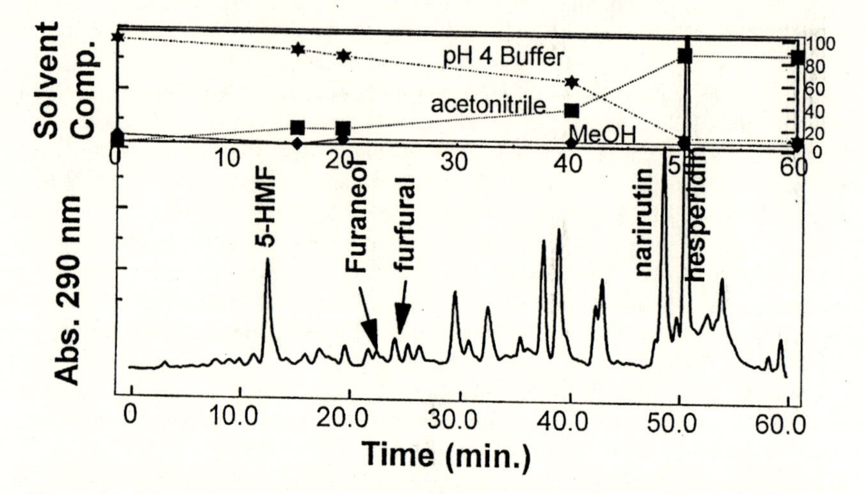

Figure 3. Orange juice stored at 40 °C for fifteen weeks. Gradient composition is indicated in the upper trace. Injection volume was 25 μL of methanol extract. Prepared using the procedure outlined in Walsh *et al.* (27).

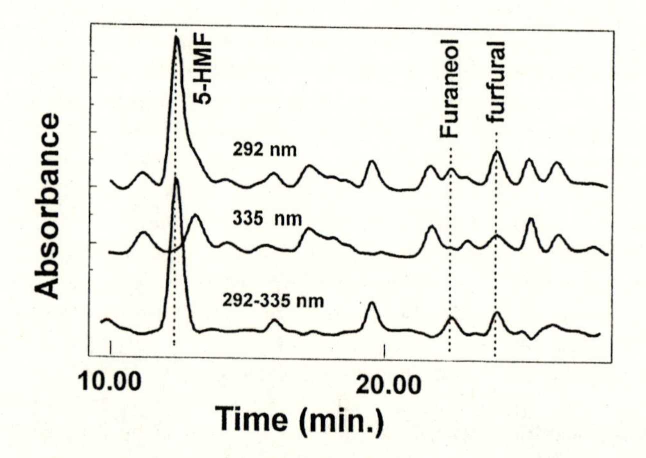

Figure 4. Expanded view of the chromatogram shown in Figure 3. Also shown are chromatograms from detection at 335 nm and a reconstructed chromatogram of the difference between the 290 nm and 335 nm chromatograms.

spectra was identical in all cases. One of the mass spectra is shown in Figure 5. Perhaps the most surprising feature is the exceptionally large molecular ion. This M^+ ion accounted for about 32% of all the ions in the spectrum. Since there was little fragmentation under 70 eV electron impact ionization, it can be said with some assurance that Furaneol is not thermally unstable in the gas phase during normal gas chromatographic analyses. We tested the thermal stability of furfural and 5-HMF in a similar manner. Just as in the case for Furaneol, there was surprisingly little fragmentation. Furfural produced two large peaks corresponding to the M^{-1} and M^+. These two ions constituted 82% of the total ions in the spectra. Compared to the other furanoids 5-HMF has a relatively small (but still appreciable) M^+ ion which corresponded to about 13% of the total ions in the spectrum. This information indicates that these compounds are thermally stable under normal GC conditions. Therefore it is unlikely that GC values for these compounds would be low due to thermal decomposition losses.

GC Chromatographic Resolution. Another possible source of error in GC furanoid analysis would be incomplete chromatographic resolution. Many if not most gas chromatograph employ flame ionization detection (FID). These detectors are not specific for any class of compounds; rather they respond to any compound with an oxidizable carbon component. Total ion chromatograms (TIC) from MS detectors coupled to GC's provide output very similar to FID detectors. Chromatograms from a canned grapefruit juice extract are shown in Figure 6. Peak 1 (furfural) is observed as a minor peak in the total ion chromatogram. If a reconstructed ion chromatogram at m/z 95 (furfural, M^{-1}) is plotted, greatly increased signal to noise ratio as well as increased selectivity is realized. Furfural (peak 2) can hardly be seen in the TIC chromatogram but is easily detected in the reconstructed ion chromatogram at m/z 128 (Furaneol, M^+). The 5-HMF peak is the largest of the three peaks of interest as it is observed as a major peak in the TIC chromatogram. The reconstructed ion chromatogram for this peak offers slightly improved selectivity but little improvement in signal/noise ratio. There does not appear to be a problem with incomplete resolution of the peaks of interest from other components in the juice. Quantification from reconstructed ion chromatograms would further reduce the chance of extraneous compounds producing erronously high values.

Conclusion

The major reason that HPLC techniques have produced greater values for compounds like Furaneol appears to be due to incomplete resolution of other compounds in citrus juices. However, a procedure was devised whereby the compounds of interest could be spectrally resolved from the interfering peaks. These interfering compounds have considerable absorption at 290 nm, but have an absorbance maximum at approximately 335 nm. It was noted that the compounds of interest have essentially no absorbance at this wavelength. By subtracting the chromatogram obtained at 335 from that obtained at 290 nm the interfering compounds could be spectrally eliminated leaving primarily the peaks of interest to be accurately quantified.

GC-MS studies have shown that Furaneol and other furanoids are not thermally

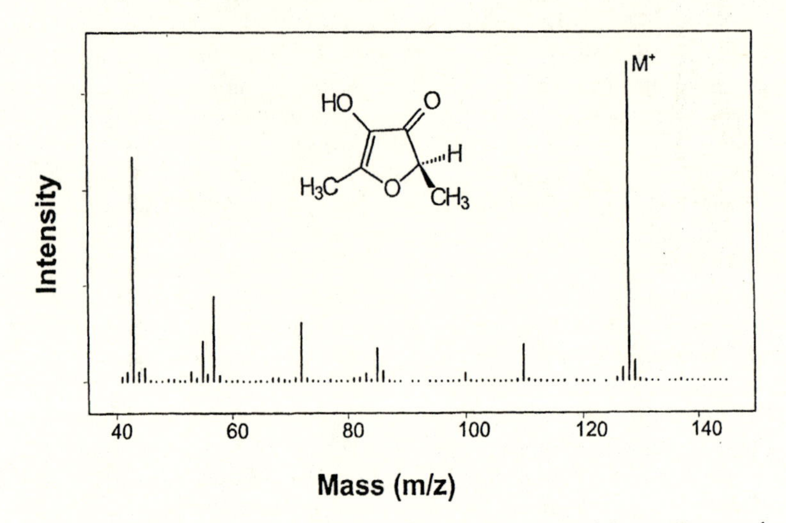

Figure 5. Background corrected mass spectrum obtained from a Furaneol standard. Note prominent molecular ion at m/z 128.

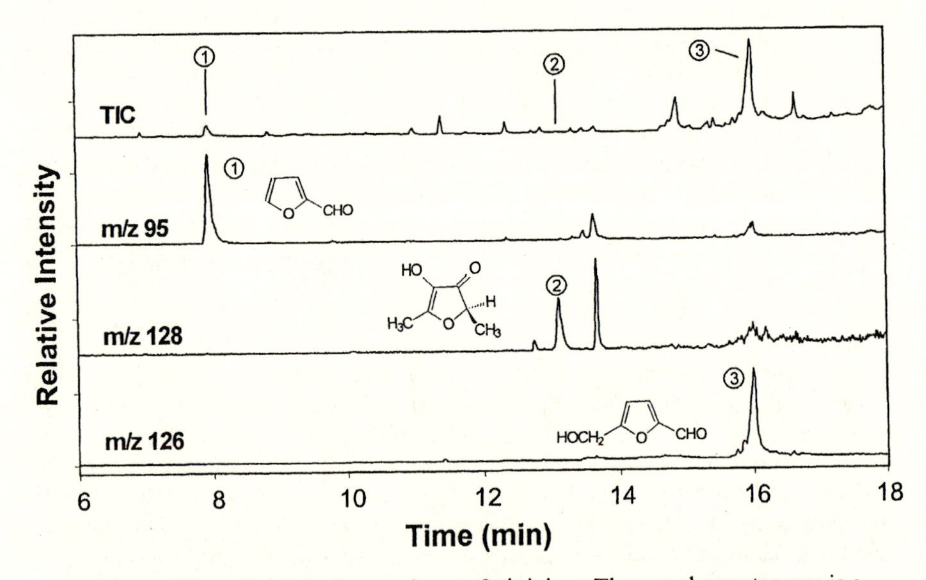

Figure 6. Thermally abused canned grapefruit juice. The top chromatogram is a total ion chromatogram with the bottom three being single ion chromatograms based on the m/z of 95, 128, and 126 respectively where (1)-furfural, (2)-Furaneol, and (3)-5HMF.

unstable under normal GC conditions. On the contrary, their mass spectra exhibit large molecular ions (M^+) or (M^+-1) which would suggest excellent thermal stability. Therefore losses due to thermal decomposition in the GC are highly unlikely.

Acknowledgments.

This research was supported in part by grant no. I-1967-91 from BARD, The United States-Israel Binational Agricultural Research and Development fund.

Literature Cited.

1. Haleva-Toledo, E.; Naim, M.; Zehavi, U.; Rouseff, R. L. *J. Agric. Food Chem.* **1997,** *45*, 1314-1319.
2. Tatum, J. H.; Nagy, S.; Berry, R. E. *J. Food Sci.* **1975,** *40*, 707-709.
3. Shaw, P. E.; Tatum, J. H.; Kew, T. J.; Wagner, C. J. J.; Berry, R. E. *J. Agric. Food Chem.* **1970,** *18*, 343-45.
4. Kallio, H. *J. Food Sci.* **1976,** *41*, 563-566.
5. Pyysalo, T. *Z. Lebensm. Unters. Forsch.* **1976,** *162*, 263-72.
6. Honkanen, E.; Kallio, H.; Pyysalo, T. *Kem. - Kemi* **1976,** *3*, 180-1.
7. Pickenhagen, W.; Velluz, A.; Passerat, J. P.; Ohloff, G. *J. Sci. Food Agric.* **1981,** *32*, 1132-1134.
8. Lee, H. S.; Nagy, S. *J. Food Sci.* **1987,** *52*, 163-165.
9. Tressl, R.; Bahri, D.; Koeppler, H.; Jensen, A. *Z. Lebensm. Unters. Forsch.* **1978,** *167*, 111-114.
10. Rapp, A.; Knipser, W.; Engel, L.; Ullemeyer, H.; Heimann, W. *Vitis* **1980,** *19*, 13-23.
11. Krammer, G. E.; Takeoka, G. R.; Buttery, R. G. *J. Agric. Food Chem.* **1994,** *42*, 1595-1597.
12. Ping, W.; Kuo, M. C.; Hartman, T. G.; Rosen, R. T.; Ho, C. T. *J. Agric. Food Chem.* **1991,** *39*, 170-172.
13. Sanz, C.; Pérez, A. G.; Richardson, D. G. *J. Food Sci.* **1994,** *59*, 139-141.
14. Rymal, K. S.; Wolford, R. W.; Ahmed, E. M.; Dennison, R. A. *Food Technol.* **1968,** *22*, 1592-95.
15. Dinsmore, H. L.; Nagy, S. *J. Food Sci.* **1972,** *37*, 768-770.
16. Mori, M.; Kaneko, K.; Wakatake, N.; Shimizu, E. *Kanzume Jiho* **1974,** *53*, 239-245.
17. Maraulja, M. D.; Blair, J. S.; Olsen, R. W.; Wenzel, F. W. *Proc. Fla. State Hort. Soc* **1973,** *86*, 270-275.
18. Winkler *Z. Lebensm. Unters. Forsch.* **1955,** *102*, 161.
19. Espinosa Mansilla, A.; Salinas, F.; Berzas Nevado, J. J. *J. of the AOAC International* **1992,** *75*, 678-684.
20. Sanz, C.; Richardson, D. G.; Perez, A. G. In *Fruit Flavors*; Rouseff, R., M., L., Eds.; American Chemical Society: Washington, D.C., 1995; Vol. 596, pp 268-75.
21. Marcy, J. E.; Rouseff, R. L. *J. Agric. Food Chem.* **1984,** *32*, 979-981.

22. Cunha Ramos, M.; Gomes, L. G. *Vinea et Vino Portugalie Documenta, Series II: Enologia* **1968,** *4*, 2-17.
23. Buttery, R. G.; Takeoka, G. R.; Krammer, G. E.; Ling, L. C. *Lebensm. Wiss. Technol.* **1994,** *27*, 592-594.
24. Ito, O.; Sakakibara, H.; Yajima, I.; Hayashi, K. In *Flavour Sci. Technol.,*; Bessiere, Y. T., Alan Francis, Ed.; Wiley Publishing: Chichester, UK., 1991, pp 69-72.
25. Paik, J. S.; Venables, A. C. *J. Chromatogr.* **1991,** *540*, 456-463.
26. Fischer, N.; Hammerschmidt, F. J. *Chem., Mikrobiol., Technol. Lebensm.* **1992,** *14*, 141-8.
27. Lee, H. S.; Rouseff, R. L.; Nagy, S. *J. Food Sci.* **1986,** *51*, 1075-1076.
28. Rouseff, R. L.; Dettweiler, G. R.; Swaine, R. M.; Naim, M.; Zehavi, U. *J. Chromatogr. Sci.* **1992,** *30*, 383-387.
29. Walsh, M.; Rouseff, R.; Naim, M.; Zehavi, U. *J. Agric. Food Chem.* **1997,** *45*, 1320-1324.
30. Mijares, R. M.; Park, G. L.; Nelson, D. B.; McIver, R. C. *J. Food Sci.* **1986,** *51*, 843-844.
31. Lee, H. S.; Rouseff, R. L.; Nagy, S. *J. Food Sci.* **1986,** *51*, 1075-1076.
32. Lee, H. S.; Nagy, S. In *Chemical Markers for Processed and Stored Foods*; American Chemical Society: Washington, D.C., 1996.
33. Schreier, P. *J. Sci. Food Agric.* **1980,** *31*, 487-94.
34. Shaw, J. J.; Burris, D.; Ho, C. T. *Perfumer & Flavorist* **1990,** *15*, 60-66.
35. Schieberle, P. In *Flavor Precursors*; Teranishi, R., Takeoka, G. R., Guentert, M., Eds.; American Chemical Society: Washington, D.C., 1992; Vol. Symposium Series No. 490, pp 164-174.

Chapter 19

Application of Microwave-Assisted Process and Py–GC–MS to the Analysis of Maillard Reaction Products

V. A. Yaylayan and A. Keyhani

Department of Food Science and Agricultural Chemistry, McGill University, 21 111 Lakeshore, Ste. Anne de Bellevue, Quebec H9X 3V9, Canada

A focused microwave system under atmospheric pressure conditions was used to synthesize and extract selected Maillard reaction products. L-Phenylalanine, glycine, or L-tryptophan / D-glucose mixtures were sequentially treated with microwave irradiation in an aqueous media to initiate the synthesis step which was followed by irradiation in an microwave transparent solvent to perform extraction of the products formed. The extracts were analyzed by GC-MS. Phenylalanine model system generated a mixture consisting of phenylacetaldehyde, 2-(5'-hydroxymethyl-2'-formylpyrrol-1'-yl)-3-phenyl propionic acid lactone, 3,5-diphenylpyridine, and an unknown compound. Tryptophan model system generated three indole derivatives. Glycine model system, on the other hand, generated a mixture consisting of 1,6-dimethyl-2(1H)-pyrazinone, 1,5,6-trimethyl-2(1H)-pyrazinone, and one major and two minor unknown compounds. Spectroscopic analysis in conjunction with labeling studies, using Py-GC-MS, have indicated that the major unknown compound is 5-hydroxy-1,3-dimethyl-2[1H]-quinoxalinone. Serine model system was used to demonstrate the effect of solvent on the product distribution.

The Maillard reaction initiated by the interaction of reducing sugars and amino acids in foods, can generate a complex mixture of products that are considered to be important contributors to the flavor and color of processed foods (*1*). The industrial significance of this reaction stems from the fact that it can also be carried out outside of the food matrix to generate what are known as reaction flavors (*2*). The complex mixtures generated through the heating of model systems consisting of reactive sugars and selected amino acids have widespread applications in the food industry. However, the ability to generate and extract specific Maillard reaction product(s) can: (a) provide the opportunity to identify unknown structures through spectroscopic means,

(b) allow the determination of their sensory properties, and (c) can serve as a means of producing Maillard reaction mixtures rich in desired components for specific applications. The ability of focused microwave irradiation under atmospheric pressure conditions, to selectively synthesize and quantitatively separate Maillard reaction product(s), was investigated using a two-stage microwave assisted process (MAP). MAP (*3*) has been applied successfully to various liquid phase and gas-phase extractions and is currently used extensively as a sample preparation tool. The first stage - microwave-assisted synthesis (MAS) - could be carried out in a microwave active solvent such as water, ethanol, or water-ethanol mixtures depending on the energy requirements of the reaction; and the second stage - microwave assisted extraction (MAE) - could be carried out in an microwave transparent solvent such as petroleum ether, hexane, or mixtures of hexane and acetone to selectively extract a minimum number of products formed in the first stage. The polarity of the solvent will determine the type of products extracted from the initial reaction mixture. After evaporation of the solvent, the residue could be analyzed to determine its composition or sensory characteristics, and if needed, be further purified by chromatography to isolate the components. Such an approach was successfully applied to synthesize and isolate Maillard reaction products (*4*) and extract selected components from complex mixtures (*5*).

Experimental

Petroleum ether (boiling range 60 - 80°C, analytical reagent) and hexane were purchased from BDH (Montréal, Canada). D-Glucose, glycine, L-serine, L-tryptophan and L-phenylalanine were purchased from Aldrich Chemical Company (Milwaukee, WI, USA). The Soxwave 100 (focused microwave extraction system at atmospheric pressure) was obtained from Prolabo (Fontenay-Sous-Bois, France). The basic apparatus consists of a command box and a microwave module operating at an emission frequency of 2450 MHz and a 300 W full power. It is equipped with a 250 mL quartz vessel, a Graham type refrigerant column (400 mm length), and a bent extraction tube.

Microwave-Assisted Synthesis and Extraction of Maillard Reaction Products from D-Glucose/Glycine Model System. A mixture of D-glucose (1.00 g, 0.005 moles) and glycine (1.25 g, 0.016 moles) was transferred into the 250 mL quartz extraction vessel of the Soxwave 100 microwave extraction system. Two mL of water was then added. The vessel was inserted inside the extraction cavity and fitted with a condenser. The irradiation was carried out in the following sequence at full power (300 W): 2 min on, 30s off, 2 min on, 30s off, 2 min on, 30s off and 2 min on. (Total irradiation time was 8 min). At the end of the irradiation sequence a dark brown and dry slurry was formed that had characteristic baked bread notes. The extraction step was carried out with a 40 mL of petroleum ether (60-80 °C) using the following sequence of irradiation: 40 s on, 30 s off, 90 s on. The solvent was decanted, dried over sodium sulfate, and analyzed by GC/MS. The resulting oil was further purified by thick layer chromatography on silica gel using ethyl acetate as the solvent (*4*).

Microwave-Assisted Synthesis and Extraction of Maillard Reaction Products from D-Glucose/L-Phenylalanine Model System. A mixture of D-glucose (0.5 g, 0.0027 moles) and L-phenylalanine (1.37 g, 0.083 moles) was transferred into the 250 mL quartz extraction vessel of the Soxwave 100 microwave extraction system. Two mL of water/methanol (1:1) was then added to solubilize L-phenylalanine. The vessel was inserted inside the extraction cavity and fitted with a condenser. The irradiation was carried out in the following sequence at full power (300 W): 2 min on, 30s off and 2 min on. (Total irradiation time was 4 min). At the end of the irradiation sequence a dark brown and dry slurry was formed that had characteristic flowery notes. The extraction step was carried out with 40 mL of petroleum ether (60-80 °C) using the following sequence of irradiation: 40 s on, 30 s off, 90 s on. The solvent was decanted, dried over sodium sulfate, and analyzed by GC/MS. Further extraction with the same solvent did not yield any additional product.

Microwave-Assisted Synthesis and Extraction of Maillard Reaction Products from D-Glucose/L-Tryptophan and from L-Serine Model Systems. The same procedure of D-glucose/glycine was followed except hexane was used as the extraction solvent.

Pyrolysis-GC-MS and GC-MS Analysis: A Hewlett-Packard GC/mass selective detector (5890 GC/5971B MSD) interfaced to a CDS pyroprobe 2000 unit was used for the Py/GC/MS analysis. Solid samples (1-4 mg) of amino acid/glucose in different ratios, were introduced inside a quartz tube (0.3 mm thickness); the tube was plugged with quartz wool and inserted inside the coil probe. The Pyroprobe was set to the desired temperature (250 °C) at a heating rate of 50 °C/ms and with a THT (total heating time) of 20 s. The GC column flow rate was 0.8 mL/min for a split ratio of 92:1 and a septum purge of 3 mL/min. The pyroprobe interface temperature was set at 250 °C. Capillary direct MS interface temperature was 180 °C; ion source temperature was 280 °C. The ionization voltage was 70 eV, and the electron multiplier was 1682 V. The mass range analyzed was 30-300 amu. The column was a fused silica DB-5 (30m length x 0.25 mm i.d. x 25 um film thickness; Supelco, Inc.). Unless otherwise specified, the column initial temperature was -5 °C for 2 min and was increased to 50 °C at a rate of 30 °C/min; immediately the temperature was further increased to 250 °C at a rate of 8 °C/min and kept at 250 °C for 5 minutes.

Results and Discussion

The Maillard reaction, under controlled experimental conditions, could become a potential source of compounds with vast molecular diversity, which would, otherwise, required complex multistep synthetic approaches for their production. A broad spectrum of flavor-active compounds is known to be formed during Maillard reaction (*1*). As such, Maillard reaction mixtures could be viewed as potential combinatorial flavor chemistry libraries. However, under reflux conditions, the complexity and the low yields of diverse products makes it, next to, impossible to isolate specific components for spectroscopic and sensory analysis. To investigate the feasibility of using simultaneous microwave-assisted synthesis and extraction for preparation of selected Maillard reaction products, we exposed model systems, consisting of D-glucose and amino acids, to focused microwave irradiation at atmospheric pressure conditions. By controlling the irradiation time and temperature (dependent on solvent mixture) during the MAS stage certain products could be formed preferentially, thus producing mixtures rich in specific products. Further selectivity could be obtained during the MAE stage, specific products formed in the MAS stage could be extracted sequentially by controlling the solvent polarity and extraction time.

L-Phenylalanine and L-Tryptophan/D-Glucose Mixtures. Phenylalanine - glucose model systems have been shown to produce complex mixtures of products under various experimental conditions, such as refluxing in a solvent, autoclaving (*6*), or pyrolysis (*7*). However, application of the MAS/MAE technique generated a relatively simple mixture (Figure 1) consisting of phenylacetaldehyde, 2-(5'-hydroxymethyl-2'-formylpyrrol-1'-yl)-3-phenylpropionic acid lactone (*6*), 3,5-diphenylpyridine and an unknown compound X_1. All of these compounds, except X_1, have been detected in the pyrolysates of both the D-glucose/L-phenylalanine mixture and the L-phenylalanine Amadori compound. Compound X_1 has only been detected during pyrolysis of L-phenylalanine Amadori compound, which indicates the possible formation of an Amadori intermediate during the synthesis stage. The synthesis stage was also carried out in a domestic microwave oven using an open vessel. This was done to compare the product distribution and the relative yields to those from MAS reflux system. The advantage of using the reflux system was that higher yields were obtained; whereas, fewer products in lower yields were generated using the open system. The more volatile component such as phenylacetaldehyde can evaporate under the open vessel conditions (Figure 1).

Application of MAS/MAE technique to the L-Tryptophan – glucose model system, again, generated a simple mixture (Figure 2). The mixture consisted of three major products: indole (40 %), 3-methyl-1H-indole (28 %) and 3-ethyl-1H-indole (10 %).

L-Serine: Effect of Solvent on Product Distribution. The amino acid L-serine was used to investigate the effect of MAS solvent composition on the product distribution. Serine degrades into a mixture that contains glycine, alanine, glyoxal and other reactive intermediates, that produce Maillard reaction products in the absence of a sugar source (Yaylayan, V. A., McGill University, Montreal, unpublished data). MAS degradation of serine (1 g) in the absence of water followed by hexane MAE yielded seven major products: ethyl pyrazine, 2-(1-pyrrolyl)ethanol, 2,6-diethyl pyrazine, and four unidentified compounds (Figure 3a). When one mL of water was added to serine during the MAS stage, the two major pyrazines disappeared from the extract, and the relative ratios of peaks designated as 6 and 7 increased (Figure 3b). Addition of two mL of water to the serine caused the disappearance of the peak designated as 5 (Figure 3c). These results demonstrate that careful manipulation of solvent

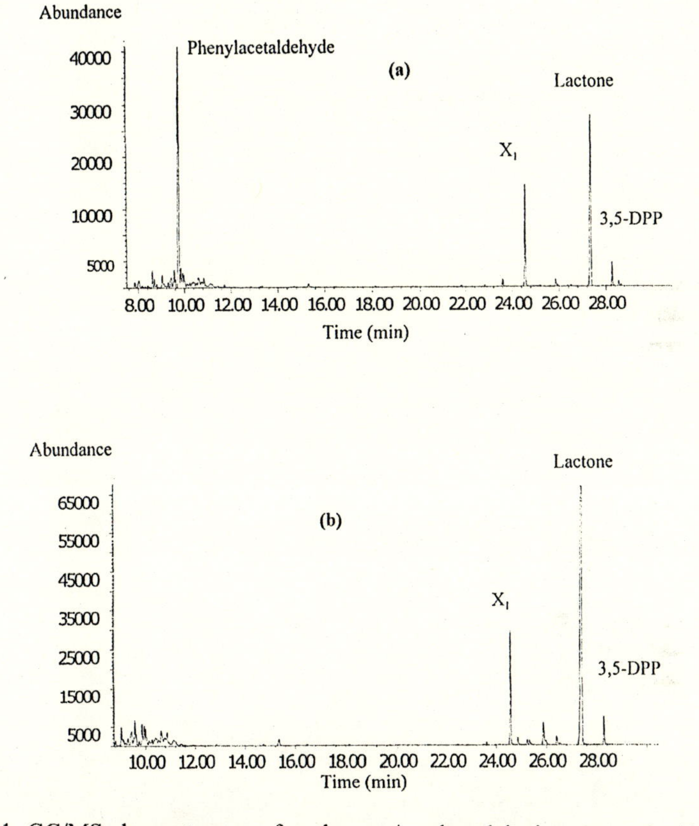

Figure 1. GC/MS chromatogram of D-glucose / L-phenylalanine extract in (a) reflux mode (in focused microwave) (b) open vessel (in domestic microwave)
Lactone = 2-(5'-Hydroxymethyl-2'-formylpyrrol-1'-yl)-3-phenylpropionic acid lactone
3,5-DPP = 3,5 Diphenylpyridine, X_1 *= unknown*

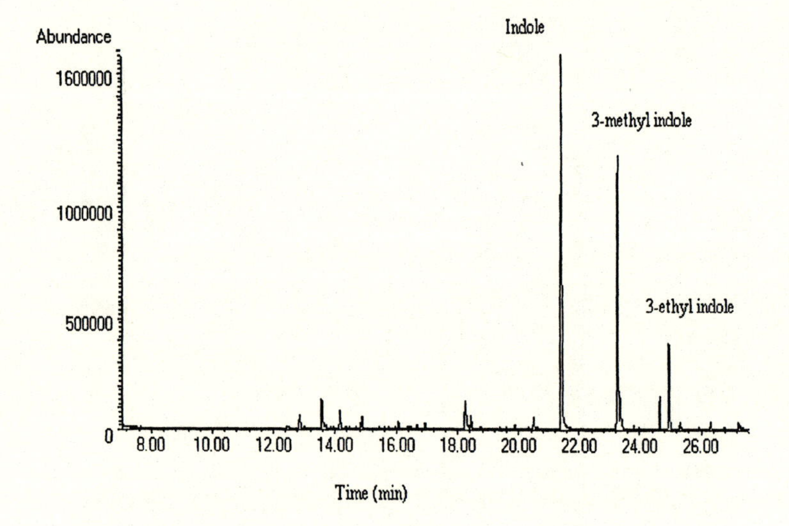

Figure 2. GC/MS chromatogram of D-glucose / L-tryptophan extract

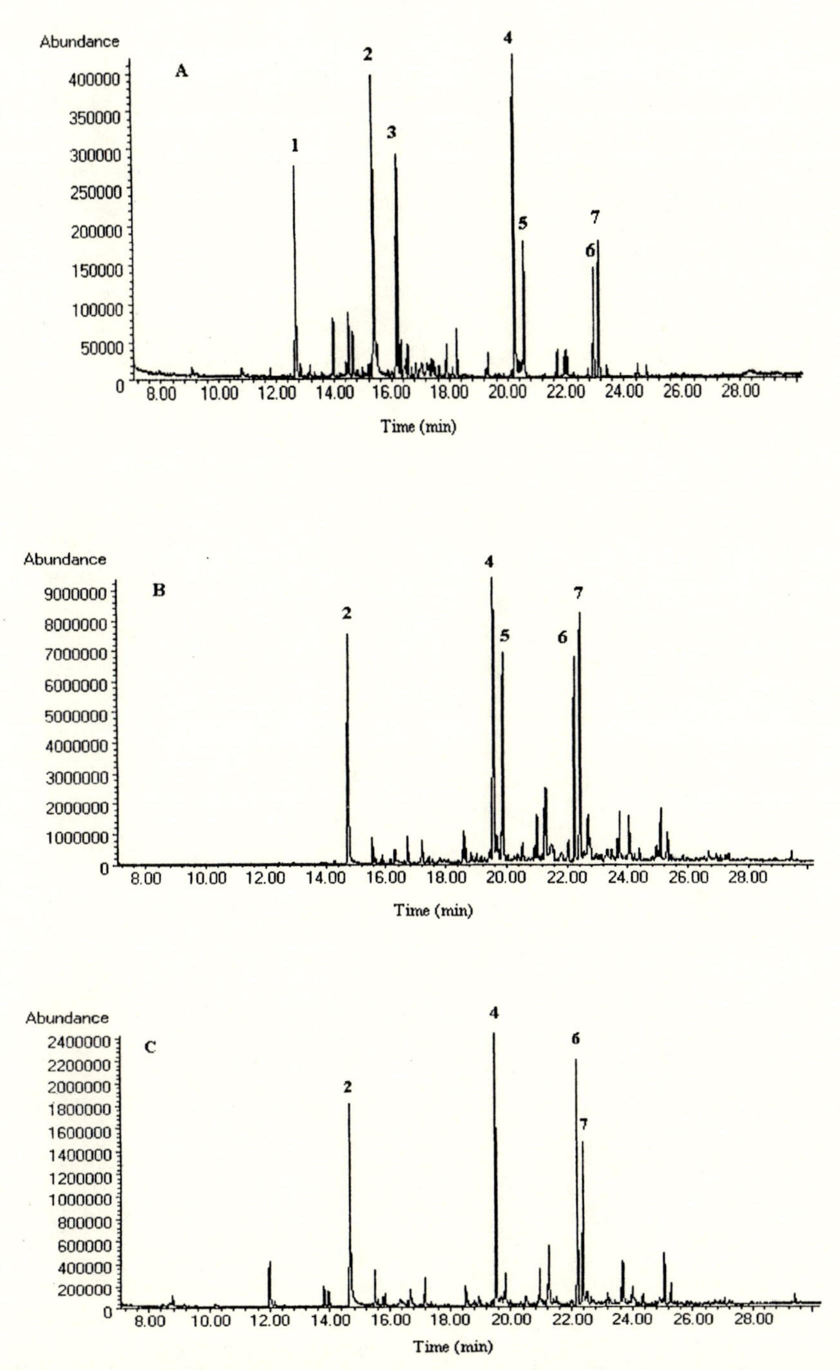

Figure 3. GC/MS chromatogram of D-glucose / L-serine extracts. (a) 0 mL water; (b) 1 mL water; (c) 2 mL water

composition during the synthesis stage can introduce some selectivity to the product profile.

Glycine/D-glucose Mixture: Application of Py-GC-MS to Identify Unknown Product. The glycine/D-glucose model system produced five major components (Figure 4), two of which (1,6-dimethyl and 1,5,6-trimethyl-2(1H)-pyrazinones) have been detected and identified previously (*8*) in glycine / D-glucose model systems. The mechanism of formation of alkyl substituted pyrazinones have been determined by ^{13}C labeling studies (*9*). Both pyrazinones can be formed from heated D-glucose glycine mixtures or from their corresponding Amadori product. Interestingly, all the components detected in the MAS/MAE extract, including the three unknowns termed Y_1, Y_2, and Y_3, were also detected in the pyrolysis mixtures of D-glucose glycine systems (*10*); Y_1 was the major product formed when three moles of glycine were reacted with one mole of D-glucose. In order to identify the structure of Y_1 (m/z 190), labeling studies were carried out using the technique of Py-GC-MS (*10*). Experiments performed with ^{13}C-labeled D-glucoses (independently labeled at each carbon atom) and ^{15}N- and ^{13}C-labeled glycines indicated the incorporation of all the six carbon atoms of the sugar and two nitrogens, one C-1, and three C-2 atoms of glycine into Y_1 (Table I). When excess glycine was reacted with synthetic 1-glycino-1-deoxy-D-fructose (Amadori product), the peak due to m/z 190 was the most intense in the pyrogram (Figure 5). In addition, due to the similarity of the glycine substitution pattern in this unknown to that of alkyl pyrazinones and intense molecular ion in the mass spectrum, it was predicted that Y_1 may possess aromatic pyrazinone structure and as such could be extracted into non-polar solvents, such as hexane. All attempts to extract the unknown from the heated D-glucose/glycine system failed to produce enough quantities for spectroscopic analysis; however, when the synthesis in water and extraction by hexane was performed by focused microwave irradiation under

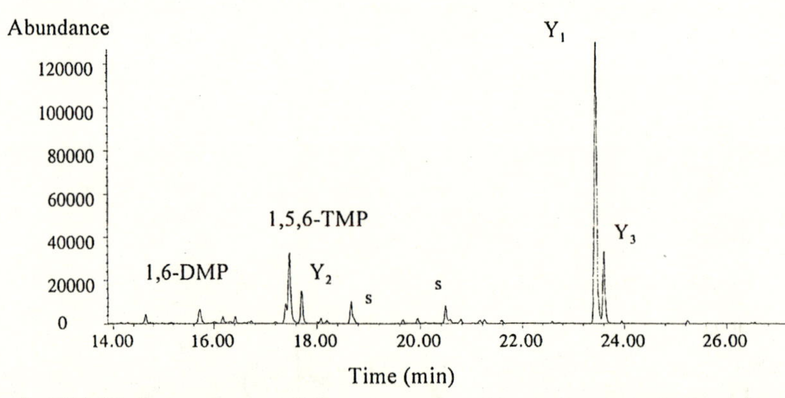

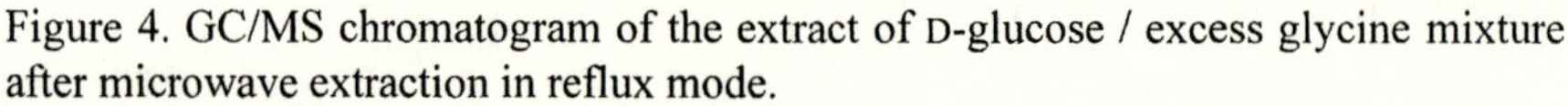

Figure 4. GC/MS chromatogram of the extract of D-glucose / excess glycine mixture after microwave extraction in reflux mode.

1,6-DMP = 1,6-dimethyl-2(1H)-pyrazinone, 1,5,6-TMP = 1,5,6-trimethyl-2(1H)-pyrazinone, Y_1 = 5-hydroxy-1,3-dimethyl-2[1H]-quinoxalinone, Y_2, Y_3 are unknown compounds, s = solvent

Table I. Percent distribution of molecular ion m/z 190 (compound Y_1) generated from labeled D-glucoses or labeled glycines*.

D-Glucose / excess glycine	m/z 190	m/z 191	m/z 192	m/z 193
D-glucose - glycine	99	1	0	0
D-[1-^{13}C]glucose - glycine	0	100	0	0
D-[2-^{13}C]glucose - glycine	0	100	0	0
D-[3-^{13}C]glucose - glycine	0	100	0	0
D-[4-^{13}C]glucose - glycine	0	100	0	0
D-[5-^{13}C]glucose - glycine	0	100	0	0
D-[6-^{13}C]glucose - glycine	0	95	5	0
D-glucose - [1-^{13}C]glycine	0	100	0	0
D-glucose - [2-^{13}C]glycine (90 % enriched	0	0	16	84
D-glucose - [^{15}N]glycine (98 % enriched)	0	2	98	0

* values are adjusted for natural abundance, compounds less than 99 % enriched are indicated

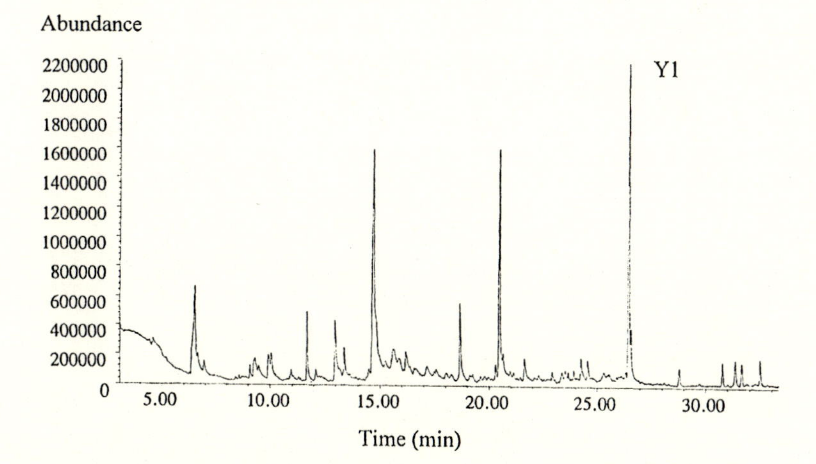

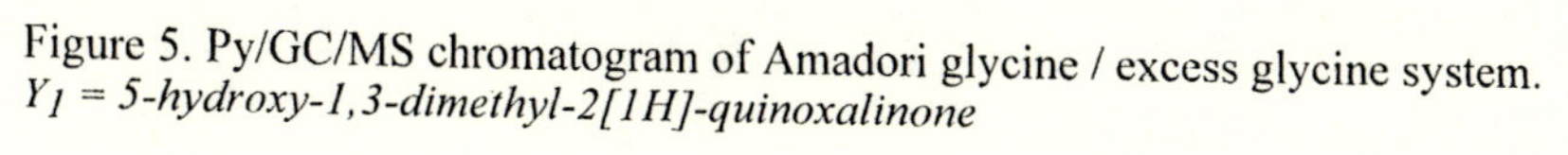
Figure 5. Py/GC/MS chromatogram of Amadori glycine / excess glycine system.
Y_1 = 5-hydroxy-1,3-dimethyl-2[1H]-quinoxalinone

atmospheric pressure conditions, a simple mixture was obtained containing the compound Y_1 (Figure 4) as the major component. Further purification by preparative chromatography yielded the pure compound which was assigned the structure of 5-hydroxy-1,3-dimethyl-2-[1H]-quinoxalinone based on spectroscopic analysis (*4, 11*).

Acknowledgments

The authors acknowledge funding for this research by Natural Sciences and Engineering Research Council (NSERC) of Canada.

Literature Cited

1. Thermally Generated Flavors: Maillard, Microwave, and Extrusion Processes. Parlimant, T. H.; Morello, M. J.; McGorrin, J., Eds.; ACS Symposium Series; No 543, American Chemical Society: Washington DC., 1994.
2. Nagodawithana, T. W. *Savory Flavors*; Esteekay Associates: Wisconsin, 1995, pp 164-224.
3. Paré J.R.J.; Bélanger J.M.R.; Stafford S.S., *Trends Anal. Chem.*, **1994**, *13*, 176-184.
4. Yaylayan, V.; Keyhani, A.; Pare, J.R.J.; Belanger, J. *Spectroscopy* **1997**, *13*, 15-21.
5. Yaylayan, V.; Matni, G.; Pare, J.R.J.; Belanger, J. *J. Agric. Food Chem.* **1997**, *45*, 149-152.
6. Kunert-Kirchhoff, J.; Baltes, W. *Z. Lebensm.-Unters.-Forsch.* **1990**, *190*, 9-13.
7. Keyhani, A.; Yaylayan, V. *J. Agric. Food Chem.* **1996**, *44*, 223-229.
8. Oh, Y-C.; Shu, C-K.; Ho, C-T. *J. Agric. Food Chem.* **1992**, *40*, 118-121.
9. Keyhani, A.; Yaylayan, V. *J. Agric. Food Chem.* **1966**, *44*, 2511-2516.
10. Yaylayan, V.; Keyhani, A. In *Contribution of low and non-volatile materials to the flavor of food.* Pickenhagen, W.; Spanier, A. M.; Ho, C-T., Eds. Allured Publishing Company: Carol Stream, Ill, 1996, pp 13-26.
11. Keyhani, A.; Yaylayan, V. *J. Agric. Food Chem.* **1997**, *45*, 697-702.

Chapter 20

Characterization of Coumarin and Psoralen Levels in California and Arizona Citrus Oils

Valerie L. Barrett and Denny B. Nelson

Sunkist Growers, Inc., P.O. Box 3720, Ontario, CA 91761

Lemon oil, because of high price, is an essential oil subject to economic adulteration. Lime oil components, such as herniarin (a coumarin) and isopimpinellin (a psoralen), thought to be unique to lime oil have been detected in lemon oils believed to be adulterated. Using GC/MSD in the SIM mode we observe lemon oil to contain low levels of herniarin, isopimpinellin, and bergapten, another psoralen identified as being absent in various studies of lemon oil. This study reports the levels of herniarin, isopimpinellin, and bergapten found naturally in cold-pressed lemon oils prepared from fruit grown in California and Arizona. Levels of these compounds found in the oils of navel orange, Valencia orange, tangerine, white grapefruit, and pink grapefruit will also be presented.

Citrus oils are important to the flavor industry, as they are used in a wide variety of food and beverage products. Because lemon is a high priced flavor oil, it is subject to economic adulteration. Detection of an adulterated lemon oil is a challenging issue because most adulteration is designed to avoid detection by the normal quality control tests. Most adulteration involves dilution of high value cold-pressed (CP) lemon oil with low value distilled lemon oil or other distilled citrus oils. Distilled oils are ultraviolet transparent, so the absorption at 315 nm, characteristic of CP lemon oil, is decreased in an adulterated lemon oil. A UV spectrometric test procedure, commonly referred to as the CD line test, was developed to detect such dilution (*1*). To mask dilution of CP lemon oil with distilled oil, materials are added that enhance UV absorption at 315 nm. Examples of such masking materials are ethyl 4-dimethylaminobenzoate (EDMAB) and solids obtained from winterized lime oil. The amorphous solids from chilled lime oil are rich in coumarins and psoralens that have strong absorption near 315 nm.

David McHale and John Sheridan of Cadbury Schweppes Research in the United Kingdom have published on detection of adulterated CP lemon oil (*2*) and on the levels of oxygen heterocyclic compounds in citrus peel oils (*3*). Their findings indicated that herniarin (7-methoxycoumarin), isopimpinellin (5,8-dimethoxypsoralen), and bergapten (5-methoxypsoralen) were unique to lime oil and not naturally present in lemon oil. Hence, any detection of these three compounds in lemon oil would infer adulteration. The method used by McHale and Sheridan was normal phase high pressure liquid chromatography (HPLC) with diode array detection. They did detect the lime oil compounds (herniarin and isopimpinellin) and EDMAB in adulterated lemon oil samples. Reverse phase HPLC work by Herta Ziegler and Gerhard Spiteller on coumarins and psoralens in Sicilian lemon oil (*4*) also supported McHale and Sheridan's conclusions.

In our laboratory we had extended the scope of our normal GC analysis for lemon oil. The amount of oil injected was increased and analyzed at a higher initial oven temperature, with a rapid temperature ramp to 300 °C. This allowed us to look at the less volatile constituents, which are normally seen as small peaks eluting after the lemon oil sesquiterpenes. We were able to use GC/FID or GC/MSD to detect the simple coumarins and psoralens (those without geranyloxy substitutions). This approach became the basis for a gas chromatographic method (*5*) to screen lemon oils by detecting EDMAB and lime oil components. The apparent presence of trace levels of herniarin, bergapten, and isopimpinellin in authentic lemon oils prompted us to establish a GC/MSD method to quantitate the levels of these compounds found in CP oils produced by Sunkist.

Experimental

The coumarin and psoralen standards were purchased from Indofine Chemical Company (Somerville, NJ). The (*R*)-limonene solvent was purchased from Aldrich (Milwaukee, WI) and biphenyl from Crescent Chemical Company (Hauppauge, NY). The GC/MSD system used was a Hewlett-Packard 5890 GC was equipped with a 7673B autoinjector, a 5971A mass selective detector, with helium carrier gas at 1 mL/min in the constant flow mode. The fused silica capillary column used was a J&W Scientific DB-5MS (30 m x 0.25 mm i.d. with 0.25 μ film). Samples of 1.0 μL were injected at 280 °C at a 50:1 split at an initial oven temperature of 120 °C, 4 minute hold, 15 °C/min ramp to 260 °C, 20 °C/min ramp to 300 °C, 5 minute hold for a 20.33 minute run time.

Solutions of each compound were prepared on a weight/weight basis with (*R*)-limonene to mimic the sample matrix. These standard solutions were analyzed with the MSD operating in the scan mode from 10-550 amu at 70 eV to determine the mass spectral pattern for each compound. Table I lists the four ions for each compound that were chosen to be acquired in the selected ion monitoring (SIM) mode for the quantitation method. Linear regression calibration curves consisted of three levels for each of the four compounds: 2, 24, & 59 ppm for herniarin and citropten, 2, 24, & 48 ppm for bergapten, and 2, 18, & 36 ppm for isopimpinellin. Citropten (5,7-dimethoxycoumarin) occurs naturally in lemon and lime oils (*2,3*), is commercially available, and functioned as an internal marker. Weekly calibration of the GC/MSD consistently yielded correlation coefficients of 1.000. Fortified samples and standard solutions were routinely analyzed, generally resulting in 90-105 % recoveries. Sample preparation consisted of adding 100 uL of a 100 ppm biphenyl internal standard (ISTD) solution to 900 uL of a citrus oil in a glass GC vial. A labeled chromatogram (Figure

1) of a Sunkist blend of coastal and desert CP lemon oil shows the ISTD and the four compounds identified. The MSD has been turned off until 6.8 minutes, therefore the early volatiles are not visible. EDMAB, if present, would elute ahead of bergapten at about 10.5 minutes.

Results and Discussion

Sunkist Growers, Inc., a leading producer of cold-pressed lemon oil, processes lemons from two major growing regions. Lemons are grown along the coast of California and in the desert areas of southeastern California and Arizona. Coastal and desert type lemon oils are usually marketed as a uniform blend. Daily production samples of CP citrus oils produced at the Sunkist processing facility in Ontario, California from August 1994 to September 1995 were analyzed.

Figure 2 is a histogram that shows the distribution of 395 samples of raw coastal CP lemon oils that were analyzed. The majority of samples had levels of 0-10 ppm herniarin, 0-10 ppm bergapten and 0-5 ppm isopimpinellin. Citropten results, which are not shown on the chart, ranged between 700-1300 ppm. These values were far above the 59 ppm citropten used for the highest calibration level. Citropten levels did agree with McHale and Sheridan results of citropten levels near 1,000 ppm (*2*). The histogram in Figure 3 shows the distribution of 89 samples of raw CP desert lemon oil. Most of the samples had herniarin levels below 10 ppm, which is similar to the coastal type CP lemon oil. But the bergapten and isopimpinellin levels were much higher than in the coastal type lemon oil. Over one third of the desert samples had bergapten levels between 50-100 ppm, with some falling between 100-350 ppm. Isopimpinellin levels in the desert oils were also higher, with almost half of the samples falling in the 35-110 ppm range. Citropten levels ranged between 700-1700 ppm. It does appear that there can be differences in the concentrations of the coumarins and psoralens in the oils cold pressed from lemons grown in different regions. The last histogram (Figure 4) shows data for 48 samples of a finished lemon oil product, which is a blend of coastal and desert lemon oils. Most of the samples contain levels of 5-10 ppm herniarin, 0-10 ppm of bergapten and isopimpinellin. Citropten results were between 900-1100 ppm. The results for the other types of raw CP citrus oils produced by Sunkist are listed in Table II.

Our limited experience with lemon oils suspected to be diluted and masked with lime oil components revealed herniarin, bergapten, and isopimpinellin levels all above 500 ppm. The concentration of lime components will depend on the level of added distilled oil these components must mask. In cases where herniarin, bergapten, and isopimpinellin are all in excess of 300 ppm, adulteration would be suggested.

Conclusion

The GC/MSD method provides a quick, reliable way to quantitate the simple coumarins and psoralens found in citrus oils. The GC/MSD method is more sensitive than the traditional HPLC analysis to detect low levels of these compounds. Detection of adulterant materials such as EDMAB can also be easily included in the method. Examination of Sunkist CP lemon oils did reveal that low levels of herniarin, bergapten,

Table I. Ions Selected for GC/MSD SIM Quantitation Method

Compound	*Target Ion*	*Qualifier Ion*	*Qualifier Ion*	*Qualifier Ion*
Biphenyl ISTD	154	153	152	76
Herniarin	176	133	148	77
Citropten	206	178	163	135
Bergapten	216	173	201	145
Isopimpinellin	231	246	175	188

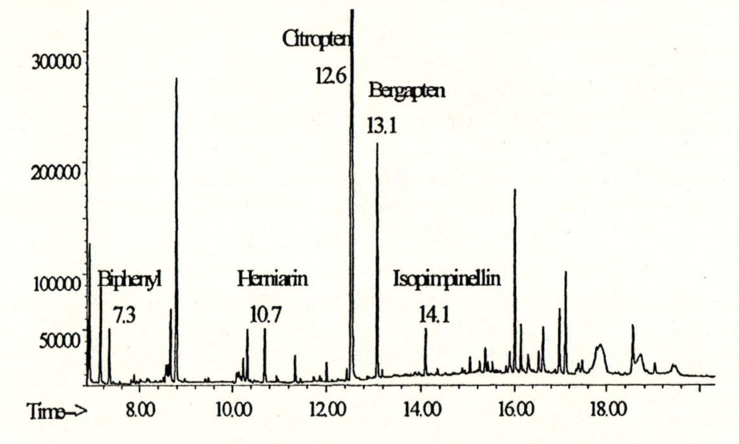

Figure 1. Chromatogram of a Blended Coastal and Desert Lemon Oil.

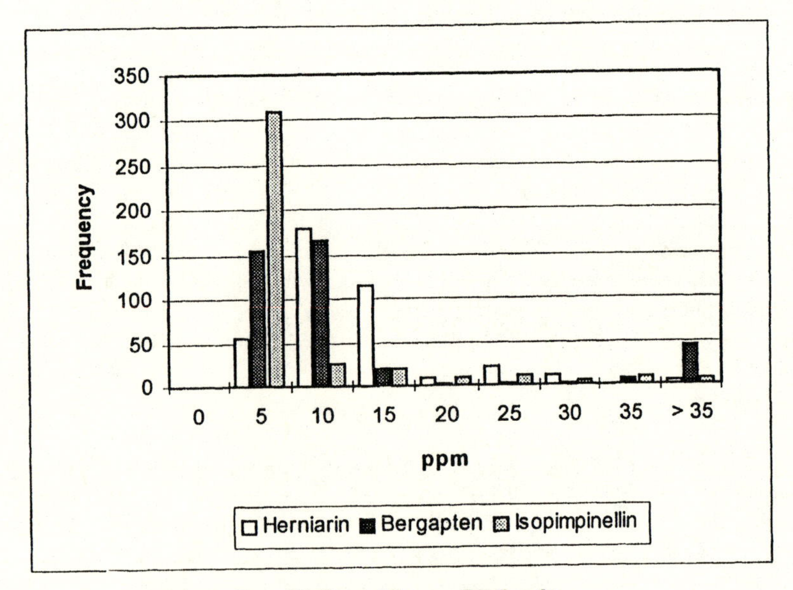

Figure 2. Raw CP Coastal Lemon Oil Results.

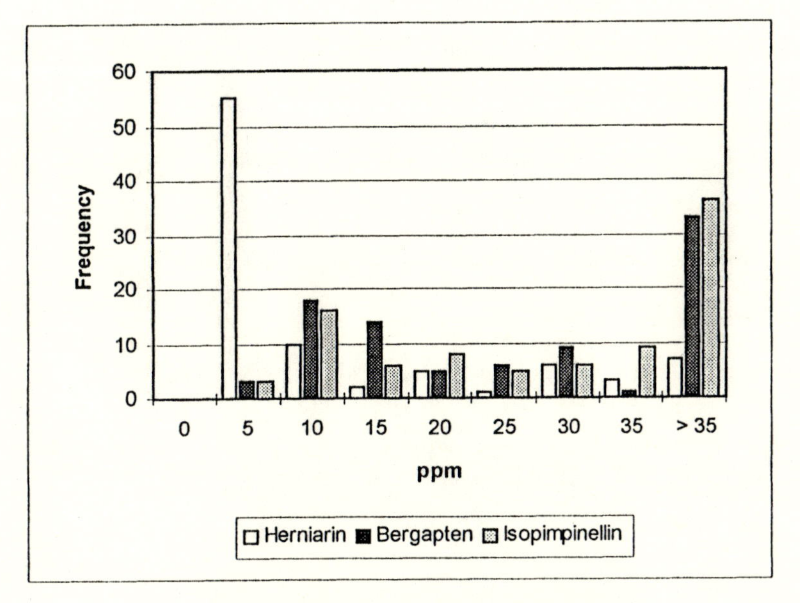

Figure 3. Raw CP Desert Lemon Oil Results.

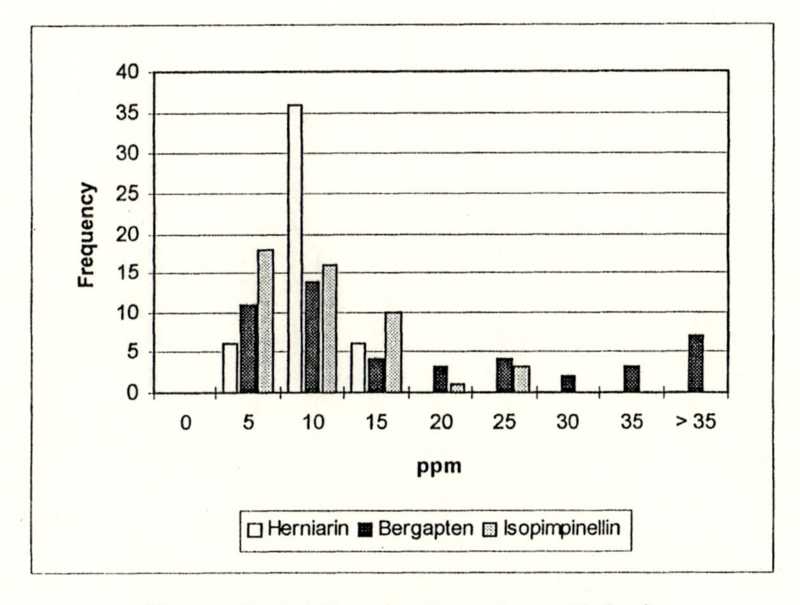

Figure 4. Blended Coastal and Desert Lemon Oil Results.

Table II. Concentrations (ppm) of Coumarins and Psoralens Present in Raw CP Citrus Oils.

Raw CP Oil Type	*Herniarin*	*Citropten*	*Bergapten*	*Isopimpinellin*	*# of Samples*
Navel Orange	< 2	20-40	<2	<2	23
Valencia Orange	<2	3-25	<2	<2	181
Pink Grapefruit	<2	6-60	20-75	5-25	169
White Grapefruit	2-5	5-50	10-40	10-40	116
Tangerine	<2	15-75	<5	<5	35

and isopimpinellin are naturally present. These findings indicate that detection of trace levels of these compounds in lemon oil does not prove adulteration.

References

1. Sales, J.W., et.al. *J. Assoc. Offic. Anal. Chem.* **1953**, *36*, 112.
2. McHale, D.; Sheridan, J.B. *Flavour and Fragrance J.* **1988**, *3*, 127.
3. McHale, D.; Sheridan, J.B. *J. Ess. Oil Res.* **1989**, *1*, 139.
4. Ziegler, H.; Spiteller, G. *Flavour and Fragrance J.* **1992**, *7*, 129.
5. Nelson, D.B.; Barrett, V.L. *Food Tech Europe*, **1994**, *1*, 106

Chapter 21

Profiling of Bioactive and Flavor-Active Natural Products by Liquid Chromatography–Atmospheric Pressure Chemical Ionization Mass Spectrometry

Kenneth J. Strassburger

Givaudan Roure Flavors, 1199 Edison Drive, Cincinnati, OH 45216

The phenolic/flavonoid/glycoside/saponin class(es) of natural products has been an important component of the human diet for centuries. As flavor active compounds they provide spice, sweetness, bitterness, and other flavor modifying effects, as well as serving as precursors to many volatile compounds. As bio-active compounds, the antioxidant, antimutagenic and therapeutic effects of these compounds are receiving significant research interest, with many herbal medicines and folk cures being re-evaluated. The literature, however, provides very little information on any uniform method(s) of analysis. We have developed simple LC gradient methods to separate a large number of these components from a variety of complex matrices, providing an efficient, and sensitive method for qualitative and quantitative analysis.

We began a program to study nonvolatile natural products and flavor precursors to better understand their effect on flavor perception, mouthfeel and balance, as well as antioxidant activity. We also wanted to identify key components in specific flavor applications and better characterize raw materials. Many food and beverage companies are interested in using these compounds in development of new, so-called "new age" products with health benefits.

Experimental

Materials. Botanicals and extracts were obtained from Hauser Chemical, Boulder, CO or Meer Corp., North Bergen, NJ. Solvents and buffers were purchased from Mallinckrodt Specialty Chemical, Paris, KY. Standards were purchased from Sigma Chemical Co., St. Louis, MO.

Liquid Chromatography. LC analyses were performed on a Waters 600-MS pump system with a 717 plus autosampler. All analyses were reverse phase using an Alltech Alltima C-18, 5 micron, 4.6 X 250 mm column. The mobile phases consisted of:

A = H_2O/1% CH_3COOH/25mM NH_4OAc
B = 90% MeOH/10% H_2O/1% CH_3COOH/25mM NH_4OAc
C = 90% CH_3CN/10% H_2O/1% CH_3COOH/25mMNH_4OAc

Glycoside and polyphenolic materials were analyzed using solvents A and B (designated as AB), applying a linear gradient from 90% A and 10% B, changing to 30% A and 70% B over 60 minutes. Saponin samples were analyzed using solvents A and C (designated as AC), also applying a linear gradient from 90% A and 10% C, changing to 30% A and 70% C over 60 minutes. The flow rate was 0.6 ml/min. The sample size was 10 micro litres.

Mass Spectrometry. Mass spectra were acquired on a Finnigan TSQ-700 equipped with an Atmospheric Pressure Chemical Ionization (APCI) interface. The vaporizer and capillary lens temperatures were set at 450 °C and 200 °C respectively. A 10V offset was applied to the octapole to suppress adduct formation. Mass analysis was performed by scanning Q_1 from 200 to 1200 amu at 2 sec/scan.

Structural Diagrams. Structures were imported using STN Express from the Chemical Abstract Service and are used with the permission of the American Chemical Society.

Results and Discussion

In the initial phase of method development, we chose two classes of compounds as important and typical of the food and flavor industry, namely saponins and polyphenolic glycosides. Both occur in a wide variety of products, can impact flavor performance, and have been difficult to characterize. Much of the previous work involved extensive pre-fractionation, collection, and subsequent analysis using off-line techniques such as FAB (fast atom bombardment) or the synthesis of volatile derivatives.

Stevioside and glycyrrhizin were selected as saponin standards and hesperidin and neo-hesperidin dihydrochalcone as polyphenolic standards because they were easily obtained in high purity. From previous LC analysis we realized that chromatography of these very polar compounds was best using an acidic mobile phase. However these conditions limited MS analysis to positive ions with a great deal of interference from basic compounds such as caffeine. The addition of ammonium acetate to the mobile phase, allowed very selective and sensitive detection in negative ion mode, with no interference from basic compounds. We were able to apply this buffer system in methanol water to the polyphenolic compounds with success. However, using methanol, the saponins were not eluted from a C-18 column; they could be very quickly eluted using acetonitrile (Figure 1.). In both examples, the spectra are characterized by simple deprotonated ions [M-H] ˉ or acetate adducts [M+OAC] ˉ as shown in Figures 2. and 3. with little background noise. We are currently working on a ternary mobile phase, $H_20/MeOH/CH_3CN$ (with buffers), to consolidate the methods. To illustrate the utility of this method results from four diverse sample types will be discussed: tea, oregano, licorice, and orange juice.

Tea. A very large amount of research and literature have been devoted to the health benefits of green tea. A number of therapeutic attributes have been ascribed to the polyphenolic catechins, they are also important sweetness and acidulent potentiators and important mouthfeel contributors. Figure 4. displays the catechin region of an LC/MS analysis of green tea; spectra of the four catechins are shown in Figure 5. While obviously important to a tea profile, we have found the catechins widely distributed in nature. For example they appear equally important to the flavor profiles of wines and cocoa (data not shown).

Oregano. As a spice, oregano has been used for centuries but has also been reported to contain potent antioxidant and antimutagenic compounds. A full characterization of oregano could not be found in the literature. Kikuzaki (1) identified 2-caffeoyloxy-3-[2-(4-hydroxybenzyl)-4,5-dihydroxy] phenyl propionic acid (**I**) in an oregano fraction having high antioxidant activity. Our LC/MS analysis in (Figure 6.) revealed (**I**) at scan 1511 (spectra and structure Figure 7.). In addition, two structurally similar compounds were found at scan numbers 1785 and 1935 (Figure 8.); both exhibit the loss of 162 amu in the only fragment ions. A loss of 162 is often indicative of a glucose or other sugar as part of the molecule; however, after a 24 hour hydrolysis in 1N HCl no sugar was present. In this case the loss of 162 amu is most probably from the common caffeoyl structural feature. Recently Kanazawa (2) identified the aglucon quercetin, in oregano as a strong desmutagen to Trp-P-2, a heterocyclic amine and potent mutagen from broiled beef. In our analysis, we identify two bound forms of quercetin: quercetin-3-L-rhamnoside and quercetin 7-D-glucoside (Figure 9.).

Licorice. Licorice is an important natural product used in the tobacco, confectionery, flavor, and pharmaceutical industries. It has been used as a medicine for centuries in the relief of rheumatic and other pain, treatment of ulcers, and recently as an anti-cavity and anti-cancer agent (Fenwick (3)). Gordan and An (4) have recently reported on the anti-oxidant activity of the flavanoid aglucons isolated in a chloroform extract of licorice.Our LC/MS analysis of a methanol extract of licorice root is shown in Figure 10. The mobile phase was H_2O/CH_3CN (AC). In addition to the saponin glycyrrhizin, four flavanoid glycosides were found (Figure 11.) These identifications are considered tentative as we could not locate a source for standards and MS/MS analysis was not attempted; however, they are consistent with FAB data in the literature.

Orange Juice. As a final example, we look at orange juice, an important commercial product whose flavor quality can vary significantly with processing. Figure 12. compares fresh juice to a canned, from concentrate, juice product. The sample

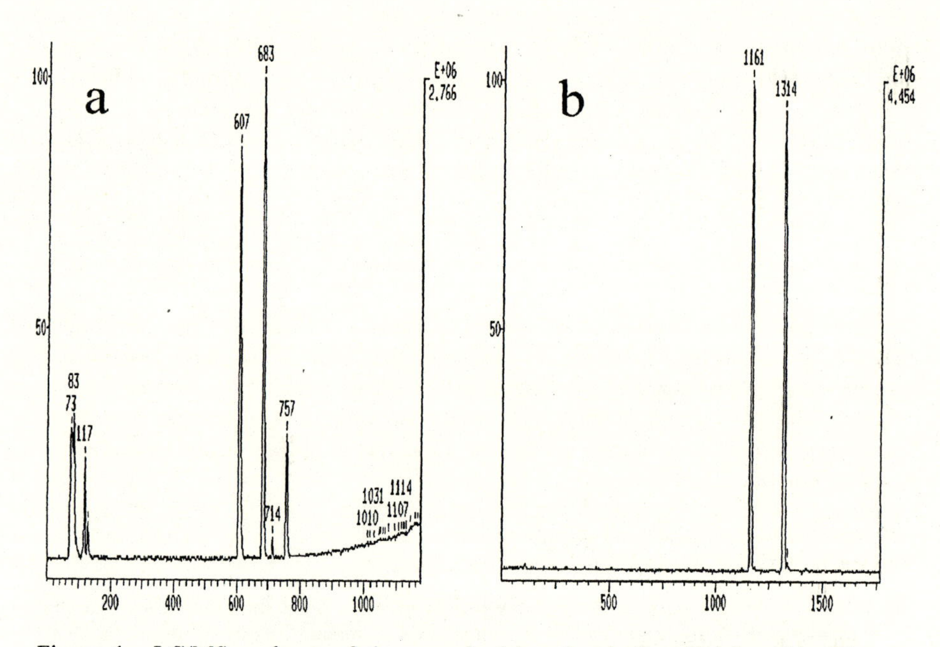

Figure 1. LC/MS analyses of the saponin (a) and polyphenolic/glycoside (b) standards. The saponin standards are: scan number 607 Stevioside [57817-89-7] and scan number 683 Glycyrrhizin [1405-86-3], using AC as the mobile phase. The polyphenolic/glycoside standards are: scan number 1161 Hesperidin [520-26-3] and scan number 1314 Neohesperidin dihydrochalcone [20702-77-6] chromatographed using AB as the mobile phase. The concentration of each standard was 1 mg/ml in methanol.

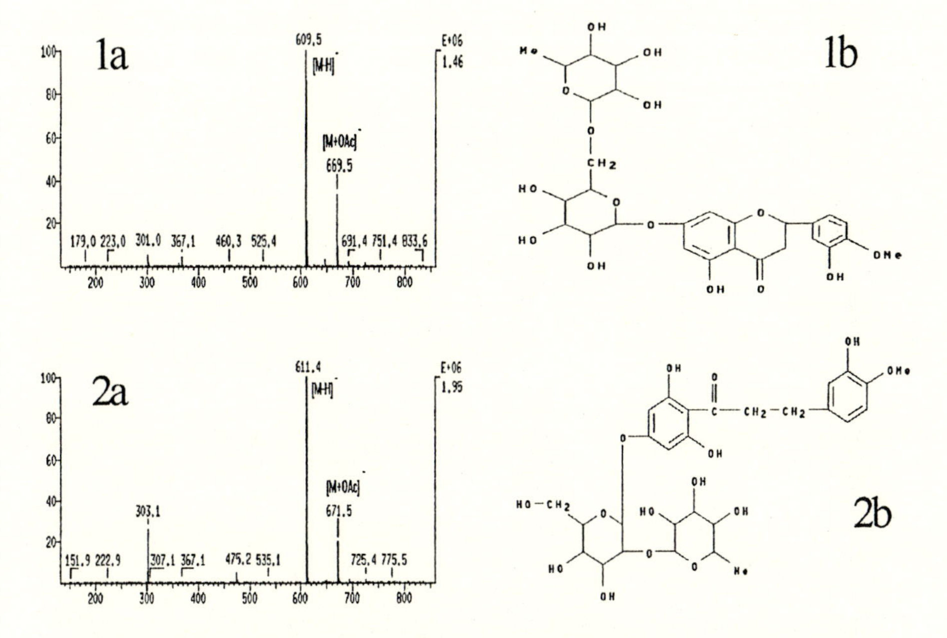

Figure 2. APCI spectra and structures of Hesperidin $C_{28}H_{34}O_{15}$=610.6 (1a&b) and Neohesperidin dihydrochalcone $C_{28}H_{36}O_{15}$=612.6 (2a&b).

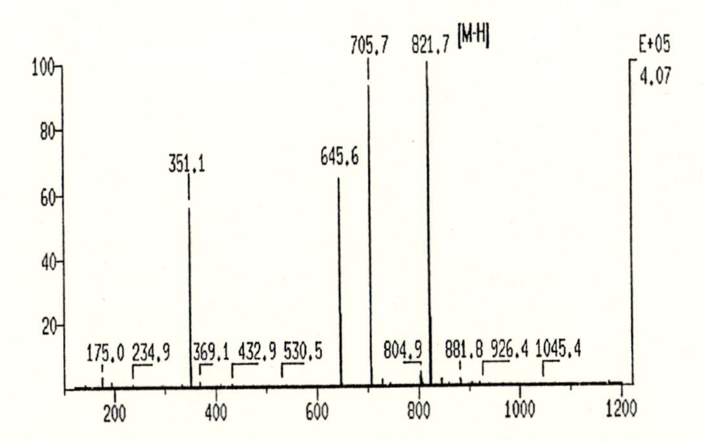

Figure 3. Spectrum of the saponin Glycyrrhizin [1405-86-3] $C_{42}H_{62}O_{16}$=822.9. The structure is shown in the analysis of Licorice Root (Figure 10.).

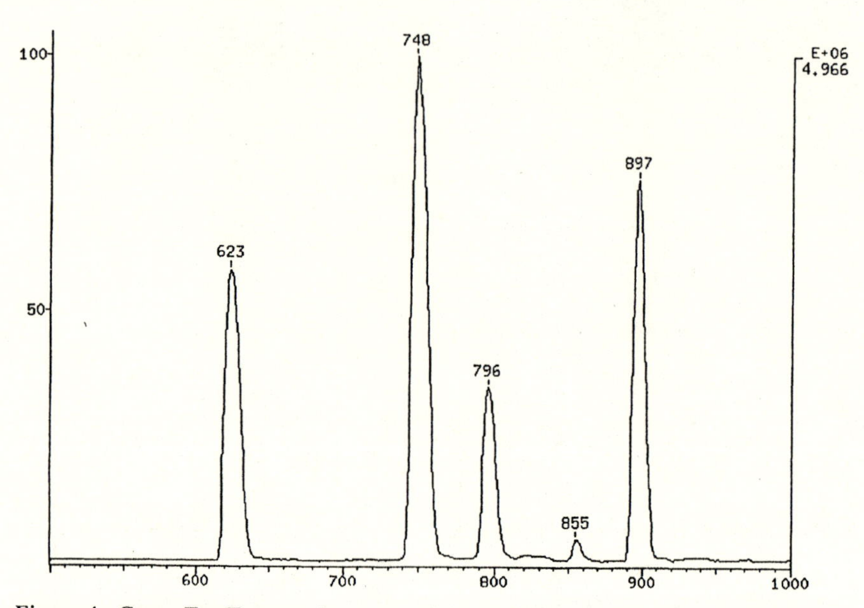

Figure 4. Green Tea Extract: A commercial sample diluted 1mg/ml methanol. LC mobile phase was AB (methanol-water).

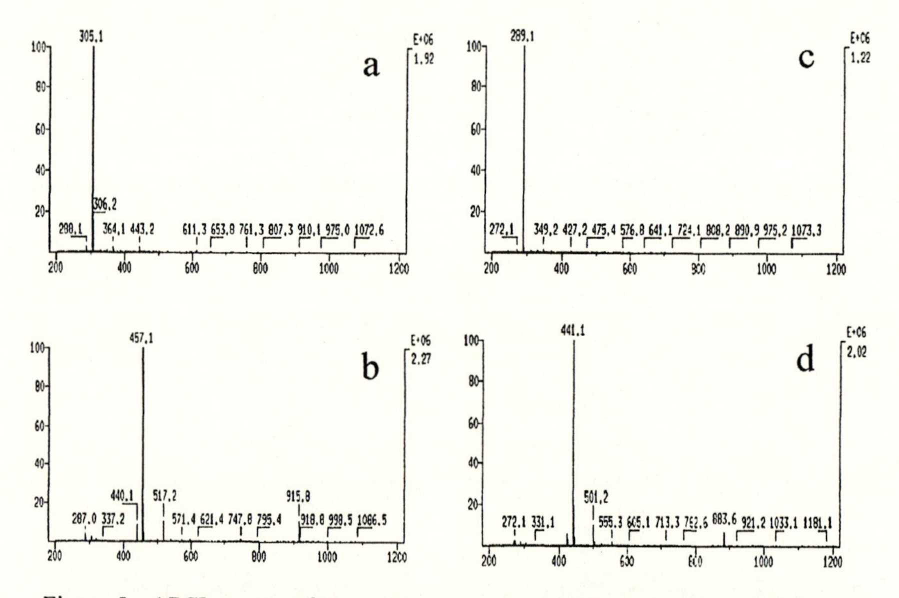

Figure 5. APCI spectra of Green Tea components (Figure 4.) identifications are a=epi-gallocatechin scan number 623, b=epi-gallocatechin-3-gallate scan number 748, c=epi-catechin scan number 796, and d=epi-catechin gallate scan number 897.

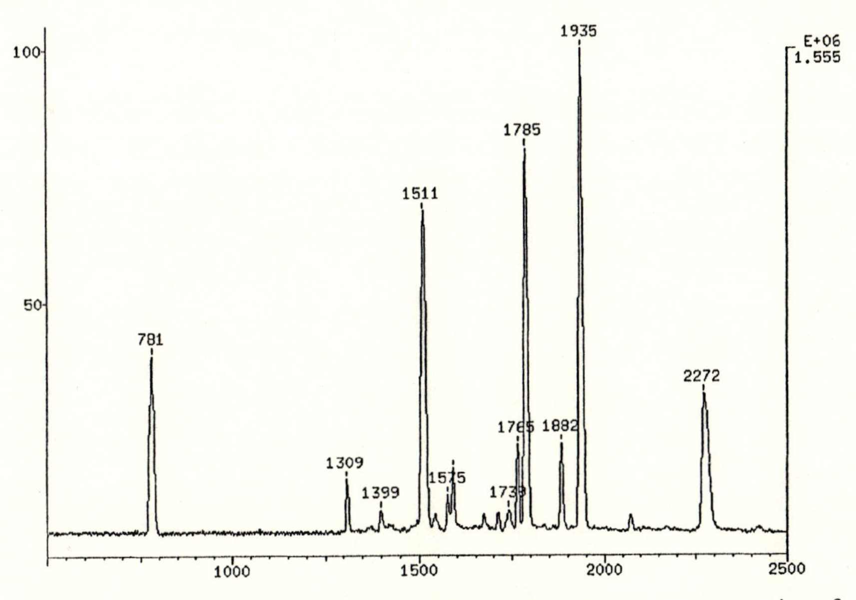

Figure 6. LC/MS analysis of a commercial Oregano extract at a concentration of 0.1026g in 2 ml of methanol. The mobile phase was AB.

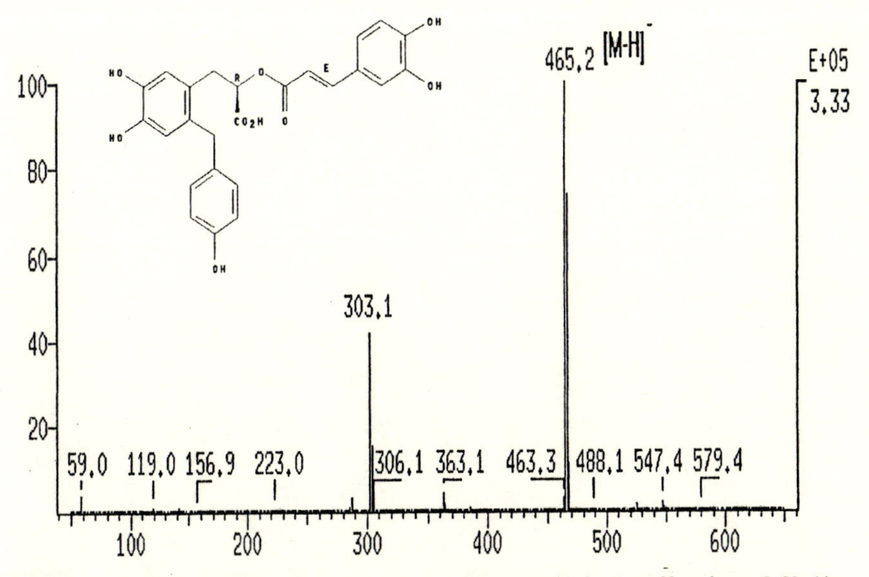

Figure 7. Scan number 1511 in the Oregano analysis 2-caffeoyloxy-3-[2-(4-hydroxybenzyl)-4,5-dihydroxy] phenyl propionic acid, was first identified by Kikuzaki (1).

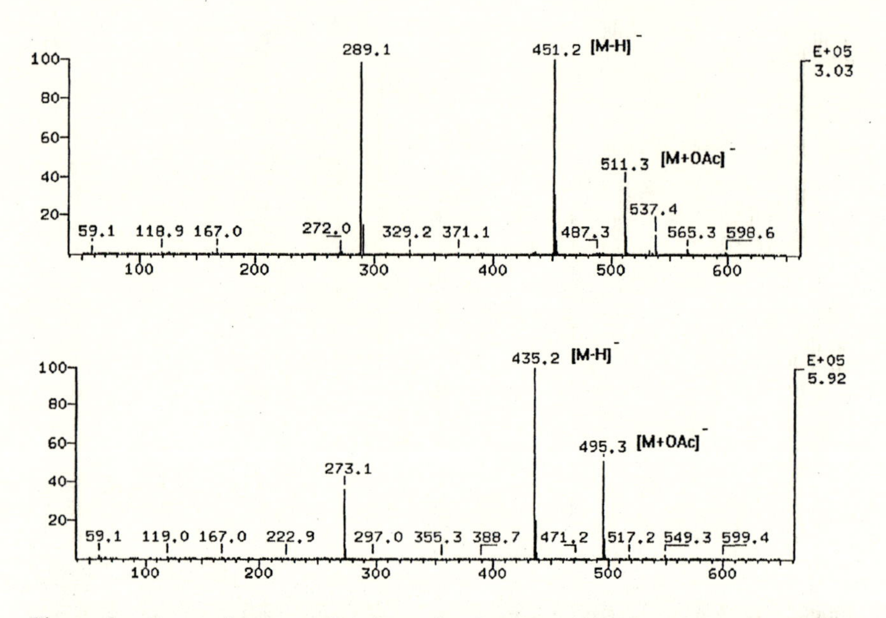

Figure 8. Two unknowns also from the Oregano analysis (Figure 6.), which appear to be structurally related to scan 1511 (Figure 7.). Scan 1785 (top) and scan 1935 (bottom) both exhibit the loss of 162 amu as the only fragmentation, consistent with scan 1511.

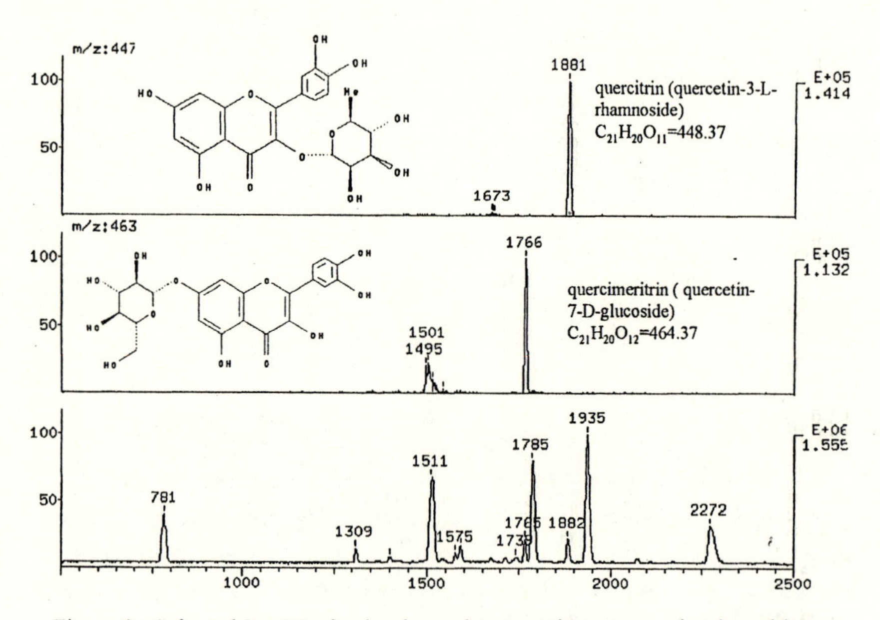

Figure 9. Selected Ion Monitoring is used to quantitate quercetin glycosides.

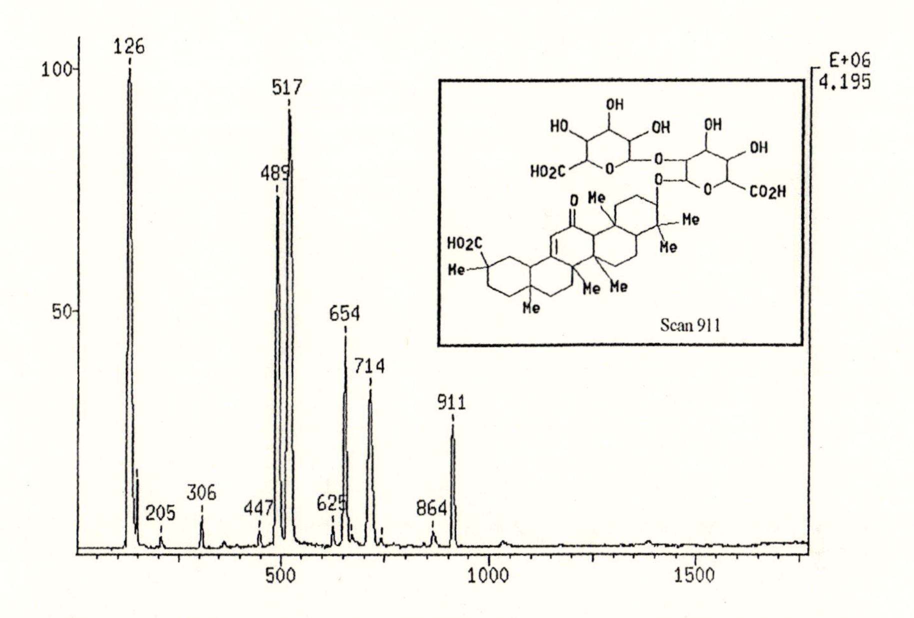

Figure 10. LC/MS analysis of a methanol extract of Licorice Root, the structure of the saponin glycyrrhizin is highlighted. The mobile phase was AC.

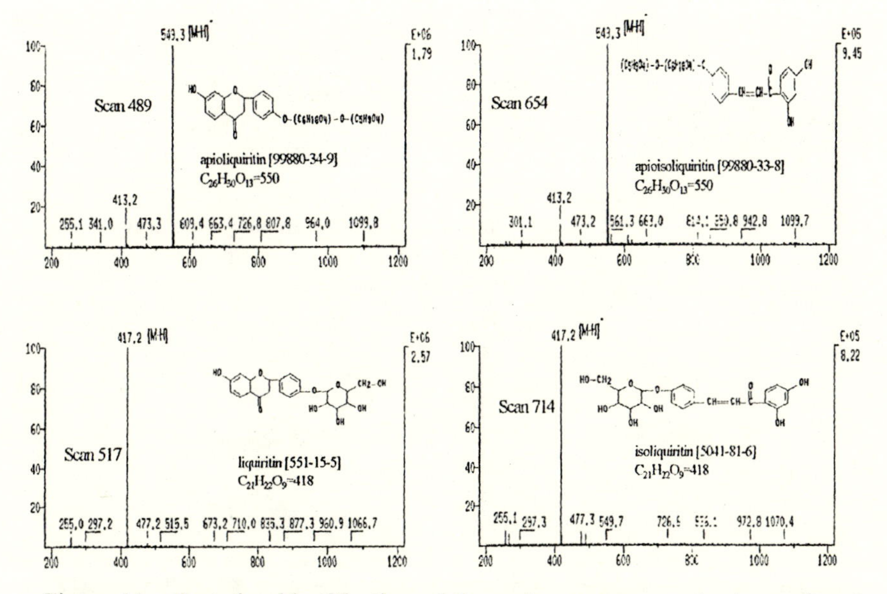

Figure 11. Tentative identification of the major components in the methanol extract of Licorice Root.

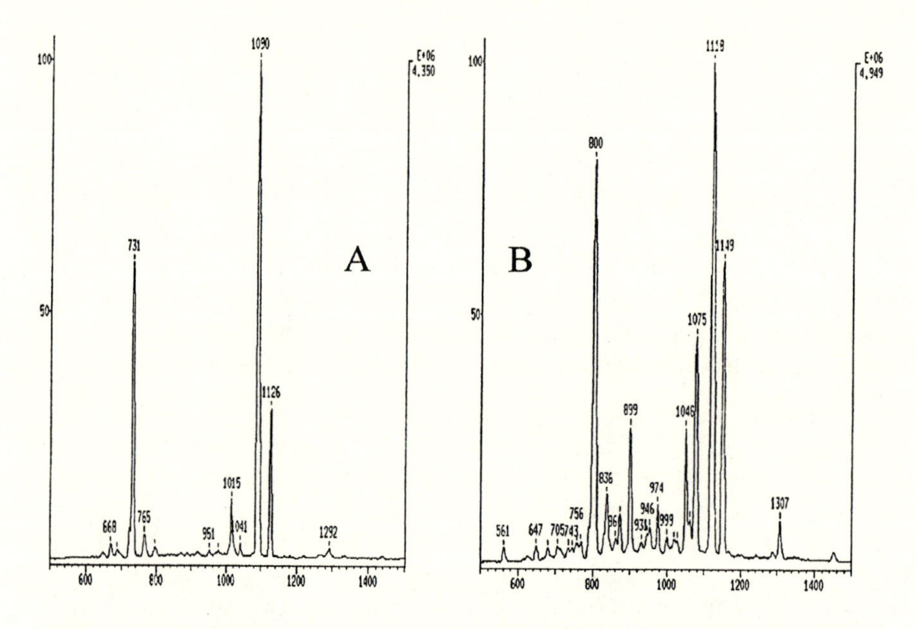

Figure 12. Fresh (A) vs. canned orange juice (B). Samples were centrifuged followed by cleanup on a C-18 SPE using 10 ml juice and eluted with 3 ml of methanol. LC solvent was AB.

preparation involved centrifuging the juice followed by cleanup on a C-18 solid phase extraction cartridge using 10 ml of juice and elution with 3 ml of methanol. The mobile phase used was water - methanol (AB). The retention times had shifted slightly in the second run (Figure 12B), however peak identities were confirmed by their mass spectra. The additional peaks in the canned juice, especially narirutin (Figure 13.), indicates the addition of pulp wash to the concentrate (Cancalon (5)).

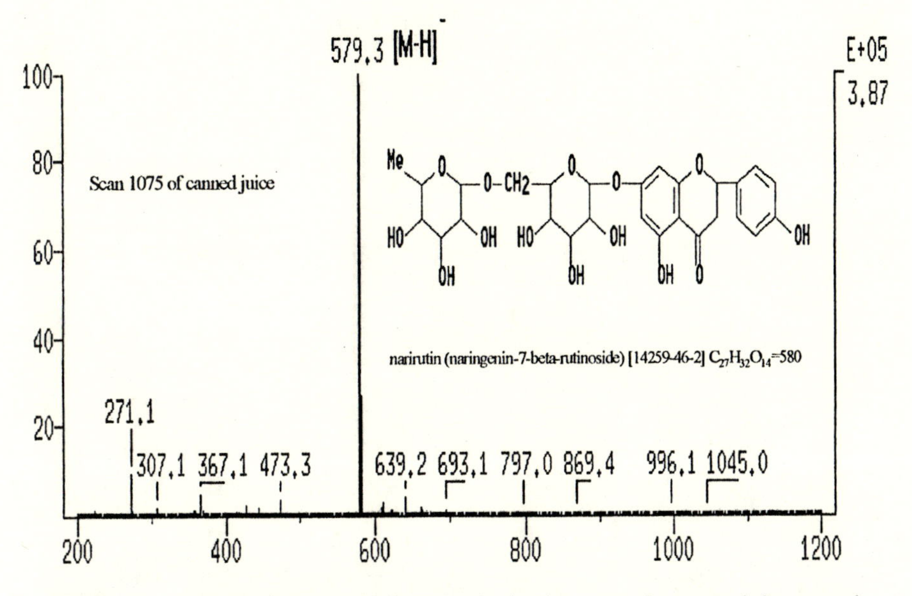

Figure 13. The identification of Narirutin in the canned orange juice sample indicates the addition of pulp wash to the concentrate.

Conclusions

The method presented has enabled us to analyze a large number of natural products. The mobile phase(s) employed produce good chromatographic separation for quantification and very specific mass spectral characterization. We are continuing to develop this methodology, including MS/MS techniques, to uniquely identify the many isomers usually present.

References

1 Kikuzaki, H; Nakatani, N. *Agric. Biol. Chem.* **1989**, 53(2) 519
2 Kanazawa, K; et al. *J. Agric. Food Chem.* **1995**, 43(2) 404
3 Fenwick, G. R.; Lutomski, J.; Nieman, C. *Food Chem.* **1990**,38(2) 119
4 Gordan, M. H.; An, J. *J. Agric. Food Chem.* **1995**, 43(7) 1784
5 Cancalon, P. *LC-GC* **1993**, 11(10) 748

Chapter 22

Quantification of Impact Odorants in Food by Isotope Dilution Assay: Strength and Limitations

Christian Milo and Imre Blank

Nestec Ltd., Nestlé Research Center, Vers-chez-les-Blanc, P.O. Box 44, 1000 Lausanne 26, Switzerland

A foods' aroma is one of the main criteria for its acceptability by the consumer. Successful product development is closely related to aroma profile and consumers' expectations. As perception of a desirable aroma depends on a subtile balance of certain odorants, understanding of a complex aroma at the molecular level means focusing on sensorially relevant odorants. Reliable quantitative data is a prerequisite for evaluating the contribution of a single odorant to a positive aroma or off-flavor and in flavor precursor studies. Therefore, improved methods are needed for accurately quantifying key odorants, particularly if they are unstable and/or occur in low concentrations. The technique of isotope dilution assay (IDA) as applied to aroma research will be presented with the aim of objectively assessing its strengths and current limitations.

Isotope dilution assay (IDA) was introduced in 1966 for the determination of glucose in plant tissues (*1*) and has been widely used in the field of pharmacological chemistry. In the '70s a theoretical basis for IDA was established (*2*) and it was proposed as a 'reference method' (*3*). The first application in aroma research was published in 1987 by Schieberle and Grosch (*4*) who quantified acetylpyrazine, 2-methyl-3-ethylpyrazine, and 2-acetyl-1-pyrroline in wheat bread crust. This research group has developed IDAs for more than 100 potent aroma compounds. Most of them are cited in a recent review article (*5*). Since about five years, there is an increasing interest in using IDA not only in academia (*6-8*), but also in industry (*9-11*, Preininger, M. In *Proceedings of the 9th International Flavor Conference*, Limnos, Greece, July 1-4, 1997 (in press)).

The method has also been proven very useful in research related to off-flavours that are often caused by a misbalance of odorants. The meaty-brothy odor defect in low fat Cheddar cheese was primarily attributed to an increase in furaneol, homofuraneol, and methional (*8*). Warmed over flavor (WOF) in precooked beef is primarily caused by hexanal and *trans*-4,5-epoxy-(*E*)-2-decenal, both odorants also occur in freshly cooked meat, but at lower concentrations (*12*).

Principle of Isotope Dilution Assay (IDA)

The analyte in the sample is quantified by means of an internal standard labeled with stable isotopes. A known amount of standard is added to the sample, preferably a liquid or a slurry of a solid food to ensure homogeneous distribution of the standard in the sample. Due to the nearly identical chemical and physical properties of the labeled standard and the analyte, losses of the analyte caused by the isolation procedure are compensated and, hence, errors in the analytical measurements are reduced to a minimum. The concentration of analyte is determined by GC-MS via the peak areas of selective mass traces according to the equation:

$$Q_a = \left[\frac{\frac{A_a}{A_s} - b}{a} \right] Q_s$$

with Q_a: Quantity of analyte in the sample
Q_s: Quantity of labeled internal standard added to the sample
A_a: Peak area of selected mass of the analyte
A_s: Peak area of selected mass of labeled internal standard
a : Slope of the calibration curve
b : Intercept of the calibration curve

The response factor R, represented by the term in brackets, is determined prior to the analysis by injecting mixtures of known amounts of analyte and standard into the GC-MS system and measuring the peak ratios versus the amount ratios (*4,13*). Consequently, both the labeled and unlabeled compounds must be available for calibrating the MS. The response factor accounts for slightly different fragmentation patterns of analyte and standard. It depends on the synthesis procedure, in particular the isotopic purity of the standard, and the number and position of labeling. Factors different from 1 have been found even for well defined molecules, such as $^{13}C_2$-labeled acetaldehyde and acetic acid.

Requirements for Labeled Standards

The isotopically labeled analogue of the analyte is used as a standard. In order to be considered stable, labeling with stable isotopes is performed in a non-exchangeable position. Deuterium (D) is preferably used because it is rather cheap compared with other isotopes. Under certain circumstances, however, ^{13}C-labeling is strongly recommended. Protons in the α-position to carbonyl-functions are known to enolize and may exchange with protons from the sample. A possible D/H exchange of a labeled standard during the clean-up procedure has to be ruled out. In some cases, the standard may be tested under conditions used for isolating the aroma components.

As an example, 4-hydroxy-2,5-dimethyl-3(2*H*)-furanone (furaneol) should be labeled with ^{13}C as shown in **c-Ia** (*14*) and **c-Ib** (*11*). By contrast, its ethyl analogue (homofuraneol) can be deuterated, but only in a well defined position, i.e. as 2(or 5)-([2,2,2-$^{2}H_3$]ethyl-4-hydroxy-5(or 2)-methyl-3(2*H*)-furanone (**d-II**) (*11,15*). As shown in Figure 1, the CH_2 in the ethyl group functions as a barrier to D/H exchange.

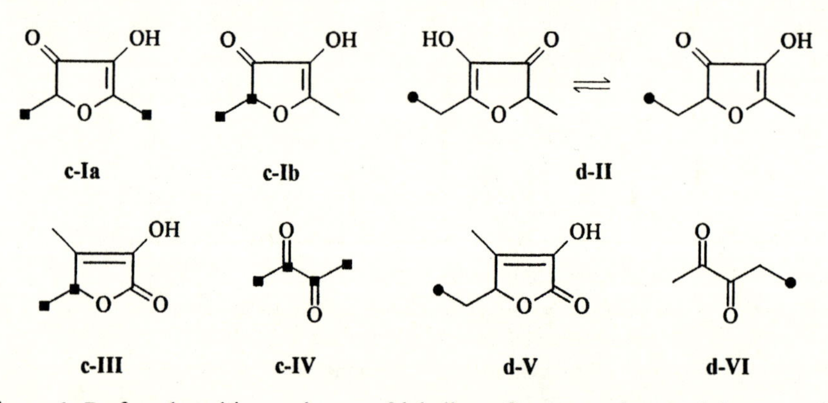

Figure 1. Preferred position and type of labeling of compounds containing an α,β-dicarbonyl moiety (■ indicates the labeling position with ^{13}C, ● with deuterium).

Further compounds which require ^{13}C-labeling are, for example, 3-hydroxy-4,5-dimethyl-2(5*H*)-furanone (sotolone, **c-III**) (*10*) and diacetyl (**c-IV**) (*16*). However, 5-ethyl-3-hydroxy-4-methyl-2(5*H*)-furanone (**V**) (*9*) and 2,3-pentanedione (**VI**) (*17*) can be deuterated provided that a CH_2 group is located between the enolizing structure and the deuterated moiety (Figure 1).

The labeling should also increase the molecular weight of the standard by at least two units to minimize interferences with masses originating from the analyte. Another source of interference may result from possible by-products from synthesis of the standard. Most isotopically labeled potent odorants are not commercially available and are synthesized in rather small quantities that prohibit extensive clean-up to avoid losses in yields. From a practical point of view, by-products may be tolerated if they do not react with the standard, are not readily converted into it, and do not interfere with any of the analytes in the sample during GC-MS measurements.

For reasons of stability the standards are kept frozen and diluted in an appropriate solvent (normally dichloromethane or methanol). Standard solutions of diacetyl (**c-IV**) and 2,3-pentandione (**d-VI**) stored at –30 °C, for example, did not change in concentration over 2 years. For quantification of odorants food samples are generally spiked with 0.1–1 ml of a standard solution corresponding to a few µg. Therefore, 100–500 mg of standard is sufficient for up to 10,000 determinations.

Accuracy and Precision of IDA

In principle, addition of an isotopically labeled odorant to a sample results in absolute quantitative measurement of this odorant. A prerequisite for absolute and correct values is fully homogeneous distribution of the standard **and** analyte in the sample. The influence of stirring time was studied by extracting a finely ground cheese with diethyl ether spiked with standards. The amounts obtained for four potent Cheddar cheese odorants are shown in Table I. No changes were found in the concentration of acetic acid, butyric acid and furaneol after 1 hour and for the lactone after 2 hours of stirring (*8*). This confirms that under the conditions chosen an equilibrium was reached between standard and analyte after 2 hours of extraction. Furthermore, these data show the high precision of the method. As an example, furaneol (**I**) was quantified with a standard deviation below 5 %.

Table I. Concentration of Selected Odorants Important to Cheddar Cheese Aroma as Affected by the Extraction Time.

ompound	*Concentration of odorants (μg/kg) after extraction for*				
	1 h	*2 h*	*3 h*	*4 h*	*6 h*
cetic acid	108900	104600	106900	108900	117000
utyric acid	11200	11100	11000	11300	11800
-Decalactone	1708	2278	2074	1989	2659
uraneol	29.8	30.1	30.4	27.0	31.5

The main advantage of IDA, however, is the high selectivity of GC-MS, particularly in selected ion monitoring (SIM) mode. Compounds that are poorly resolved by GC may not interfere in GC-MS due to different fragmentation patterns. Therefore, less sample clean-up is needed. Another way to enhance sensitivity and selectivity for certain compounds depends on the MS ionisation mode. As shown in Figure 2, a 20-40 fold increase in abundance was obtained for *trans*-4,5-epoxy-(*E*)-2-decenal (**VII**) by using negative CI compared to positive CI. The increased selectivity of negative CI is another interesting feature for the quantitation of certain odorants in complex mixtures of volatile compounds.

Application of IDA in Foods

On the basis of its high accuracy and selectivity, IDA is the method of choice for the quantitation of odorants that are either not stable, like thiols, (*Z*)-3-alkenals, and epoxy-compounds or very polar and, hence, difficult to isolate quantitatively, e.g. furaneol and its derivatives.

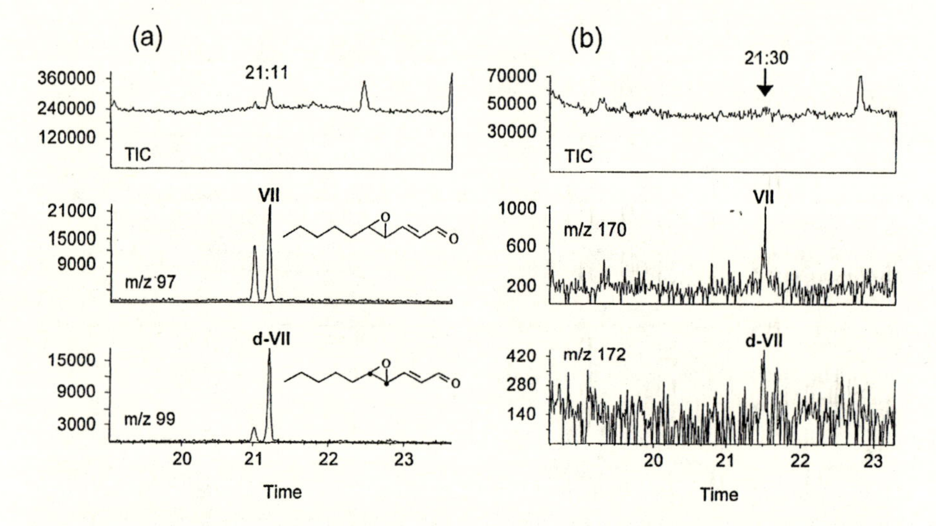

Figure 2. Detection of *trans*-4,5-epoxy-(*E*)-2-decenal **(VII)** and **d-VII** by GC-MS in the (a) negative CI and (b) positive CI mode using ammonia as reagent gas (J. Lin, L. B. Fay, I. Blank, unpublished results).

Accurate quantification of thiols is a challenging task because of their tendency to degrade and their relatively low concentration in foods. Compounds such as 2-furanmethanethiol (**VIII**) and 2-methyl-3-furanthiol (**IX**) possess very low odor thresholds in water, i.e. 0.01 μg/kg (*18*) and 0.007 μg/kg (*19*), respectively. They significantly contribute to the aroma of many foods like cooked meat and roasted coffee (Figure 3).

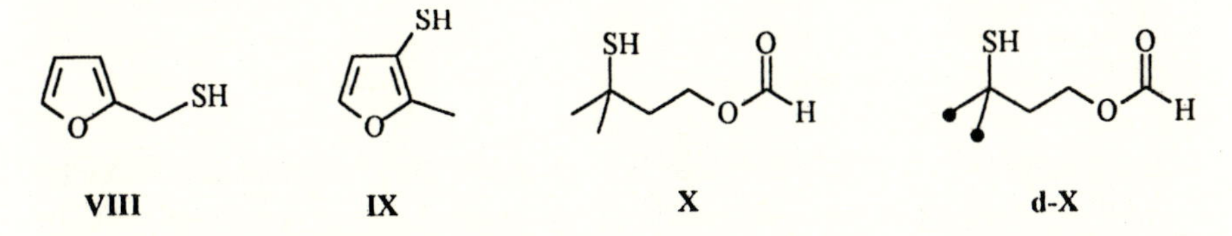

Figure 3. Sensorially relevant thiols contributing to the aroma of roast coffee and cooked meat (explanation in text).

3-Mercapto-3-methylbutyl formate (**X**) is more specific to coffee (*20*). About 50 g of roast and ground (R&G) coffee were required for identification of **X** by GC-MS (*21*) and 20 kg for obtaining a ^{1}H-NMR spectrum (*22*). The concentration of this compound in R&G coffee (50 g) and coffee brew (1 L) was assessed for the first time by Grosch and coworkers on the basis of IDA using GC-MS in the CI mode, i.e. ~100 μg/kg in R&G coffee (*18*) and ~5 μg/kg in the brew (*23*). Because of the low amounts of **X**, quantification can be complicated by interference from matrix constituents. A simplified and rapid method for quantitation of **X** from only 10 g R&G coffee using [$^2H_6$]-3-mercapto-3-methylbutyl formate (**d-X**) as internal standard (*24*) will be briefly described. It is based on selective clean-up of thiols (*23, 25*).

Ten g coffee was extracted with dichloromethane (50 mL) in the presence of 7.4 μg labeled standard. After 3 h of stirring, the sample was filtered and the coffee reextracted with dichloromethane (50 mL) overnight. The combined extracts were concentrated to ~40 mL and then passed through a preconditioned Affi-Gel 501 column (Bio-Rad Laboratories) which reversibly binds the thiols. Thiols were eluted with dithiothreitol (10 mMol/L) in pentane/dichloromethane (2/1, v/v, 50 mL). Excess dithiothreitol was removed by precipitation through concentration on a Vigreux column and freezing. The supernatant was further concentrated to ~0.5 mL and an aliquot of 1 μL analyzed by GC-MS in the electron ionisation (EI) mode at 70 eV. The quantitation was performed on a Hewlett-Packard HP-5971 GC-MS connected to an HP-5890 gas chromatograph equipped with an HP-7673 autosampler. The interface was kept at 300 °C and the ion source was held at 180 °C. A DB-Wax capillary column was used: 30 m x 0.25 mm, film thickness 0.25 μm. The carrier gas was helium (8 psi). The GC was held at 20 °C for 0.5 min, then the temperature was raised at 70 °C/min to 60 °C, and then at 6 °C/min to 180 °C.

Figure 4 shows an excerpt of the total ion chromatogram of a coffee aroma extract 'enriched' in thiols and the MS(EI) spectra of **X** and **d-X** obtained. The concentration of the odorant **X** was determined using the selective mass 75 for **d-X** and 69 for **X**, resulting in 120 µg/kg R&G coffee.

Limitations of IDA in Flavor Research

The use of isotope labeled standards for the quantification of key odorants is still limited because, unfortunately, most aroma compounds are not yet commercially available. There are only a few exceptions, e.g. ^{13}C-labeled acetaldehyde, guaiacol and vanillin as well as some deuterated fatty acids and dimethylsulfide which can be obtained from Cambridge Isotope Laboratories (Andover, MA, USA) or Aldrich. In-house synthesis can be time-consuming. Indeed many flavor and food companies do not have organic synthesis facilities, but depend on external vendors.

Extensive references for the synthesis of labeled odorants have been published (*5*). The synthetic routes described in the literature, however, are not always fully documented in terms of yield for each step, purity of the final product, and analytical characterisation including NMR. Optimisation of the synthesis from a yield and cost perspective is needed to attract laboratories to provide such standards. Some recommendations are given below.

It is advantageous to develop synthesis procedures that can be applied to a series of homologous standards (*10*). They may even be prepared in one batch, like C_6, C_8, C_{10} and C_{12}-δ-lactones or several α-dicarbonyl compounds. Labeling should be done at the end of a reaction sequence to avoid loss of precious intermediates. The type of isotope and the position of labeling will depend on stability considerations (see Requirements of Labeled Standards), but also on the MS detection mode.

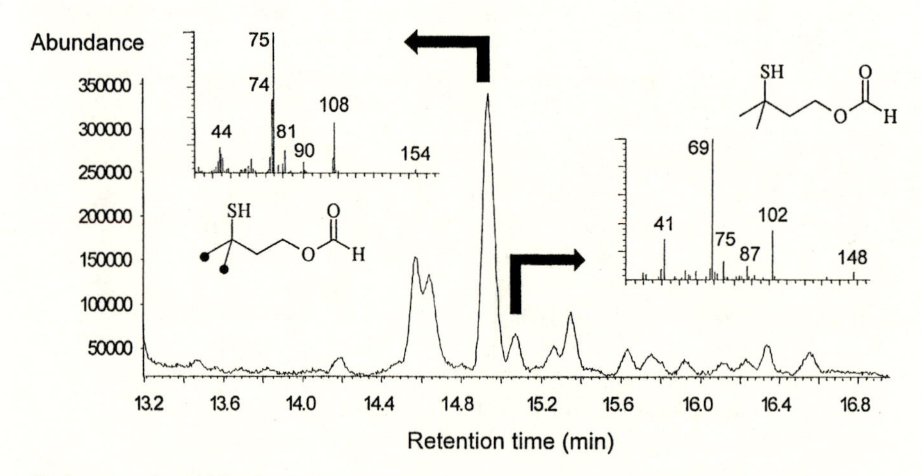

Figure 4. Quantification of 3-mercapto-3-methylbutyl formate (**X**) in a roast and ground coffee aroma extract enriched in thiols by IDA using **d-X** as internal standard.

As an example, quantitation of δ-decalactone was described using IDA and GC-MS(CI) (*26*). However, δ-lactones can also be quantified by IDA using GC-MS(EI). In the latter, special labeling is required taking into consideration the fragmentation pattern of δ-lactones. As outlined in Figure 5, the synthesis procedure described by Schieberle et al. (*26*) was modified to introduce deuterium not only in the side chain of the molecules, but also in the lactone ring.

With this modification, various δ-lactones can be monitored by detecting the fragment m/z 102 for the standards, containing 3 deuterium atoms, and m/z 99 for the analytes (Figure 6). This example shows that standards may not be universally useable, but rather custom-tailored for certain detection methods.

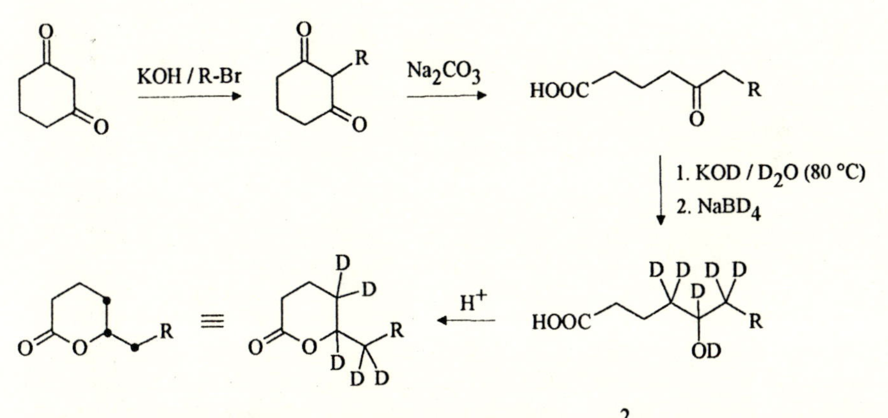

Figure 5. General synthetic route for the preparation of [$^{2}H_5$]-δ-lactones (D stands for deuterium; • indicates the labeling position).

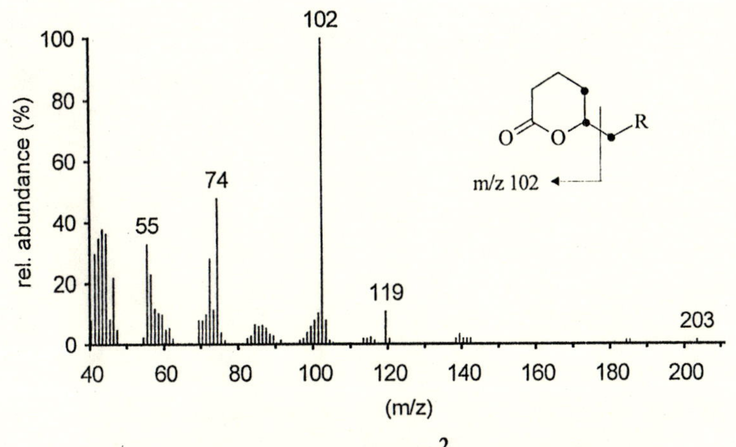

Figure 6. Mass spectrum (EI) of [$^{2}H_5$]-δ-dodecalactone.

Conclusion

Isotope dilution assay is the most accurate method currently available for quantification of aroma compounds, particularly if they are labile and occur in low concentrations. IDA does not require quantitative isolation of odorants from the matrix provided that the internal standard is homogeneously distributed. After equilibrium is reached between the labeled standard and the corresponding analyte, the concentration determined by GC-MS isotope ratio measurements reflects the absolute amount of analyte in the sample. Its high selectivity and sensitivity, both also dependent on the MS ionisation technique, make IDA a powerful tool for evaluating the contribution of sensorially relevant compounds to complex aromas.

Acknowledgement

We are grateful to Mrs J. Lin for synthesizing the unlabeled and deuterated *trans*-4,5-epoxy-(*E*)-2-decenal (**VII**), Dr. L. Fay for performing the GC-MS measurements of **VII** and **d-VII** in the PCI and NCI mode, and Dr. E. Prior for linguistic proofreading.

References

1. Sweeley, C. C.; Elliot, W. H.; Fries, I.; Rhyage, R. *Anal. Chem.* **1966**, *38*, 1549.
2. Pickup, J. F.; McPherson, K. *Anal. Chem.* **1976**, *48*, 1885.
3. Björkhem, I.; Blomstrand, R.; Lantto, O.; Svensson, L.; Ohman, G. *Clin. Chem.* **1976**, *22*, 1789.
4. Schieberle, P.; Grosch W. *J. Agric. Food Chem.* **1987**, *35*, 252.
5. Schieberle, P. In *Characterization of Food: Emerging Methods*, Gaonkar A. G., Ed; Elsevier Science, Amsterdam, Netherlands, 1995, pp. 403-431.
6. Dollmann, B.; Wichmann, D.; Schmitt, A.; Koehler, H.; Schreier, P. *J. AOAC Int.* **1996**, *79*, 583.
7. Aubry, V.; Giniès, C.; Henry, R.; Etiévant, P. In *Flavour Science – Recent Developments*, Taylor, A. J.; Mottram, D. S., Eds.; Spec. Publ. - Roy. Soc. Chem., Cambridge, UK, 1996, pp. 331-334.
8. Milo, C.; Reineccius, G. *J. Agric. Food Chem.* **1997**, *45*, 3590.
9. Blank, I.; Schieberle, P.; Grosch, W. In *Progress in Flavour and Precursor Studies*, Schreier, P.; Winterhalter, P., Eds; Allur. Publ., Wheaton, USA, 1993, pp. 103-109.
10. Blank, I.; Lin, J.; Fumeaux, R.; Welti, D. H.; Fay, L. B. *J. Agric. Food Chem.* **1996**, *44*, 1851.
11. Blank, I.; Fay, L. B.; Lakner, F. J.; Schlosser, M. *J. Agric. Food Chem.* **1997**, *45*, 2642.
12. Kerler, J.; Grosch, W. *J. Food Sci.* **1996**, *61*, 1271.
13. Guth, H.; Grosch, W. *Lebensm. Wiss. Technol.* **1990**, *23*, 513.
14. Sen A.; Schieberle, P; Grosch, W. *Lebensm. Wiss. Technol.* **1991**, *24*, 364.
15. Preininger, M.; Grosch, W. *Lebensm. Wiss. Technol.* **1994**, *27*, 237.
16. Schieberle, P.; Hofmann, P. *J. Agric. Food Chem.* **1997**, *45*, 227.

17. Milo, C.; Grosch, W. *J. Agric. Food Chem.* **1993**, *41*, 2076.
18. Semmelroch, P.; Laskawy, G.; Blank, I.; Grosch, W. *Flavour Fragrance J.* **1995**, *10*, 1.
19. Schieberle, P.; Hofmann, T. *Lebensmittelchemie* **1996**, *50*, 105.
20. Holscher, W.; Vitzthum, O. G.; Steinhart, H. *Café Cacao Thé* **1990**, *34*, 205.
21. Blank, I.; Sen, A.; Grosch, W. *Z. Lebensm. Unters. Forsch.* **1992**, *195*, 239.
22. Holscher, W.; Vitzthum, O. G.; Steinhart, H. *J. Agric. Food Chem.* **1992**, *40*, 655.
23. Semmelroch, P.; Grosch, W. *J. Agric. Food Chem.* **1996**, *44*, 537.
24. Masanetz, C.; Blank, I.; Grosch, W. *Flavour Fragrance J.* **1995**, *10*, 9.
25. Full, G.; Schreier, P. *Lebensmittelchemie* **1994**, *48*, 1.
26. Schieberle, P.; Gassenmeier, K.; Guth, H.; Sen, A.; Grosch, W. *Lebensm. Wiss. Technol.* **1993**, *26*, 347.

Chapter 23

Carbon Isotope Composition of Selected Flavoring Compounds for the Determination of Natural Origin by Gas Chromatography/Isotope Ratio Mass Spectrometer

R. A. Culp, J. M. Legato, and E. Otero

Center for Applied Isotope Studies, University of Georgia, Athens, GA 30602–4702

Capillary gas chromatography coupled to an on-line Isotope Ratio Mass Spectrometer (GC/IRMS) is used to determine individual compound $\delta^{13}C$ values for selected flavoring mixtures. These mixtures include citrus oils from geographical regions including China, Argentina, Italy, and the United States; peppermint oils from Oregon; and vanilla extracted from beans grown in geographical regions including Madagascar, Java, Indonesia, and Mexico. The resultant data can be used to generate a fingerprint pattern of the major and ancillary compounds found in these specific flavors. Comparison of these patterns to those of synthetically derived flavors or compounds can be applied to establish the authenticity or natural origin of these flavors.

The analytical techniques gas chromatography, thin layer chromatography and atomic absorption spectroscopy have provided researchers the tools to determine authenticity of the component flavors in various foods and beverages (*1*). More recently, by examining the nuclear structure of the flavor molecules using such techniques as isotope ratio mass spectrometry (*2,3*), liquid scintillation counting (*4-6*) and nuclear magnetic resonance spectroscopy (*7*) more complex flavor adulteration has been revealed. Each of these techniques has advantages and limitations for authenticity testing. A technique may have high precision and use extremely small samples but lack specificity. When techniques are used in combination, they can be very effective for verifying the true naturalness of flavoring material.

The radionuclides ^{14}C and ^{3}H can yield undeniable evidence of a flavor compound's synthetic origin when their activities fall below the normal range for botanically derived materials (*8-10*). The stable isotopes of carbon and hydrogen have been used to substantiate and characterize specific biochemical processes and sources by which botanical products can be defined (*11-16*).

The demand for these techniques has grown from the need to reveal and expose the sophisticated manipulation of synthetic sources to mimic those of natural ones. The economic incentive is present for such manipulation by virtue of the extreme discrepancy between the production costs of natural materials versus their synthetic counterparts and the fact that a large consumer demand for the natural materials exists.

Two techniques have recently moved to the forefront by virtue of their ability to detect flavor adulteration. One is compound specific isotope analysis, sometimes abbreviated CSIA or GC/IRMS. This technique combines gas chromatography and isotope ratio mass spectrometry (*17-21*). The other technique uses nuclear magnetic resonance spectroscopy (*22,23*). Both of these techniques can reveal the isotopic abundances of individual compounds or their components in a complex mixture or on certain functional groups on purified compounds. This study seeks to investigate and define the accuracy of the GC/IRMS or compound specific isotope analysis technique for three selected flavoring materials: citrus oils, peppermint oils, and vanilla extracts from natural origins. The vanilla extracts were processed as underivatized and derivatized components to assess and compare the derivatization technique. Derivatization has been shown to be favorable for chromatographic separations in GC/IRMS for compounds such as amino acids, porphyrins, and alkaloids (*24-26*). The stable carbon isotope ratios and concentrations of the primary constituents from each flavor material were measured and compared to the total isotopic ratio from each. This information helps define the accuracy of the technique and develop a range of isotopic values that are characteristic of the flavor materials true origin or process of manufacture.

Previous studies have relied heavily on the isotopic abundance of the bulk material where isotopic abundances of both the radiometric and stable isotopes have been defined. These have been determined on specific isolated compounds, prepared independently, or on the total mixture of compounds making up the flavor or oil. Though the present study looks only at the stable isotopes of carbon, ^{13}C and ^{12}C, the potential for the other stable isotopes including deuterium, hydrogen, ^{15}N, and ^{14}N to be measured by similar techniques is evident. By developing a fingerprint or multi-component stable isotope picture based on the primary constituents of a flavor, we hope to better define the characteristics of natural and synthetic flavors.

Methodology

Preparation of Samples. The samples were supplied by the Flavor and Extract Manufacturer's Association of the United States Isotopic Studies Committee and reported as being natural oils and extracts from their respective botanical source. The citrus and peppermint oils were received as neat solutions and were diluted to 2% with 100% residue-analyzed ethyl alcohol (Baker Chemical Co.). This allowed for better chromatography and analyte separation. The vanilla extracts were received as one-fold solutions so no further dilution was necessary. They were, however, extracted with diethyl ether (Baker Chemical Co.) (WARNING: use in a hood away from open flames!) 10 ml of vanilla extract were extracted three times with 10 ml of ether. The combined extracts were dried over anhydrous sodium sulfate and concentrated under a stream of nitrogen to a 5 ml volume. A one milliliter aliquot was removed for analysis.

Validation of Naturalness. The peppermint and citrus oil samples had their ^{14}C activities measured to confirm their modern ^{14}C activity. These samples were measured using a Packard 1150 liquid scintillation counter (Packard Instrument Co.) (*27*). Calibration was performed using NIST reference oxalic acid standard. The measured activity levels ranged from 15.01 to 15.39 disintegrations per minute per gram carbon or 0.25 to 0.26 Bequerels which is equivalent to approximately 98% of the natural level of ^{14}C activity photosynthesized by plants today.

Quantitative Analysis. To determine the components in complex flavorants and ascertain which are best for component specific isotope analysis the flavor material was first analyzed by gas chromatography/mass spectrometry (GC/MS). The resulting mass spectra yielded not only qualitative analysis but quantitative data based on known standards of identity. The percent concentration of each of the components of interest for each of the flavor materials studied are tabulated and listed in Tables I-III. The chromatographic conditions were manipulated so that the optimum separation of the components of interest was attained. Lastly, depending on the components found versus those known to exist in a particular flavor material, a determination was made as to whether derivatizing it to enhance chromatographic separation was necessary. A Hewlett Packard 6890 gas chromatograph with a 5973 mass selective detector was used to identify and quantify the components used in this study. A NIST 75K spectral library (Hewlett Packard Corp.) was used for spectral identification. A 30 meter, 0.32 millimeter o.d. DB5MS capillary column was used for separation of the components. The temperature program for the citrus and peppermint oils was 60°C for 5 minutes then ramped to 180°C at 3°C per minute, held for 10 minutes, then ramped to 220°C at 5°C per minute. For the vanilla extract the temperature program was 100°C for 5 minutes then ramped to 180°C at 3°C per minute, held for 10 minutes, then ramped to 220°C at 5°C per minute. Citrus and peppermint oils were run as the 2% ethanolic solutions prepared as described above. Vanillin extracts in ether were taken to dryness and rehydrated with 0.5 ml HPLC grade acetonitrile (Baker Chemical Co.). For each analysis, a 0.2 microliter injection was made with a split ratio of 100:1, allowing only 1% of the injected volume to enter the chromatographic column. The analyses were run at constant flow to closely resemble conditions expected in the gas chromatograph/isotope ratio mass spectrometer.

Total Stable Carbon Isotope Ratio Analysis. The total $\delta^{13}C$ value was determined for each of the twelve citrus oil samples, sixteen peppermint oil samples and fifteen vanilla extracts. For each of the citrus and peppermint oils, one microliter of neat sample was loaded into pyrex ampules. For the vanilla extracts a small aliquot of the extract previously extracted with ether was taken to dryness under a stream of nitrogen. This was extracted with pentane (WARNING: use in a hood away from open flames!) and allowed to form vanillin crystals on a warm watch glass. An amount approximating one milligram was scraped off the watch glass and placed into an ampule. To each of the ampules was added 0.5 grams pre-burned (800°C) copper oxide wire. The ampules were connected to a diffusion pumped vacuum line and evacuated to less than 10 millitorr for 10 minutes. Those samples exhibiting some volatility at low pressure were frozen with

Table I. Citrus oil GC/MS component retention times and concentrations (%)

Retention Time (min.)	alpha-thujene	alpha-pinene	beta-thujene	beta-pinene	beta-myrcene	alpha-phellandrene	alpha-terpinene	para-cymene	limonene	gamma-terpinene	terpinolene	linalool	caryophyllene	total
	5.80	6.03	7.51	7.62	8.25	8.75	9.28	9.64	9.88	11.18	12.51	13.10	26.80	
Source														
Florida Tangerine-5825	0.10	0.88	0.13	0.20	1.48	0.12	0.05	0.27	93.21	2.98	0.11	0.33		99.86
Chinese Tangerine-5826	0.20	1.10	0.12	0.35	1.43	0.11	0.10	0.43	89.74	4.97	0.21	1.03		99.78
Argentinian Tangerine-5827	0.16	0.93	0.18	0.35	1.29	0.10	0.04	0.52	91.33	4.57	0.17	0.24		99.89
Clementine Mandarin-5828		0.59	0.90	0.16	1.32	0.12	0.08	0.11	96.00	0.37		0.21		99.84
Italian Mandarin-5829	0.91	2.47	0.22	1.51	1.19	0.13	0.52	1.02	67.35	22.86	0.96	0.05	0.67	99.84
Tangerine-5830	0.15	0.95	0.19	0.32	1.41	0.11	0.07	0.42	91.53	4.01	0.16	0.51		99.83
California Tangelo-5831	0.10	0.88	0.19	0.21	1.56	0.06	0.05	0.32	93.52	2.68	0.10	0.19		99.85
Mandarin-6502	1.98	5.56	0.34	2.80	1.61	0.12	0.33	1.52	41.86	37.35	2.58	0.21	2.17	98.43
Mandarin-6503	0.89	2.48	0.24	1.66	1.43	0.12	0.50	2.11	65.16	22.99	1.01	0.06	0.90	99.56
Tangerine-6504	0.23	2.13	1.69	0.36	1.78	0.13		0.92	84.95	3.74	0.26	2.60		98.77
Tangerine-6505	0.11	0.84	0.40	0.21	1.49	0.09	0.03	0.76	92.64	2.46	0.09	0.69		99.79

Blanks represent below detectable level

Table II. Peppermint oil GC/MS component retention times and concentrations (%)

Retention Time (min.)	alpha-pinene	beta-thujene	beta-pinene	cymene	limonene	cineol	menthone	menthol	pulegone	menthol acetate	caryophyllene	germacrene	total
	6.02	7.49	7.60	9.62	9.80	9.90	15.56	16.55	19.47	22.05	27.29	29.83	
Source													
Yakima-5908	0.45	0.26	0.59	0.37	1.46	5.04	18.13	46.96	1.59	5.76	1.92	0.96	83.49
Yakima-5909						0.18	17.86	53.00	1.91	6.62	2.28	1.39	83.24
Yakima-5910	0.71	0.33	0.76	0.18	1.48	5.73	18.37	43.83	1.26	5.61	1.85	1.54	81.65
Yakima-5911	0.50	0.25	0.61	0.29	1.46	4.65	20.87	47.02	1.11	4.46	2.09	0.47	83.78
Oregon-5912							22.39	53.01	1.61	5.26	2.11	0.60	84.98
Oregon-5913				0.07	0.13	0.66	20.28	53.40	1.03	6.28	2.28	1.01	85.14
Oregon-5914	0.61	0.32	0.68	0.22	1.45	4.43	20.78	44.20	1.50	4.55	2.17	2.11	83.02
Yakima-6167	0.41	0.25	0.58	0.25	1.36	5.12	17.77	47.04	1.45	4.95	2.07	1.06	82.31
Yakima-6168							20.28	50.15	1.46	6.42	2.10	1.44	81.85
Yakima-6169	0.69	0.34	0.73	0.16	1.53	5.57	18.35	44.31	1.28	5.65	1.85	1.58	82.04
Yakima-6170	0.50	0.26	0.62	0.33	1.47	4.65	21.66	47.19	1.12	4.54	1.70	1.20	85.24
Oregon-6171							22.54	52.89	1.28	5.11	2.02	1.56	85.40
Oregon-6172						0.16	20.93	55.09	1.31	5.35	2.03	1.24	86.11
Oregon-6173	0.71	0.33	0.79	0.17	1.48	4.84	22.52	43.69	1.03	4.92	1.95	1.94	84.37

Blanks represent below detectable level

Table III. Vanilla extract GC/MS component retention times and concentrations (%)

	p-OH benzaldehyde	ortho-vanillin	vanillin	syringaldehyde	p-OH benzoic acid	vanillic acid	total
Retention Time (min.)	12.1	12.4	13.7	16.2	17.6	19.9	
Source							
Madagascar-6227	4.08	0.56	93.60	0.10	0.12	1.03	98.46
Madagascar-6228	5.24	0.43	91.40		0.14	1.49	97.21
Indonesia-6229	3.57	2.17	91.85	0.33	0.17	1.33	98.09
Indonesia-6230	4.35	1.11	92.97	0.15		0.50	98.58
Comore-6231	5.17	1.85	92.16	0.12	0.08	0.54	99.38
Comore-6232	4.18	0.81	88.85	0.37	0.23	1.71	94.43
Mexico-6233	4.78	1.18	90.41	0.25	0.20	2.15	96.82
Mexico-6234	3.95	0.55	92.65	0.07	0.13	1.28	97.35
Tonga-5269	3.84	0.61	91.45	0.25	0.05	0.22	96.19
Madagascar-5270	4.95	0.24	88.33		0.05	0.64	93.58
Comore-5271	4.15	1.60	88.17	0.23	0.11	0.90	94.25
Java-5272	3.29	3.56	77.35	4.00	0.12	1.07	88.31
Bali-5273	4.55	10.58	51.67	1.11	0.40	1.75	68.31
Bourdon (Bali)-5274	4.38	3.05	84.01	2.43	0.35	2.91	94.21
Madagascar-5275	4.42	4.48	82.66	0.64	0.11	0.67	92.30

Blanks represent below detectable level

liquid nitrogen (WARNING: Use gloves and eyeglasses and do not contact skin!) just prior to evacuating. The ampules were flame sealed and their contents combusted at 575°C for 8 hours (*28,29*). Duplicates of each sample were prepared in this manner. The resulting carbon dioxide was separated and isolated from other gases cryogenically and sealed in sample bottles to await analysis. The bottles were sequentially analyzed on a Finnigan MAT 251 isotope ratio mass spectrometer. The resulting average carbon isotope ratios for each of the flavor samples are listed under the column marked measured in Table IV.

Compound Specific Isotopic Analysis. GC/IRMS analysis was performed using a Finnigan MAT 252 isotope ratio mass spectrometer coupled to a Varian 3400 gas chromatograph. The two instruments were coupled using a unique interface which allowed for the direct combustion of the analyte peaks as they exited the capillary column. Combustion to carbon dioxide was performed by passing the effluent from the capillary column through a tubular furnace 200 millimeters long by 0.5 millimeters i.d. containing a single copper oxide and platinum wire heated to 850°C The resulting carbon dioxide entered the mass spectrometer where the ion currents for the respective masses 46, 45, and 44 amu were determined. These corresponded to the mass of carbon dioxide as $^{12}C^{18}O^{16}O$, $^{13}C^{16}O^{16}O$, and $^{12}C^{16}O^{16}O$ respectively. From this information the $\delta^{13}C$ value was computed by comparison to a standard CO_2 reference gas, admitted into the mass spectrometer at predefined intervals, whose isotopic ratio is well defined. The standard CO_2 reference gas is maintained at a pressure below 750 psi to prevent formation of liquid CO_2 which may effect its isotopic value (*30*). The Varian gas chromatograph used the same column and conditions as previously described for the HP6890/5973 GC/MS. Citrus and peppermint oils were analyzed as 2% ethanolic solutions seven times each. Injection and split ratios varied depending on whether the major and minor constituents were being analyzed. By adjusting for higher injection volumes and lower split ratios, more sample could be placed onto the chromatographic column with subsequent greater precision with the isotope ratio mass spectrometer. Likewise for major component analysis less sensitivity was required and injection volumes were decreased and split ratios increased. The resultant average stable carbon isotopic ratios for the components of interest for the three flavor materials; citrus oils, peppermint oils, and vanilla extracts, underivatized and derivatized, are listed in Tables V, VI, and VII respectively. The one-sigma standard deviations for the same are listed in Tables VIII, IX, and X respectively.

Derivatization of Vanillin Samples. To enhance separation of components during gas chromatography, the vanilla extracts were exposed to the derivatizing reagent N,O-Bis(trimethylsilyl)trifluoroacetamide (BSTFA) containing 1% Trimethyl chlorosilane (TMCS) (Alltech Corp., Deerfield, IL part no. 18089). This is the derivatizing reagent of choice for active hydrogens, carboxylic acids, and phenolic compounds. Derivatization was completed in 15 minutes at 100°C by combining 50 microliters of the vanilla extract, previously extracted with ether and dried over anhydrous sodium sulfate then taken up in HPLC grade acetonitrile (Baker Chemical Co.), with 50 microliters of BSTFA (1%TMCS) reagent (*31*).

Table IV. Citrus oil, peppermint oil, and vanilla extract comparison between measured 13C/12C values (o/oo PDB) via off-line preparation and calculated 13C/12C values (o/oo PDB) based on recombined GC/IRMS component 13C/12C values (o/oo PDB

Citrus oils				Peppermint oils				Vanilla extracts			
Source	Measured	Calculated	Difference	Source	Measured	Calculated	Difference	Source	Measured	Calculated	Difference
Florida Tangerine-5825	-29.14	-29.32	-0.18	Yakima-5908	-27.12	-26.96	0.16	Madagascar-6227	-20.31	-20.46	-0.15
Chinese Tangerine-5826	-29.74	-29.55	0.19	Yakima-5909	-27.29	-27.08	0.21	Madagascar-6228	-20.11	-19.74	0.37
Argentinian Tangerine-582	-29.37	-29.48	-0.11	Yakima-5910	-27.50	-27.06	0.44	Indonesia-6229	-19.74	-19.82	-0.08
Clementine Mandarin-5828	-27.15	-27.08	0.07	Yakima-5911	-27.29	-27.30	-0.01	Indonesia-6230	-20.20	-20.20	0.00
Italian Mandarin-5829	-29.11	-28.85	0.26	Oregon-5912	-27.23	-27.04	0.19	Comore-6231	-18.62	-19.18	-0.56
Tangerine-5830	-29.13	-29.26	-0.13	Oregon-5913	-27.12	-26.80	0.32	Comore-6232	-18.24	-19.08	-0.84
California Tangelo-5831	-27.56	-28.26	-0.70	Oregon-5914	-27.20	-27.06	0.14	Mexico-6233	-18.76	-19.77	-1.01
Mandarin-6502	-29.04	-29.02	0.02	Yakima-6167	-27.12	-27.03	0.09	Mexico-6234	-19.02	-20.11	-1.09
Mandarin-6503	-28.35	-28.72	-0.37	Yakima-6168	-27.17	-27.10	0.07	Tonga-5269	-20.90	-19.73	1.17
Tangerine-6504	-30.20	-30.21	-0.01	Yakima-6169	-27.11	-27.05	0.06	Madagascar-5270	-21.52	-20.84	0.68
Tangerine-6505	-29.32	-29.17	0.15	Yakima-6170	-27.41	-27.33	0.08	Comore-5271	-21.33	-20.51	0.82
				Oregon-6171	-27.40	-27.10	0.30	Java-5272	-19.90	-20.22	-0.32
				Oregon-6172	-27.48	-27.41	0.07	Bali-5273	-19.51	-19.91	-0.40
				Oregon-6173	-27.27	-27.07	0.20	Bourdon (Bali)-52	-21.24	-19.94	1.30
								Madagascar-5275	-21.04	-19.94	1.10

Table V. Citrus oil GC/IRMS component retention times and average 13C/12C values (o/oo PDB)

	alpha-thujene	alpha-pinene	beta-thujene	beta-pinene	beta-myrcene	alpha-phellandrene	alpha-terpinene	para-cymene	limonene	gamma-terpinene	terpinolene	linalool	caryophyllene
Retention Time (sec.)	910	942	1076	1100	1120	1170	1228	1256	1285	1386	1480	1515	2586
Source													
Florida Tangerine-5825	-27.17	-31.06	-28.63	-30.09	-26.99	-30.83	-29.89	-28.72	-29.34	-29.45	-28.64	-29.82	-30.28
Chinese Tangerine-5826	-28.05	-31.76	-29.31	-30.86	-27.27	-32.14	-30.81	-30.34	-29.50	-30.70	-29.25	-28.99	-30.37
Argentinian Tangerine-5827	-28.03	-31.44	-29.29	-29.82	-26.93	-30.78	-30.88	-29.68	-29.45	-30.27	-28.71	-30.54	-31.94
Clementine Mandarin-5828		-29.02	-26.22	-25.85	-24.24	-28.14		-25.27	-27.13	-29.60	-23.18	-25.59	-32.80
Italian Mandarin-5829	-27.17	-30.12	-27.46	-29.64	-26.73	-31.76	-29.84	-29.24	-28.64	-29.39	-28.56	-30.61	-29.43
Tangerine-5830	-27.64	-31.72	-28.45	-29.70	-26.77	-30.39	-30.50	-29.67	-29.26	-29.58	-27.87	-28.70	-30.93
California Tangelo-5831	-25.33	-29.96	-27.57	-28.26	-25.51	-31.10	-28.91	-27.24	-28.26	-29.45	-26.97	-27.51	-29.00
Mandarin-6502	-26.33	-29.93	-25.89	-28.30	-25.58	-30.28	-27.71	-30.60	-28.65	-29.67	-28.56	-29.85	-28.50
Mandarin-6503	-26.75	-30.12	-26.83	-29.05	-26.21	-29.10	-28.62	-29.28	-28.72	-28.72	-28.91	-30.83	-28.47
Tangerine-6504	-28.12	-31.30	-29.25	-30.20	-26.04	-29.77		-31.43	-30.29	-30.73	-28.89	-29.35	-30.65
Tangerine-6505	-27.84	-31.84	-28.95	-30.93	-26.56	-29.86	-29.03	-29.23	-29.18	-29.68	-27.14	-28.40	-32.29

Blanks represent below detectable level

Table VI. Peppermint oil GC/IRMS component retention times and average 13C/12C values (o/oo PDB)

	alpha-pinene	beta-thujene	beta-pinene	cymene	limonene	cineol	menthone	menthol	pulegone	menthol acetate	caryophyllene	germacrene
Retention Time (sec.)	942	1076	1100	1256	1285	1300	1717	1785	2000	2203	2585	2638
Source												
Yakima-5908	-29.69	-25.59	-27.95	-26.81	-25.87	-29.01	-25.94	-27.07	-25.37	-28.61	-24.86	-27.57
Yakima-5909						-31.47	-26.52	-27.18	-25.62	-28.84	-24.67	-27.82
Yakima-5910	-28.97	-26.43	-28.44	-29.99	-26.60	-29.06	-26.03	-27.10	-26.01	-28.82	-24.81	-26.36
Yakima-5911	-29.69	-27.10	-28.76	-30.33	-27.49	-29.51	-26.39	-27.27	-25.81	-29.94	-25.35	-28.52
Oregon-5912							-26.47	-27.17	-25.27	-29.74	-24.54	-26.53
Oregon-5913					-27.40	-30.18	-26.13	-26.92	-25.36	-29.18	-24.05	-26.29
Oregon-5914	-27.82	-25.37	-29.73	-26.78	-26.61	-29.25	-26.19	-27.15	-25.70	-29.83	-24.70	-26.06
Yakima-6167	-29.56	-27.81	-28.95	-27.62	-26.06	-29.14	-26.10	-27.08	-25.04	-29.01	-25.48	-26.08
Yakima-6168						-33.83	-26.32	-27.32	-25.98	-28.83	-24.99	-26.91
Yakima-6169	-29.39	-26.72	-28.83	-29.98	-25.83	-28.96	-26.09	-27.11	-26.01	-28.95	-24.95	-25.56
Yakima-6170	-28.59	-26.30	-30.56	-31.01	-26.32	-29.29	-26.63	-27.29	-25.97	-29.81	-25.06	-27.55
Oregon-6171							-26.53	-27.20	-26.22	-29.90	-24.83	-26.55
Oregon-6172						-31.56	-26.96	-27.39	-25.62	-30.19	-25.06	-29.19
Oregon-6173	-28.61	-26.88	-29.03	-28.62	-25.56	-29.31	-26.26	-27.17	-25.51	-29.77	-24.74	-24.73

Blanks represent below detectable level

Table VII. Vanilla extract GC/IRMS underivatized and derivatized component retention times and average 13C/12C values (o/oo PDB)

	underivatized compounds						derivatized compounds					
	p-OH benzaldehyde	ortho-vanillin	vanillin	syringaldehyde	p-OH benzoic acid	vanillic acid	p-OH benzaldehyde	ortho-vanillin	vanillin	syringaldehyde	p-OH benzoic acid	vanillic acid
Retention Time (sec.)	1560	1570	1660	1792	1990	2150	1566	1697	2014	2190	2215	2688
Source												
Madagascar-6227	-18.65	-20.67	-20.49	-23.20	-20.31	-24.73	-27.14	-29.02	-27.57	-35.68	-32.51	-35.11
Madagascar-6228	-17.26	-20.94	-19.83	-24.02	-19.80	-24.21	-27.00	-30.15	-27.92	-34.21	-33.49	-34.07
Indonesia-6229	-18.04	-22.30	-19.76	-25.61	-19.14	-23.56	-27.05	-30.43	-27.98	-38.01	-34.78	-34.34
Indonesia-6230	-17.82	-25.22	-20.22	-26.38	-20.21	-23.13	-27.72	-30.73	-28.56	-38.80	-34.03	-34.26
Comore-6231	-19.43		-19.53	-25.06	-20.14	-21.90	-26.61	-30.30	-27.40	-37.66	-34.83	-35.87
Comore-6232	-16.67	-21.38	-19.10	-23.16	-18.26	-22.01	-26.06	-29.87	-27.53		-29.71	-33.66
Mexico-6233	-17.49	-23.45	-19.75	-25.03	-19.72	-24.87	-26.68	-30.13	-27.74	-38.40	-33.85	-34.05
Mexico-6234	-16.97	-21.30	-20.21	-26.38	-20.27	-22.09	-26.37	-29.00	-27.87		-28.55	-36.08
Tonga-5269	-19.97	-20.72	-19.67	-31.80	-24.87	-26.80	-28.47	-27.60	-27.06	-32.11	-37.07	-30.78
Madagascar-5270	-20.87	-22.31	-20.83		-25.68	-20.49	-27.61	-25.60	-27.57	-30.73	-37.73	-33.63
Comore-5271	-20.41	-20.29	-20.40	-30.68	-24.52	-28.54	-27.29	-29.18	-26.44	-29.79	-36.59	-33.46
Java-5272	-21.12	-20.66	-19.84	-23.20	-23.67	-32.06	-27.57	-29.91	-26.46	-29.85	-32.47	-33.04
Bali-5273	-21.13	-20.07	-19.35	-23.76	-20.21	-29.80	-28.71	-29.74	-26.32	-30.81	-34.00	-30.89
Bourdon (Bali)-5274	-19.38	-18.65	-19.69	-22.67	-22.00	-27.07	-28.75	-28.32	-26.73	-31.35	-33.29	-32.48
Madagascar-5275	-19.98	-20.06	-19.80	-23.96	-24.10	-30.94	-28.10	-28.86	-27.20	-25.95	-38.96	-32.23

Blanks represent below detectable level

Table VIII. Citrus oil GC/IRMS component retention times and 13C/12C values (o/oo PDB) standard deviations

	alpha-thujene	alpha-pinene	beta-thujene	beta-pinene	beta-myrcene	alpha-phellandrene	alpha-terpinene	para-cymene	limonene	gamma-terpinene	terpinolene	linalool	caryophyllene
Retention Time (sec.)	910	942	1076	1100	1120	1170	1228	1256	1285	1386	1480	1515	2586
Source													
Florida Tangerine-5825	0.76	0.82	0.85	0.27	0.41	0.35	1.74	0.14	0.21	0.92	2.61	4.87	
Chinese Tangerine-5826	1.13	0.49	0.36	0.49	0.36	1.04	0.78	0.10	0.68	0.31	0.45	0.21	
Argentinian Tangerine-582	0.74	0.35	0.68	0.45	0.20	0.87	1.00	0.08		0.48	0.60	4.80	0.04
Clementine Mandarin-5828		0.28	0.31	0.72	0.31	0.49		0.57		1.41	0.25	0.95	1.09
Italian Mandarin-5829	0.13	1.40	0.32	0.19	0.23	0.43	0.28	1.01	0.50	0.12	0.65	3.13	0.13
Tangerine-5830	0.19	0.61	0.17	0.13	0.13	0.86	1.25	0.76		2.21	0.82	0.31	0.30
California Tangelo-5831	1.12	0.38	0.57	0.18	0.18	2.01	1.09	0.00		3.08	0.98	0.07	
Mandarin-6502	0.28	0.19	0.41	0.25	0.33	0.72	0.44	0.59	0.35	0.19	0.54	3.50	1.12
Mandarin-6503	0.28	0.74	0.55	0.27	0.45	2.53	0.15	0.84	0.41	0.41	1.51	5.39	1.01
Tangerine-6504	1.31	0.36	0.14	0.32	0.55	0.84		1.10	0.09	0.65	0.35	0.12	
Tangerine-6505	1.57	0.70	0.41	0.57	1.31	1.06	1.55	1.75	0.34	1.35	0.53	0.16	

Blanks represent below detectable level

Table IX. Peppermint oil GC/IRMS component retention times and 13C/12C values (o/oo PDB) standard deviations

	alpha-pinene	beta-thujene	beta-pinene	cymene	limonene	cineol	menthone	menthol	pulegone	menthol acetate	caryophyllene	germacrene
Retention Time (sec.)	942	1076	1100	1256	1285	1300	1717	1785	2000	2203	2585	2638
Source												
Yakima-5908	2.15	1.58	1.29	0.42	0.50	0.19	0.17	0.10	0.51	0.39	0.88	4.12
Yakima-5909						0.59	0.08	0.20	0.57	0.17	0.45	3.67
Yakima-5910	0.63	0.69	1.21	5.10	1.47	0.28	0.14	0.16	1.18	0.40	0.45	3.06
Yakima-5911	2.65	0.34	0.33	1.06	0.84	0.64	0.18	0.20	1.00	0.38	0.55	2.05
Oregon-5912							0.20	0.33	0.96	0.74	0.92	3.08
Oregon-5913					2.31	1.90	0.15	0.14	0.43	0.28	0.82	3.71
Oregon-5914	3.14	1.80	2.30	1.74	1.50	0.51	0.24	0.30	0.85	0.33	0.60	2.69
Yakima-6167	1.31	0.20	0.95	0.07	0.50	0.20	0.22	0.21	1.00	0.63	1.30	2.90
Yakima-6168							0.19	0.17	0.75	0.22	0.99	2.49
Yakima-6169	1.67	0.47	1.16	3.86	0.28	0.39	0.36	0.62	1.06	0.49	1.13	1.81
Yakima-6170	2.85	1.63	2.02	0.77	0.66	0.49	0.27	0.37	0.82	0.51	0.79	3.70
Oregon-6171							0.21	0.20	1.57	0.46	0.60	3.89
Oregon-6172						2.45	0.20	0.45	0.62	0.70	0.89	4.65
Oregon-6173	1.64	0.52	0.95	2.36	0.90	0.60	0.25	0.40	0.65	0.09	1.00	1.56

Blanks represent below detectable level

Table X. Vanilla extract GC/IRMS underivatized and derivatized component retention times and 13C/12C values (o/oo PDB) standard deviations

	underivatized compounds						derivatized compounds					
	p-OH benzaldehyde	ortho-vanillin	vanillin	syringaldehyde	p-OH benzoic acid	vanillic acid	p-OH benzaldehyde	ortho-vanillin	vanillin	syringaldehyde	p-OH benzoic acid	vanillic acid
Retention Time (sec.)	1560	1570	1660	1792	1990	2150	1566	1697	2014	2190	2215	2688
Source												
Madagascar-6227	0.17	3.60	0.12	1.39	1.29	2.75	0.32	0.21	0.35	4.50	3.18	0.10
Madagascar-6228	0.86	1.77	0.12	3.94	1.56	2.65	0.54	0.62	0.56	6.84	0.74	0.74
Indonesia-6229	1.81	0.92	0.19	1.21	0.17	2.31	0.42	0.32	0.41	0.96	0.58	0.48
Indonesia-6230	1.76	1.73	0.20	1.58	1.01	1.97	0.52	0.49	0.25	2.31	0.01	0.18
Comore-6231	0.31		0.24	3.23	1.03	0.91	0.27	0.06	0.04			0.46
Comore-6232	1.62	0.54	0.39	0.37	1.12	2.15	0.22	0.25	0.03		1.24	0.27
Mexico-6233	2.16	2.47	0.13	1.18	2.82	3.28	0.47	0.16	0.08	4.17	0.20	0.45
Mexico-6234	1.36	2.55	0.12	1.39	0.55	2.10	0.49	0.64	0.10		0.40	0.24
Tonga-5269	2.47	1.95	0.62	4.20	2.46	4.31	0.44	0.57	0.15	2.05	2.21	2.24
Madagascar-5270	2.85	1.05	0.16		2.50		0.39	0.43	0.16	1.60	2.31	0.44
Comore-5271	1.25	0.68	0.41	5.13	3.48	4.29	0.20	0.16	0.37	1.22	0.53	1.85
Java-5272	1.40		0.23	0.86	3.13		0.37	0.20	0.17	0.70	0.67	1.55
Bali-5273	1.02	1.87	0.42	0.99	3.57		0.42	0.19	0.25	3.20	0.26	1.11
Bourdon (Bali)-5274	1.07	1.03	0.35	0.75	3.31	9.43	0.38	0.36	0.25	2.00	0.63	0.75
Madagascar-5275	0.79	1.09	0.28	1.27	3.74	1.61	0.32	0.38	1.22	1.10	1.48	0.77

Blanks represent below detectable level

Terminology. Isotopic abundances are typically expressed in delta notation (δ) and written relative to the heavier mass isotope. The δ value is defined as the per mil (^{o}/oo) deviation of the sample isotopic ratio versus that of the PDB standard: the Cretaceous belemnite, *Belemnitilla americana*, which represents the zero point for the $\delta^{13}C$ scale (*32*). The δ value is expressed by

$$^{o}/oo = [(^{13}C/^{12}C)_{sample}/(^{13}C/^{12}C)_{standard} - 1] \times 10^{3}$$

Results and Discussion

To achieve an accurate comparison of the total isotopic signature of a flavor oil or extract and its' chemical components, one must first be able to separate the components adequately to minimize overlap and interferences of their respective isotopic values. Figures 1-4 show typical GC traces for the natural citrus oils, peppermint oils, and vanilla extracts; underivatized and derivatized respectively. The chemical components separated and quantified, for their content and isotopic abundance, in the citrus oils were α- and β-pinene, α- and β-thujene, β-myrcene,α-phellandrene, α- and γ-terpinene, cymene, limonene, terpinolene, linalool, and caryophyllene. For the peppermint oils α- and β-pinene, β-thujene, cymene, limonene, cineol, menthone, menthol, pulegone, menthyl acetate, caryophyllene, and germacrene were measured. In the vanilla extracts p-hydroxy benzaldehyde, p-hydroxy benzoic acid, vanillin, vanillic acid, o-vanillin and syringaldehyde were determined.

After manipulation of the chromatographic conditions, good separation of the components were achieved. Quantitation was performed based on known standards. Relative percent concentrations of the components identified are listed for the three flavors materials in Tables I-III. As seen in these tables, each oil or extract contians a major component: limonene in the citrus oils, menthol in the peppermint oils, and vanillin in the vanilla extracts. These major components range in concentration between 17 and 96 percent, while minor or trace components are typically less than 6%.

Once the chromatographic conditions were optimized, they were transferred to the GC/IRMS instrument for similar component separation. Figures 5-8 portray the chromatographic separation of the components of interest using the Finnigan 252 GC/IRMS for the natural citrus oils, peppermint oils, and vanilla extracts; underivatized and derivatized respectively. The major components aforementioned presented a problem for their measurement along with the measurement of the trace constituents in the flavor. The GC/IRMS features a back-flushing, capability which prevents the large solvent peak from reaching and exhausting the combustion interface reagents. This feature can also be initiated during the run to remove a large peak. However, we found that during the onset and termination of back-flushing the baseline would fluctuate due to the flow variation and inaccurate isotopic ratios could be observed. We therefore ran our chromatograms with only solvent back-flushing but ran the methods in a normal mode and an enhanced mode of operation. The normal mode brought the major peaks into a normal range of voltage for accurate isotopic measurement, and the enhanced mode increased sensitivity 10 to 100 fold by reducing split ratios and increasing injection volumes. Even though the major peaks were unmeasurable due to saturation of the

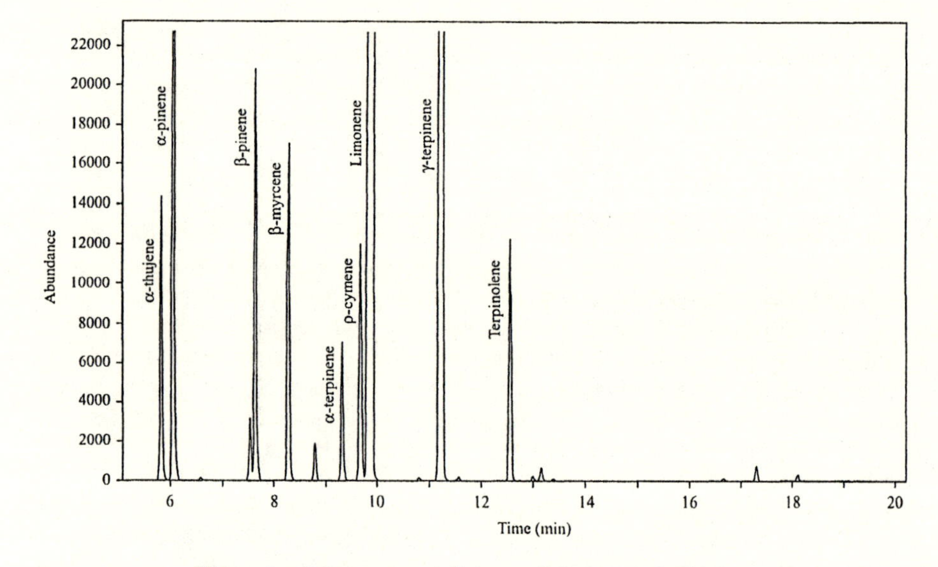

Figure 1. GC spectrum of citrus oil from HP GC/MS

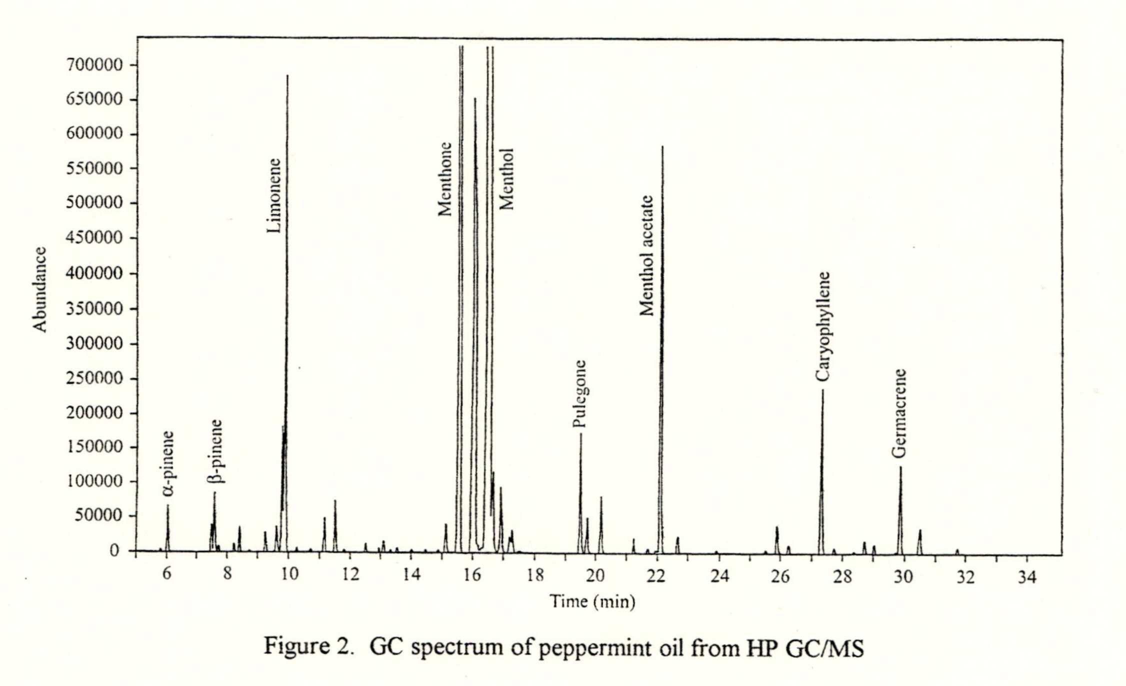

Figure 2. GC spectrum of peppermint oil from HP GC/MS

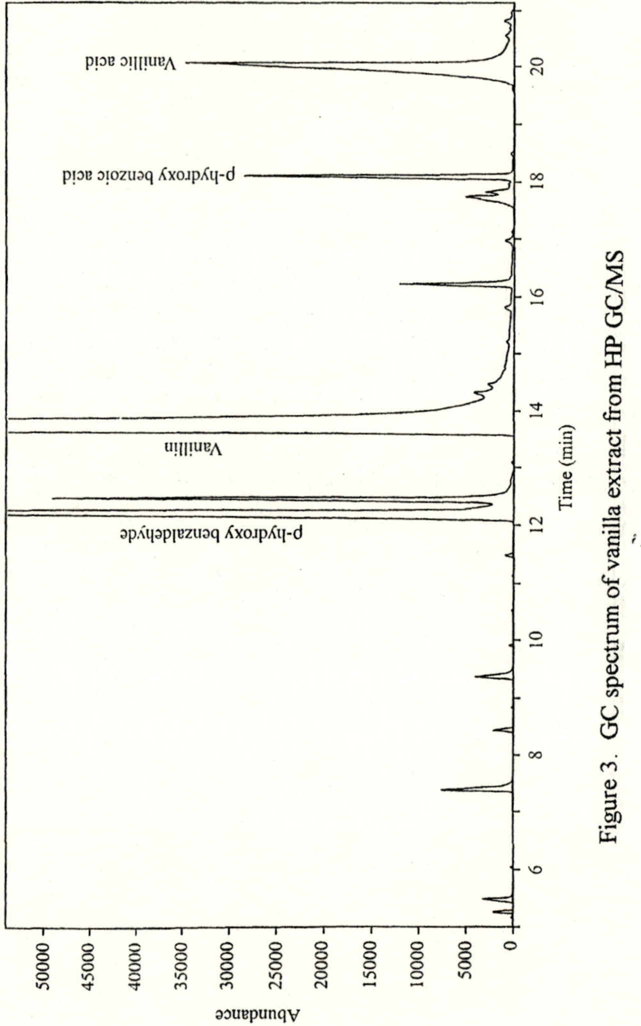

Figure 3. GC spectrum of vanilla extract from HP GC/MS

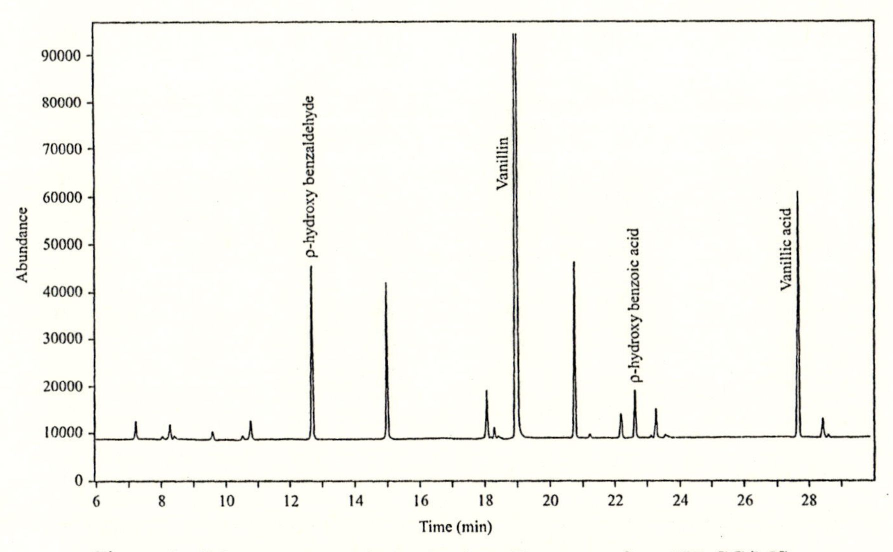

Figure 4. GC spectrum of derivatized vanilla extract from HP GC/MS

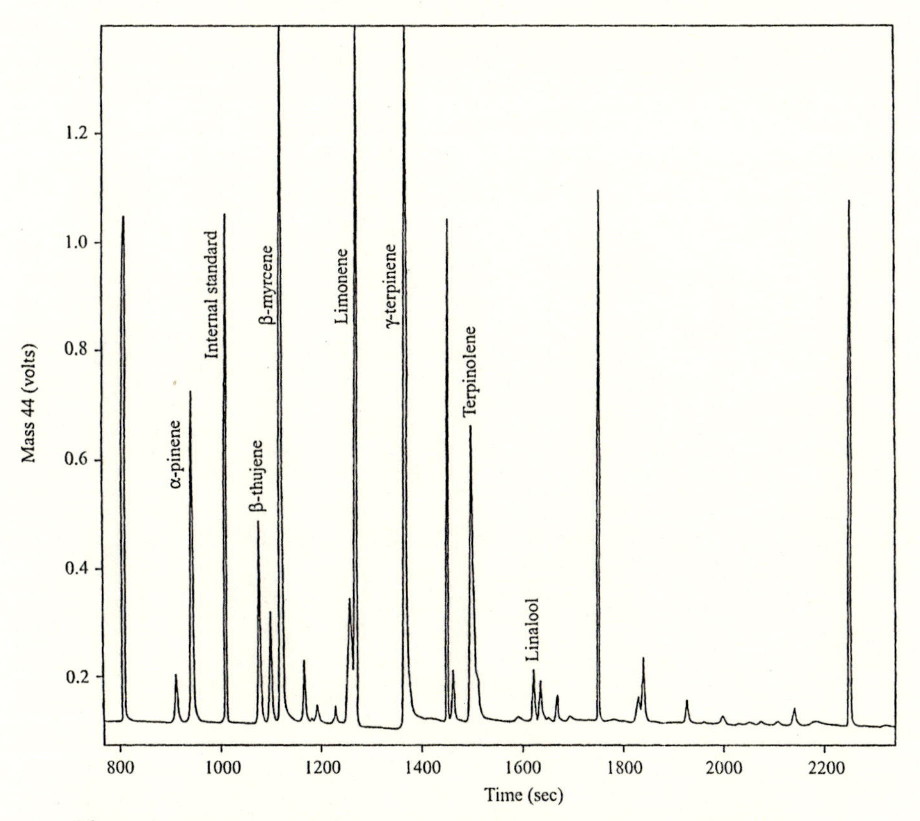

Figure 5. GC spectrum of citrus oil from Finnigan MAT 252 GC/IRMS

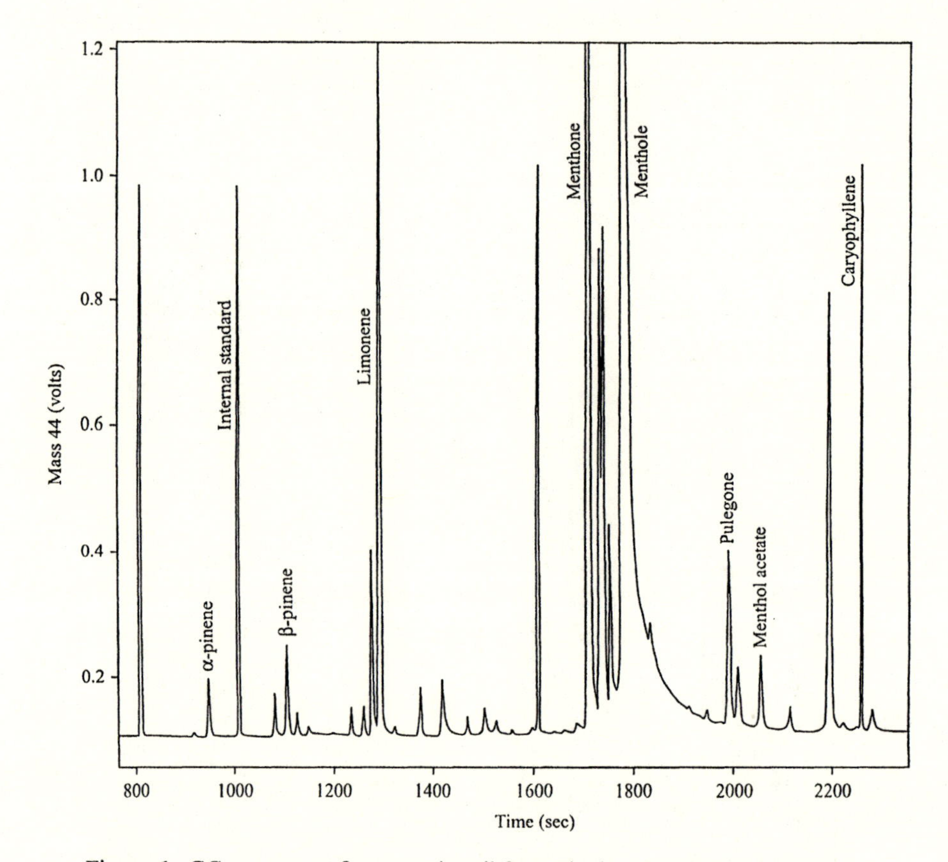

Figure 6. GC spectrum of peppermint oil from Finnigan MAT 252 GC/IRMS

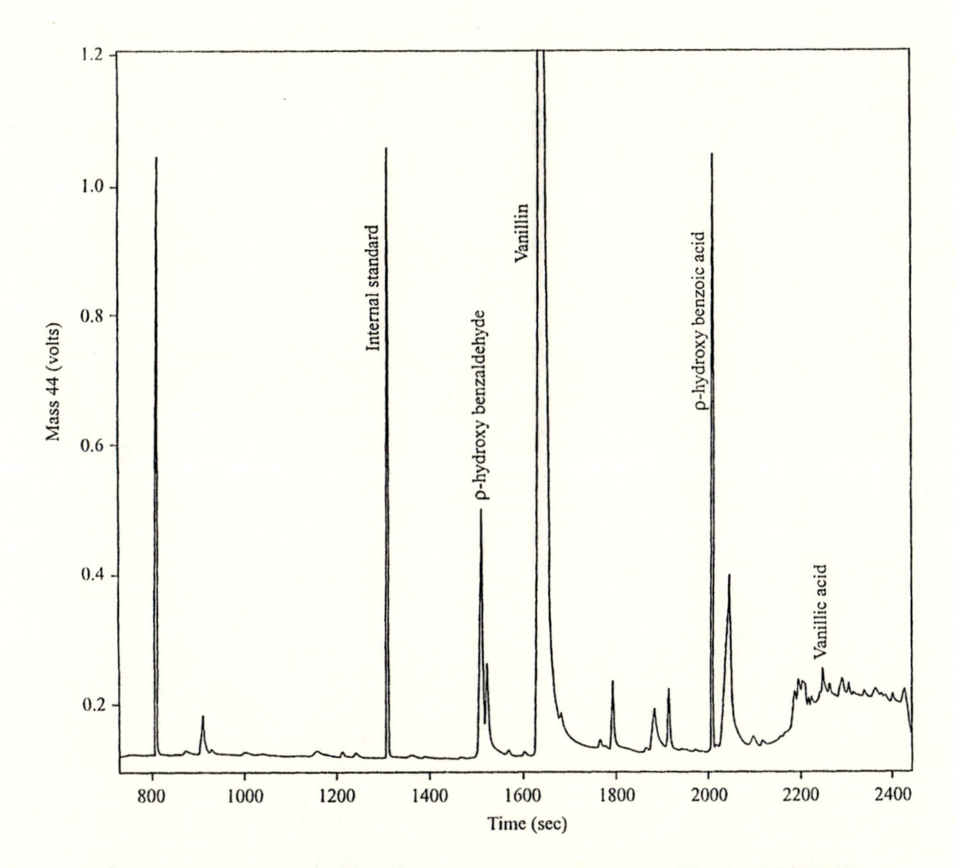

Figure 7. GC spectrum of vanila extract from Finnigan MAT 252 GC/IRMS

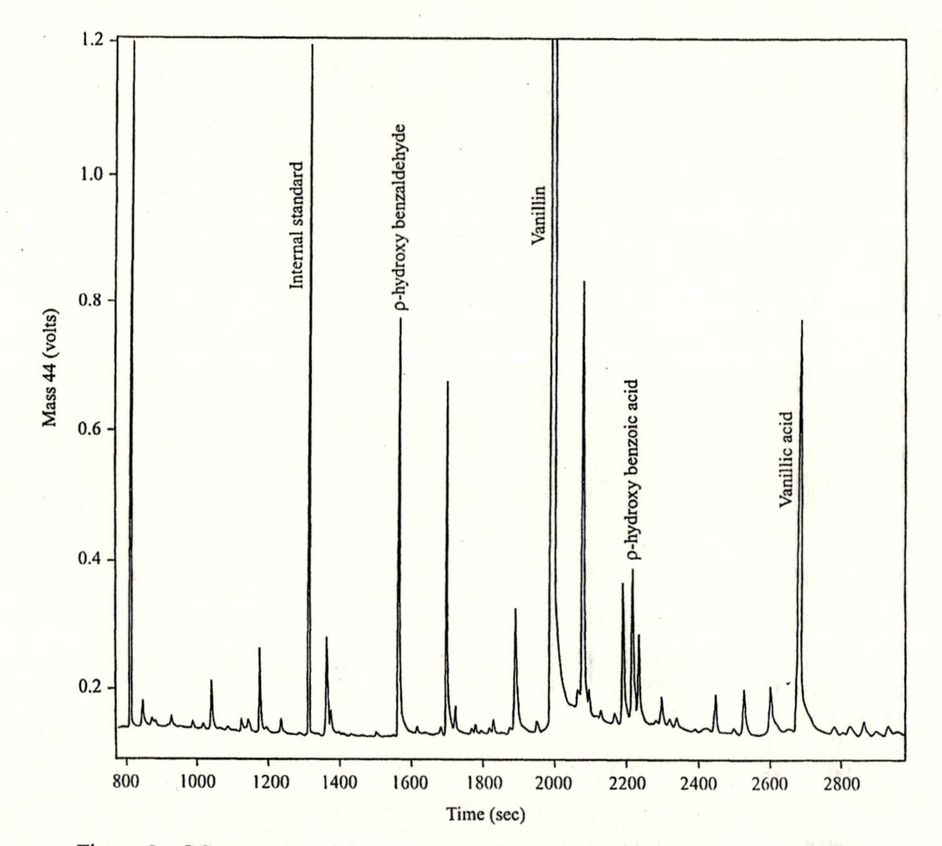

Figure 8. GC spectrum of derivatized vanilla extract from MAT 252 GC/IRMS

Faraday collectors, the minor peaks were measured with reasonable accuracy and precision. Tables V, VI, and VII list the average $\delta^{13}C$ values from a minimum of seven replicate analyses using the GC/IRMS technique. The minor and major analytes are grouped together within each flavor group even though they were measured separately. Blanks appear where the component of interest was below detectable levels.

The derivatization of the vanilla extracts was necessary because of the difficulty of chromatographing some of the known components in vanilla extract. Even though we were able to separate the components of interest without derivatization, a comparison of the methods was made. Derivatization enhanced the chromatography by yielding better resolution of the components both in peak height and separation. Acidic components such as vanillic acid and p-hydroxy benzoic acid were visibly greater in concentration when derivatized. The major drawback to derivatization is the correction for additional carbon atoms to each of the components. Based on the structural information provided by the GC/MS analysis, we were able to decipher which components had single or doubly substituted trimethylsilyl functionalities. Based on the original carbon content, the post-derivatization carbon content, and the stable isotopic value of the BSTFA derivatizing reagent a corrective algorithm to calculate the origin carbon isotopic ratio value was derived. The right hand columns of Table VII indicate the more depleted $\delta^{13}C$ values expected for the derivatized components of the vanilla extracts. The left hand columns of Table XI list the back calculated $\delta^{13}C$ values based on the derivatized samples. The right hand columns of Table XI are the differences between the actual measured component specific $\delta^{13}C$ for the vanilla extracts and those computed from the derivatized samples. As expected, there is good agreement between the underivatized and derivatized values for components of higher concentration. Vanillin and para-hydroxy benzaldehyde, the two largest components in the vanilla extracts, have differences usually less than 1 $^{o}/oo$ and only one sample exceeds 2 $^{o}/oo$. The trace components exhibit deviations greater than 5 $^{o}/oo$. Hence the applicability of this technique for trace components is questionable. The large deviations most likely are due to variation in the trimethylsilyl substitutions in the extremely low concentration components. We observed in some instances that incomplete substitution of the hydroxy groups, on some components, occurred despite a long derivatization time. These trace components may be masked or sterically hindered by the higher concentration major components within the extract.

Once the concentration of each flavor component was established and its stable carbon isotope ratio accurately measured, we used the technique of mass balance to arrive at a calculated $\delta^{13}C$ value based on the GC/MS and GC/IRMS data. This technique recombined the isotopic values, on a carbon equivalent basis, back into their origin mixture. This data is tabulated in the column heading calculated in Table IV for each of the oils or extracts and their respective sample entries. A comparison of the 251 method data, generated by off-line combustion and clean-up of the total extract, and the 252 method data which is a re-combination of the component specific isotopic data show very good agreement. The citrus oils show an average difference of approximately 0.2 $^{o}/oo$ between the two methods, and the peppermint oils exhibit an average difference of less than 0.18 $^{o}/oo$ between them. This is extremely good agreement for isotopic analysis which can sometimes reveal differences this great between two measurement of

Table XI. Vanilla extract GC/IRMS calculated 13C/12C values (o/oo PDB) from derivatized compounds versus measured 13C/12C values (o/oo PDB) from underivatized compounds

	calculated from derivatized compounds						difference of calculated and measured					
	p-OH benzaldehyde	ortho-vanillin	vanillin	syringaldehyde	p-OH benzoic acid	vanillic acid	p-OH benzaldehyde	ortho-vanillin	vanillin	syringaldehyde	p-OH benzoic acid	vanillic acid
Retention Time (sec.)	1560	1570	1660	1792	1990	2150	1566	1697	2014	2190	2215	2688
Source												
Madagascar-6227	-19.08	-21.76	-19.68	-26.86	-20.98	-25.80	0.43	1.10	-0.81	3.66	0.67	1.07
Madagascar-6228	-18.87	-23.37	-20.19	-24.13	-22.81	-23.88	1.61	2.43	0.36	0.11	3.01	-0.33
Indonesia-6229	-18.94	-23.77	-20.28	-31.20	-25.20	-24.39	0.90	1.48	0.52	5.59	6.06	0.82
Indonesia-6230	-19.91	-24.21	-21.10	-32.65	-23.80	-24.23	2.09	-1.01	0.88	6.27	3.60	1.10
Comore-6231	-18.32	-23.59	-19.45	-30.55	-25.29	-27.21	-1.11		-0.08	5.49	5.15	5.31
Comore-6232	-17.53	-22.97	-19.63		-15.78	-23.12	0.86	1.59	0.53		-2.47	1.11
Mexico-6233	-18.42	-23.35	-19.93	-31.92	-23.47	-23.84	0.92	-0.10	0.18	6.89	3.75	-1.03
Mexico-6234	-17.97	-21.74	-20.12		-13.62	-27.61	1.00	0.44	-0.09		-6.65	5.52
Tonga-5269	-20.97	-19.73	-18.95	-20.25	-29.46	-17.77	1.00	-0.99	-0.71	-11.55	4.59	-9.02
Madagascar-5270	-19.74	-16.87	-19.69	-17.68	-30.68	-23.06	-1.13	-5.44	-1.14		5.00	2.57
Comore-5271	-19.29	-21.98	-18.07	-15.93	-28.55	-22.75	-1.12	1.69	-2.33	-14.75	4.03	-5.79
Java-5272	-19.69	-23.03	-18.10	-16.05	-20.90	-21.97	-1.43	2.37	-1.74	-7.15	-2.76	-10.09
Bali-5273	-21.32	-22.78	-17.90	-17.82	-23.75	-17.97	0.19	2.72	-1.45	-5.94	3.54	-11.83
Bourdon (Bali)-5274	-21.37	-20.76	-18.49	-18.83	-22.44	-20.92	1.99	2.12	-1.20	-3.84	0.44	-6.15
Madagascar-5275	-20.44	-21.52	-19.16	-8.80	-32.96	-20.46	0.47	1.46	-0.64	-15.16	8.86	-10.48

Blanks represent below detectable level

the same sample by off-line preparative methods. The vanilla extract samples, though not nearly as close in comparison of the two methods, do exhibit differences less than 0.7 $^{o}/oo$ on average. This may be a consequence of the difference in the preparative methods used to isolate crystalline vanillin which are not used in GC/IRMS component analysis. Generally, the comparison of these methods of isotopic analysis confirm the ability of the GC/IRMS technique to accurately define single component $\delta^{13}C$ values.

Natural Profiles. As the second goal of this study, it was hoped that a carbon isotope ratio profile of the principle constituents for the natural flavor materials studied could be developed. Tables V, VI, and VII present the average $\delta^{13}C$ values for each of the flavor constituents separated from each of the naturally derived samples. This data is graphically displayed in Figures 9-11. The data is presented as series of lines with sample components along the x-axis and their respective $\delta^{13}C$ values along the z-axis for each of the samples within a flavor group. Samples are identified as to their source and identification number. The range of $\delta^{13}C$ values between samples and their respective components is well within the range expected for botanically derived materials. This data gives some indication of the natural isotopic variation existing in nature for these flavor materials based on the differences in some of the major biochemical components such as lipids and carbohydrates (*33*) and fatty acids (*34*). The isotopic abundance of these individual constituents also indicate the consistency within a particular oil or extract as specific compounds are biosynthesized. As indicated in Figure 9, citrus oils derived from various natural sources of considerable geographical variation portray a consistency across the thirteen components analyzed that could be used in defining natural origin. The trend is similar in Figure 10, although only two distinct geographic origins are indicated and these are very proximal to one another. As indicated, both the major and minor components here have similar trends in their respective isotopic abundance. The component menthyl acetate was found to be depleted in its $\delta^{13}C$ value as indicated by previous investigations of peppermint oils (*35*). For the vanilla extracts, the trends are not quite as consistent as for the oils as indicated in Figure 11. Source variation and very low concentrations of the minor components can account for some of the variation in these natural vanilla extract profiles. Though the profiles vary more than the citrus and peppermint profiles from one sample to another, the closeness to the total $\delta^{13}C$ value and good agreement with previous research (*36*) indicate the potential for use in defining a natural isotopic profile for vanilla derived from natural sources.

Conclusion

This study examined the accuracy of the hyphenated technique of gas chromatography and isotope ratio mass spectrometry to define individual component isotopic abundances which could be used to develop a pattern of isotopic values for specific flavor materials. Once defined these could be used as patterns of acceptance similar to total isotopic ratios now used for specific flavor compounds to judge their naturalness. To the first goal, the data presented here substantiates the capability of the GC/IRMS technique for accurately portraying the carbon isotopic ratios for selected components found in natural citrus oil, peppermint oil, and vanilla extracts. The comparison of two methods, one measuring the

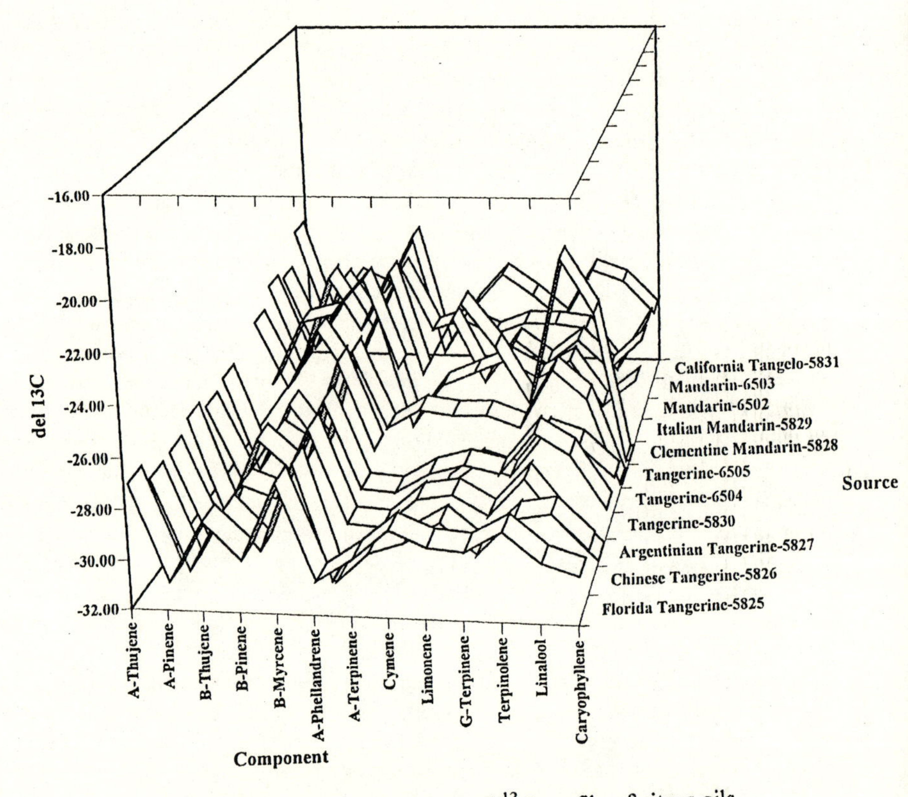

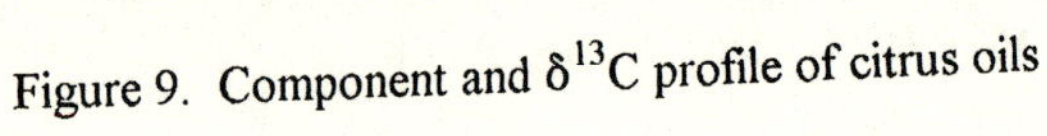

Figure 9. Component and $\delta^{13}C$ profile of citrus oils

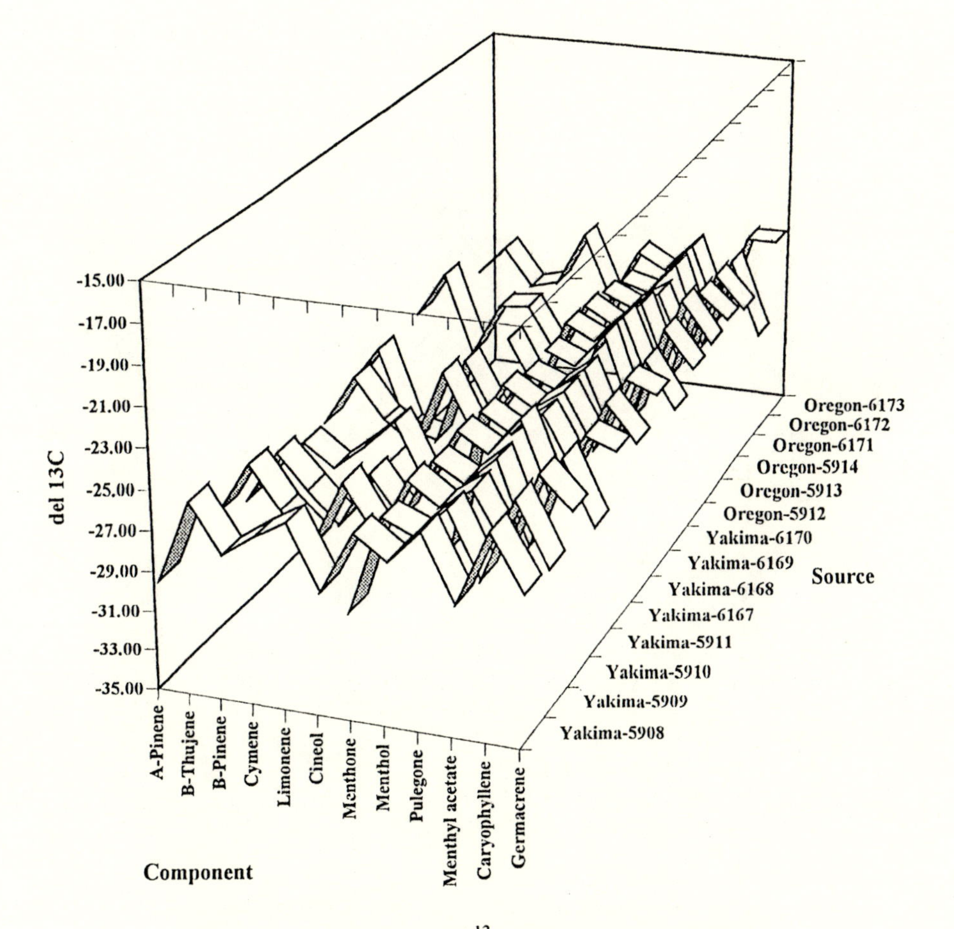

Figure 10. Component and $\delta^{13}C$ profile of peppermint oils

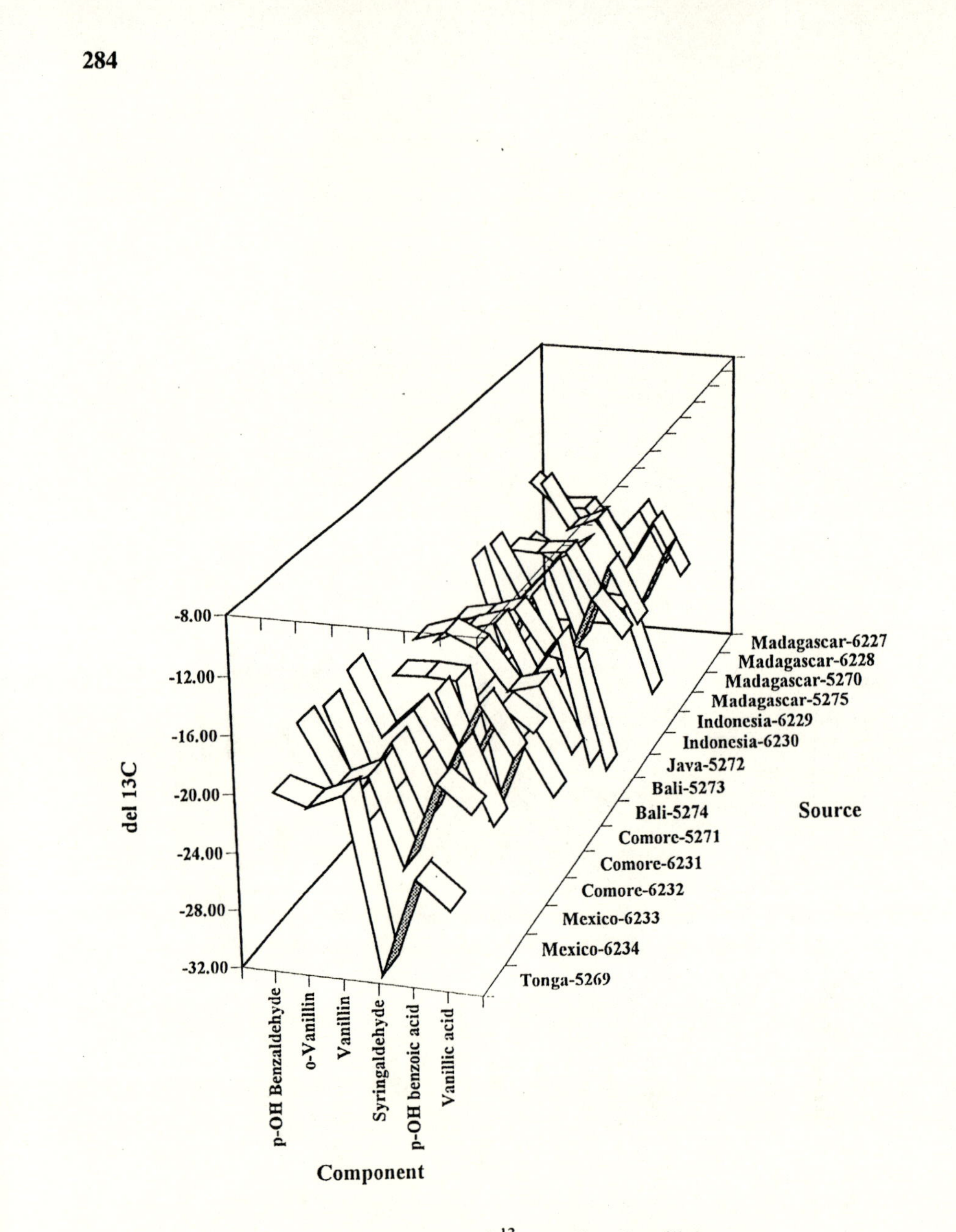

Figure 11. Component and $\delta^{13}C$ profile of vanilla extracts

total carbon isotopic value by off-line preparation and measurement on a Finnigan 251 isotope ratio mass spectrometer (IRMS), while the other performed gas chromatographic component separation with subsequent isotope ratio measurement, on the Finnigan 252 GC/IRMS, resulted in a variance of 0.7 ^{o}/oo or less.

Each of the flavor materials tested had primary components which could be expected to generate precise data based on their high concentrations. However, replicate analyses performed on the GC/IRMS indicate that even those components found at less than 1% concentration exhibit very low standard deviations and portray their isotopic ratios accurately. This indicates that by this technique we can demonstrate an increased capability to detect adulteration of flavors. The profiles shown in Figures 9 through 11 portray an isotopic fingerprint to confirm the naturalness of these flavor materials. The samples measured here were all of natural origin so only natural profiles or fingerprints can be revealed here. However, it is with considerable certainty that the profiles, of these same flavor materials, from synthetic origins, will differ significantly. This is based on numerous measurements of the total carbon isotopic ratios of synthetic flavor compounds which differ significantly from those of natural origin (*37,12*). It is anticipated that the individual component carbon isotope ratios will differ not only isotopically but also in relative concentration of the components typically found. This leads to the potential for applying multi-component isotope ratios to chemometric analysis. This should help to develop criteria best suited for different flavor materials to authenticate naturalness and uncover adulteration.

The future potential for the method of gas chromatography/isotope ratio mass spectrometry lies in its applicability to the broad range of flavors and flavor compounds that exist in our foods and beverages. Because of the minute quantities required for determination of component specific isotope ratios, typically nanogram to picograms measured, the possibility exists for direct on-line analysis of shelf products using the new techniques of sample preparation and clean-up. Though this study was limited to measurement of carbon isotopes, nitrogen isotopes can presently be measured by the same technique and research is already underway to determine the hydrogen isotopes following chromatographic separation (*38-40*).

The capability to reveal adulteration is presently available by applying both stable and radiometric analysis to bulk samples and isolates. The added capability to measure $^{13}C/^{12}C$ and soon, D/H ratios, as well as confirm the ^{14}C or ^{3}H activity of individual components by chromatographic separation will allow unprecedented confirmation of natural flavors and flavor ingredients.

Literature Cited

1. Martin, G.E.; Noakes, J.E.; Culp, R.A. unpublished data.
2. Koziet, J.; Rossman, A.; Martin, G.J.; Ashurst, P.R. *Anal. Chim. Acta.* **1993**, *271*, 31-38.
3. White, J.W.; Winters, K. *J. Assoc. Off. Anal. Chem.* **1989**, *72*, 907-911.
4. Hoffman, P.G.; Salb, M. *J. Assoc. Off. Anal. Chem.* **1980**, *63*, 1181-1183.
5. Martin, G.E.; Noakes, J.E.; Alfonso, F.C.; Figert, D.M. *J. Assoc. Off. Anal. Chem.* **1981**, *64*, 1142-1144.

6. Culp, R.A.; Noakes, J.E. *J. Agric. Food Chem.* **1990**, *38*, 1249-1255.
7. Martin, G.G.; Hanote, V.; Lees, M.; Martin, Y.L. *J. A.O.A.C. Intl.* **1996**, *79*, 62-72.
8. Allen, A. B. *J. Agric. Food Chem.* **1961**, *9*, 294-295.
9. Martin, G.E.; Krueger, H.W.; Burggraff, J.M. *J. Assoc. Off. Anal. Chem.* **1985**, *68*, 440-452.
10. Neary, M.; Spaulding, J.D.; Noakes, J.E.; Culp, R.A. *J. Agric. Food Chem.* **1997**, *45*, 2153-2157.
11. Bricout, J.; Koziet, J. *J. Agric. Food Chem.* **1987**, *35*, 758-760.
12. Bricout, J.; Koziet, J. In *Flavors of Foods and Beverages, Chemistry and Technology*; Charalambous, G.; Inglett, F.E.; Eds.; Academic Press: New York, 1978; 199-208.
13. Byrne, B.; Wengenroth, K.J.; Krueger, D.A. *J. Agric. Food Chem.* **1986**, *34*, 736-738.
14. Doner, L.W.; Bills, D.D. *J. Assoc. Off. Anal. Chem.* **1982**, *65*, 608-610.
15. Martin, G.J.; Guillou, C.; Martin, M.L.; Cabanis, M.T.; Tep, Y.; Aerny, J. *J. Agric. Food Chem.* **1988**, *36*, 316-322.
16. Hoffman, P.G.; Salb, M. *J. Agric. Food Chem.* **1979**, *27*, 352-355.
17. Santrock, J.; Studley, S.; Hayes, J.M. *Anal. Chem.* **1985**, *57*, 1444-1448.
18. Brand, W.A.; Rautenschlein, M.; Habfast, K.; Ricci, M.; Freeman, K.L.; Hayes, J. Finnigan Corp. Publ. **1989**.
19. Rautenschlein, M.; Habfast, K; Brand, W. In *Stable Isotopes in Pediatric Nutritional and Metabolic Research*, Proceedings, **1989**, 1-14.
20. Schmidt, H.L.; Butzenlechner, M.; Rossman, A.; Schwarz, S.; Kexel, H.; Kempe, K. *Z. Lebensm. Unters. Forsch.* **1993**, *196*, 105-110.
21. Caer, V.; Trierweiler, M.; Martin, G.J.; Martin, M.L. *Anal. Chem.* **1991**, *63*, 2306-2313.
22. Martin, G.J.; Remaud, G. *Flav.Frag. J.* **1993**, *8*, 97-107.
23. Martin, G.J.; Hanneguelle, S.; Remaud, G. *Ital. J. Food Sci.* **1993**, *3*, 191-213.
24. Silfer, J.A.; Engel, M.H.; Macko, S.A.; Jumeau, E.J. *Anal. Chem.* **1991**, *63*, 370-374.
25. Hayes, J.M.; Freeman, K.H.; Popp, B.N.; Hoham, C.H. *Org. Geochem.* **1990**, 1-32.
26. Weilacher, T.; Gleixner, G.; Schmidt, H.L. *Phytochem.* **1996**, *41*, 1073-1077.
27. Noakes, J.E.; Kim, S.M.; Stipp, J.J. In *Sixth International Symposium on Radiocarbon Dating and Tritium*, Proceedings of USAEC Conference 650-652, **1965**; 68-92.
28. Culp, R.A.; Noakes, J.E. *J. Agric. Food Chem.* **1992**, *40*, 1892-1897.
29. Boutton, T.W.; Wong, W.W.; Hachey, D.L.; Lee, L.S.; Cabrera, M.P.; Klein, P. *Anal. Chem.* **1983**, *55*, 1832-1833.
30. Coplen, T; Kendall, C. *Anal. Chem.* **1982**, 2611-2612.
31. Goni, M.A.; Eglinton, T.I. *Org. Geochem.* **1996**, *24*, 601-615.
32. Craig, H. *Geochim. Cosmochim. Acta* **1957**, *12*, 133-149.
33. DeNiro, M.J.; Epstein, S. *Science* **1977**, *197*, 261-263.
34. Monson, K.D.; Hayes, J.M. *J. Biol. Chem.* **1982**, *257*, 5568-5575.
35. Faber, B.; Krause, B.; Dietrich, A. Mosandl, A. *J. Essent. Oil Res.* **1995**, *7*, 123-131.

36. Kaunzinger, A.; Juchelka, D.; Mosandl, A. *J. Agric. Food Chem.* **1997**, *45*, 1752-1757.
37. Bricout, J. In *Stable Isotopes, Conference Procedures*; Schmidt, H.L., Forstel, H., Heinzinger, K., Eds.; Elsevier: Amsterdam, **1982**; 483-494.
38. Tobias, H.J.; Brenna, J.T. *Anal. Chem.* **1996**, *68*, 3002-2007.
39. Prosser, S.J.; Scrimgeour, C.M. *Anal. Chem.* **1996**, *67*, 1992-1997.
40. Tobias, H.J.; Goodman, K.J.; Blacken, C.E.; Brenna, J.T. *Anal.Chem.* **1995**, *67*, 2486-2492.

Sensory Characterization

Chapter 24

Validation of Gas Chromatography Olfactometry Results for 'Gala' Apples by Evaluation of Aroma-Active Compound Mixtures

Anne Plotto[1], James P. Mattheis[2], David S. Lundahl[1], and Mina R. McDaniel[1]

[1]Department of Food Science and Technology, Oregon State University, Wiegand Hall, Corvallis, OR 97331
[2]Tree Fruit Research Laboratory, Agricultural Research Service, U.S. Department of Agriculture, Wenatchee, WA 98801

'Gala' is an early maturing apple variety with a distinctive aroma and flavor. Previous research has determined 'Gala's aroma-active compounds by using Osme, a gas chromatography olfactometry method that records subjects' olfactory response on a time-intensity scale. Sixteen of those compounds were combined in mixtures in water solutions at concentrations determined by analyzing apple headspace. Sixteen panelists compared aromas of the solutions with fresh apples and rated degree of difference for aroma. In a pilot study, mixture solutions were prepared by combining compounds based on their intensities as perceived by Osme; results showed a large variability between panelists for perception of the solutions. Another experiment used a statistical screening design. Hexyl acetate, hexanal, butyl acetate, 2-methyl butyl acetate, and methyl-2-methyl butyrate contributed to the least difference between mixtures and apples; while pentyl acetate, hexyl-2-methyl butyrate, butyl hexanoate, and 4-allyl anisole contributed to the largest difference. Further experiments using statistical designs will be necessary to determine interactions between compounds.

Smelling gas chromatograph effluents to determine the odor characteristic of a compound has been practiced in flavor research chemistry since the development of gas chromatography in the 1950's; it has been formalized and is now known as gas chromatography olfactometry (GCO) (*1,2*). However, without any quantification of the chemical stimuli and of the subjects' responses, GCO is limited to screening odor-active volatiles among those present in a complex sample. Potent odorants are often near or beyond the limit of detectability by GC analysis (*3,4*). Patton and Josephson (*5*) introduced the idea of relating a compound's concentration to its odor threshold in order to assess its odor significance. This concept was named "aroma value" by Rothe and Thomas (*6*), "unit flavor base" by Keith and Powers (*7*), and is now used as the "odor unit" (*8*), and the "Odor Activity Value" (OAV) (*9*). The concepts of odor activity, odor potency, and odor threshold of a compound have been further developed with the use of dilution techniques in GCO analysis and named CharmAnalysis (*10*) and Aroma Extract Dilution Analysis (AEDA) (*11*). Using these techniques, the compounds that are perceived at the highest dilution level are deemed the most potent in the sample. In other words, the odor potency of a compound is determined by the quantity necessary to give a response: the smaller the concentration, the more potent the compound. Both CharmAnalysis and AEDA assume that the response to an odorous stimulus is linear and that all compounds have identical response slopes with increasing concentration. In contrast, psychophysical events are based on the principles of Stevens' law, which states that the response to a stimulus follows a power function, and that the exponent of the function is between 0.3 and 0.8 for odorants (*12,13*). Another GCO technique, Osme, is in agreement with Stevens' law of psychophysics as it measures the response to odorants on a time-intensity scale (*14,15*). Osme produces an odor profile, and comparisons between samples can be made by either comparing sample profiles (*16-18*), or by statistical analysis (*19*).

All GCO techniques are useful for determining the odor activity, quality, and potency of compounds in foods, and thus allow for sample comparisons. However, the limitation inherent to GC techniques is that the information is obtained for individual compounds, which are presented to the nose outside of the food matrix. Also, the different GCO methods and their data analyses may lead to different conclusions as to which compounds are most important in a sample (*20,16*). Validations of GCO by aroma reconstitution are required. Confirmation of GCO results by sensory comparison of mixtures with the original samples has been demonstrated for butter (*21*), strawberry juice (*22,23*), cheddar cheese (*24*), and apple (*25*). In references 21 – 24, authors combined AEDA results with the OAV concept to compare reconstituted aroma mixtures with the original samples. Schieberle and colleagues added odor-active compounds in a neutral substrate of oil for butter aroma (*21*) and pectin, sugars, and acids for strawberry juice (*23*). They determined the sensory importance for each compound by omitting them from the model solution one by one, and they compared the mixtures to the original samples. Mixtures most similar to the sample were those containing all the compounds with an OAV above one. Dacremont and Vickers (*24*) combined 15 compounds in two fractionated factorial designs and matched the resulting odors with cheddar cheeses. They narrowed the number of optimum compounds to six, and matched the mixtures' odors to 15 cheeses to determine which of the cheeses had a cheddar note. Young and co-workers (*25*) used the four most potent compounds found by CharmAnalysis, combined them in four concentration levels, and used sensory descriptive analysis to measure differences for attributes generated by the mixture's aroma and flavor. They found that the combination of 2-methyl butyl acetate, hexyl acetate, and butanol approached most closely the "Red apple" attribute associated with 'Gala' apple flavor. Synthetic tomato aroma was also made by using the odor unit values concept (but without previous determination of compounds odor potency by GCO) and reported to have a tomato odor by a sensory panel (*26,27*).

'Gala' apple (*Malus domestica* Borkh) is an early ripening cultivar that resulted from a cross between 'Kidd's Orange' ('Cox's Orange Pippin' X 'Red Delicious') and 'Golden Delicious' (*28*). 'Gala' has a sweet and perfumey aroma and flavor, which distinguishes it from other cultivars (*29*). Storage techniques such as controlled atmosphere (CA) maintain the apple fruit firmness and acidity for up to seven to eight months, but a significant aroma decrease is generally observed (*30*). Determination of compounds contributing to the aroma of fresh harvested 'Gala' aroma would assist further research aimed at maintaining 'Gala' flavor in storage. The odorants used in this study were previously determined by GCO and Osme to contribute to 'Gala' aroma (*31*). This paper explores two methods to validate Osme data by comparing 'Gala' apples with model solutions. One method is based on the results found by Osme on compounds' perceived intensities. This approach is similar to the odor unit concept, and results are discussed by comparing Osme and odor unit values. The other method explored model solutions prepared with compounds at the same concentrations as found in apples. The solutions were prepared following a statistical screening design, but ignored odor units or odor intensities.

Experimental

Materials. Volatiles emitted by 'Gala' apples previously stored in air at 1 °C were analyzed one week before sensory analysis as previously described (*32*). Briefly, headspace of ca. 1 kg apples was trapped onto 50 mg of Tenax TA traps by using a dynamic flowthrough system with purified air at 100 mL/min. Samples consisted of 100 mL of headspace, and traps were thermally desorbed. Compounds were analyzed on a HP 5890A-5971A GC-MSD system. Previous work on 'Gala' using Osme had identified 26 compounds which had various levels of odor activity (*31*). Fifteen chemically identified compounds that were perceived consistently by all three panelists using Osme were used to construct model solutions (Table I). Hexanal was also used in the experimental mixtures, even though it was not present in the Osme analysis of samples prepared from charcoal traps and eluted with CS_2. It was reported as present in the samples heat desorbed from Tenax traps, and was previously found to contribute to apple odor with a green apple descriptor (*33*). The headspace concentrations measured in the 100 mL sample used in this study were converted to concentrations in water by using Henry's law to calculate K, the air-water partition coefficient (*34-37*) (Table I). The low solubility of 4-allyl anisole did not follow the ideality assumptions necessary to calculate its air-water partition coefficient K at 25 °C (Table I). Therefore, 4-allyl anisole was used in the mixtures in the same proportions as found in the apple headspace. Experimental solutions were first prepared by mixing the compounds as calculated for the theoretical concentration in water (Table I). However, the odor intensities of these solutions were too weak to be compared with apples; therefore, concentrations in water were increased until the overall aroma could be compared with apples while keeping the same relative ratios between compounds.

All compounds were purchased from Aldrich Flavors and Fragrances (Milwaukee, WI) and were food grade. Compound purity was verified by GC and by sniffing the GC effluent of a preparation of standards in the same concentrations as found in apple headspace. Compounds were mixed in odor-free double distilled water (Milli-Q) according to the design described below.

Experimental Designs.

Pilot Study. Based on the compounds' odor intensities from Osme analysis (Table II), sixteen mixtures were prepared by sequentially adding compounds in an incremental manner. The first solution was only hexyl acetate in water, the second solution was butyl acetate added to hexyl acetate, the third solution was made by adding 2-methyl butyl acetate to butyl acetate and hexyl acetate, and so on; the final solution was the mixture of the 16 compounds in water. Two replicates of each mixture were prepared in 50 mL water at the concentrations shown in Table I and presented in 120 mL glass jars with Teflon- lined screw caps. Means separation between mixtures for degree of difference from apple was performed with the least significant difference (LSD) test, with panelist as a random effect.

Screening Design. Because all possible combinations of the 16 compounds would generate too many samples to evaluate, a screening design was used. These designs are often used by food developers to identify which among many ingredients in a sample are the most important to achieve a product characteristic; for example, which sugars and acids are necessary to combine in a fruit beverage to have a determined level of sweetness. The 16 compounds tested in the pilot study were mixed in 30 mL water following combinations computed by the ECHIP v.6.1.2 (Hockessin, DE) statistical package. A linear D-optimal screening design was used, with 16 variables (the 16 compounds) and 8 replicates (Table III). The design resulted in 25 combinations containing 6 to 10 compounds (and 16 for the combination containing all compounds). Eight combinations were replicated, as the design dictated. Therefore, a total of 33 samples were prepared for each panelist in the same jars as described above. Concentrations and sample headspace used for this experiment were adjusted based on panelists' comments during the pilot study, without altering their relative proportions (Table I). This experiment was repeated once. Each experiment, the pilot study and the two replications of the screening design, was conducted on a different day.

Sensory Analysis Procedure. Sixteen panelists participated in the testing. Procedures were discussed with the panelists for one hour before the beginning of the study. For both pilot study and screening design, the 33 samples were evaluated in three sets of 11 jars containing the compound mixtures according to a complete randomized block design across sets. The first experiment had 32 jars (two replicates of 16 samples), but the first jar was triplicated to give the total of 33 samples and was not used in the statistical analysis. Twenty 'Gala' apples were used for each testing day. They were put in four 4 L glass jars (5 apples, ca. 1 kg, per jar) and presented randomly to the 16 panelists, one jar for 4 panelists. When the testing began, one apple jar was covered with aluminum foil and presented to the panelist with the sample

Table I. Concentration of Apple Headspace Compounds, Air/Water Partition Coefficient, Theoretical Concentration in Water, and Compound Concentrations Used in the Sensory Experiments

Compound in Apple	*Apple Headspace* [a] *(μg/L)*	*Partition Coefficient* [b] *(K)*	*Theoretical Concentration* [c] *(μg/L)*	*Pilot Study Solution (mg/L)*	*Screening Design Solution (mg/L)*
Butyl Acetate	7.044	0.0133	529.60	30.00	50.00
Hexyl Acetate	1.160	0.0241	48.12	2.73	4.54
2-Methyl Butyl Acetate	0.567	0.025	22.66	1.28	2.14
Butyl Hexanoate	0.549	0.04	13.72	0.78	1.29
Hexyl 2-Methyl Butyrate	0.531	0.05	10.62	0.60	1.00
Butyl Butyrate	0.234	0.029	8.07	0.46	0.76
Butyl 2-Methyl Butyrate	0.228	0.04	5.71	0.32	0.54
Hexyl Butyrate	0.218	0.0394	5.53	0.31	0.52
Butyl Propionate	0.184	0.0197	9.34	0.53	0.88
Methyl-2-Methyl Butyrate	0.093	0.028	3.30	0.19	0.31
Pentyl Acetate	0.088	0.0211	4.18	0.24	0.39
2-Methyl Propyl Acetate	0.049	0.0221	2.22	0.13	0.21
Hexyl Propionate	0.045	0.0277	1.64	0.09	0.15
Ethyl 2-Methyl Butyrate	0.014	0.030	0.46	0.03	0.04
Hexanal	0.011	0.0087	1.26	0.07	0.12
4-Allyl Anisole	0.010	n.a.[d]	n.a.	0.042[e]	0.07[e]

[a] 100 mL of dynamic headspace

[b] Henry's law air/water partition coefficient; from Buttery et al., 1971; Jordan, 1954; Lyman et al., 1982; Pierotti et al., 1959

[c] Calculated by dividing column 1 by column 2

[d] No published vapor pressure found for this compound at 25 °C. [e] Added in the same proportions as found in apple headspace

Table II. A) Apple Compounds Sorted by Decreasing Osme Intensity[a], Corresponding Descriptors and Perceived Osme Intensity B) Apple Compounds Sorted by Decreasing Odor Units, Concentrations in Water Calculated from Headspace, Published Odor Thresholds, and Calculated Odor Units

A) Apple compounds sorted by decreasing Osme intensity	*Descriptor*	*Osme Intensity* [a]	*B) Apple compounds sorted by decreasing odor unit*	*Theoretical Concentration (μg/L)*	*Odor Threshold (μg/L)*	*Odor Unit* [b] *(C/T)*
Hexyl Acetate	gala, ripe, pear	12.26	Hexyl Acetate	48.12	2.00 [c,d]	24.06
Butyl Acetate	nail polish	9.72	Methyl-2-Methyl Butyrate	3.30	0.25 [e]	13.20
2-Methyl Butyl Acetate	solventy	8.56	Butyl Acetate	529.60	66.00 [c,d]	8.02
Methyl-2-Methyl Butyrate	sweet fruity	7.36	Ethyl-2-Methyl Butyrate	0.46	0.10 [d]	4.60
Ethyl-2-Methyl Butyrate	sweet strawberr	7.28	2-Methyl Butyl Acetate	22.66	5.00 [d]	4.53
4-Allyl Anisole	anise, licorice	6.48	Pentyl Acetate	4.18	5.00 [d]	0.84
Pentyl Acetate	gala	5.82	Hexyl-2-Methyl Butyrate	10.62	22.00 [c]	0.48
Hexyl-2-Methyl Butyrate	apple, grapefruit	5.74	2-Methyl Propyl Acetate	2.22	5.00 [d]	0.44
Butyl-2-Methyl Butyrate	fruity, apple	5.64	Butyl Propionate	9.34	25.00 [d]	0.37
Butyl Propionate	fruity, apple	5.24	Butyl-2-Methyl Butyrate	5.71	17.00[c]	0.34
2-Methyl Propyl Acetate	tea, leaves	4.22	Hexanal	1.26	4.50 [c,d]	0.28
Hexanal	n.a.	n.a.	Hexyl Propionate	1.64	8.00[c]	0.21
Hexyl Propionate	apple	2.97	Butyl Butyrate	8.07	100.00 [c]	0.08
Butyl Butyrate	rotten apple	2.43	Hexyl Butyrate	5.53	250.00 [c]	0.02
Hexyl Butyrate	apple	2.02	Butyl Hexanoate	13.72	700.00 [c]	0.02
Butyl Hexanoate	green apple	2.02	4-Allyl Anisole	0.74	35.00 [f]	0.02

[a] From Plotto et al., 1995; intensity on a 16-point category scale: 0 = none, 7 = moderate, 15 = extreme

[b] Teranishi et al., 1991. [c] Takeoka et al., 1990. [d] Flath et al., 1967. [e] Takeoka et al., 1989. [f] Williams et al., 1977.

n.a.: non applicable. Hexanal was not present in the samples used for Osme analysis

Table III. Combinations of Compounds for the Mixture Solutions as Computed by ECHIP Statistical Software[a]

	Solution Number																								
	1	*2*	*3*	*4*	*5*	*6*	*7*	*8*	*9*	*10*	*11*	*12*	*13*	*14*	*15*	*16*	*17*	*18*	*19*	*20*	*21*	*22*	*23*	*24*	*25*
Hexyl Acetate	X			X	X		X			X				X		X		X	X	X	X			X	X
Butyl Acetate	X		X			X	X		X						X		X		X	X	X	X			X
2-Methyl Butyl Acetate		X	X		X			X	X		X					X		X		X	X	X	X		X
Methyl-2-Methyl Butyrate		X		X	X		X			X	X	X					X		X		X	X	X	X	X
Ethyl-2-Methyl Butyrate	X			X		X	X		X			X	X					X		X		X	X	X	X
4-Allyl Anisole	X		X			X		X	X		X		X	X					X		X		X	X	X
Pentyl Acetate	X		X		X			X		X	X	X		X	X					X		X		X	X
Hexyl-2-Methyl Butyrate	X		X		X		X			X		X	X		X	X					X		X		X
Butyl-2-Methyl Butyrate		X	X		X		X		X				X	X		X	X					X		X	X
Butyl Propionate	X			X	X		X		X		X			X	X		X	X					X		X
2-Methyl Propyl Acetate		X	X			X	X		X		X	X			X	X		X	X					X	X
Hexanal	X			X	X			X	X		X	X	X			X	X		X	X					X
Hexyl Propionate		X	X			X	X			X	X	X	X	X			X	X		X	X				X
Butyl Butyrate		X		X	X			X	X			X	X	X	X			X	X		X	X			X
Hexyl Butyrate		X		X		X	X			X	X		X	X	X	X			X	X		X	X		X
Butyl Hexanoate		X		X		X		X	X			X		X	X	X	X			X	X		X	X	X
Total compounds	8	8	8	8	9	7	10	6	10	6	9	9	8	9	8	8	7	7	8	9	9	8	8	8	16

[a] "X" indicates presence

mixtures. The apple jar remained covered during the testing. Panelists were asked to lift the 4 L jar cover, smell the apples, close the lid, open the sample containing the mixture solution, smell it, and rate degree of difference between the mixture and the apples on a 16-point category scale (0 = no difference, 15 = extremely different). Panelists could also comment on the quality of the mixture. Panelists were asked to rest one minute after the first five samples, and take a 10 minute break between sets. They were only allowed to smell the samples once. All samples and apples were presented at room temperature. Panelists were seated in individual testing booths equipped with PCs and Compusense Five, v. 2.2 (Guelph, Ontario) software for data recording.

Results and Discussion

Pilot Study: Mixtures Prepared from Osme Odor Intensity Values. The differences in degree of difference ratings between solutions were small (Table IV).

Table IV. Degree of Difference Between Odorant Mixtures and Apples (n = 32 observations)

Number of Compounds in Solution[y]	*Average Difference from Apples*[z]
1	7.56 [a]
2	5.81 [bc]
3	5.84 [bc]
4	6.44 [ab]
5	6.41 [ab]
6	6.29 [ab]
7	6.69 [ab]
8	5.81 [bc]
9	5.59 [bc]
10	6.06 [bc]
11	6.22 [bc]
12	6.13 [bc]
13	5.03 [c]
14	5.06 [c]
15	5.56 [bc]
16	5.81 [bc]

[y] Compounds were added incrementally in the order shown in Table I

[z] Difference from apple: 0 = no difference, 15 = extremely different. Means followed by the same letter were not significantly different by the LSD test ($P < 0.05$), with panelist as random variable

Average difference ratings ranged from 5.03 to 7.56 (slightly to moderately different). The largest average difference ratings was given to the solution containing hexyl acetate alone and the solutions containing four, five, six, and seven compounds; the least differences were found for the solutions containing 13 and 14 compounds. Based on previous research (*21, 23*), a decrease in the degree of differences from apples as more compounds were present in the solutions was expected for the first five compounds with an odor unit above one, but this trend was not observed. Variability between panelists' perception of the sample aromas and their comparison with apples was considerable. Some mixtures were found to be very close to the apples by some panelists, and rather different for others. Some of this variability may have been due to variation in apples used as reference. A variation of 20% is not uncommon in apple headspace (*38*) and was observed by sampling 'Gala' headspace with Tenax traps (unpublished results). Additionally, apples produce volatile compounds continuously, and it is possible that, within the few hours in which the experiment took place, headspaces were different from jar to jar when presented to panelists. Another source of variability was the lack of training for this specific task although panelists had been trained for other types of sensory analysis. Finally, different perceptual response between panelists is usually expected.

The experimental design in this pilot study was based on assumptions about the relative odor activity of the 16 compounds. The combination of aroma-active compounds was determined from the data obtained by GCO of 'Gala' apples where odorant peak intensities were rated on a 16-point category scale (*31*). To relate Osme data to the odor unit concept, ranking of compounds by decreasing odor intensity was compared to the ranking of compounds by decreasing calculated odor units (Table II). Odor units were calculated by using odor threshold values published by the U.S.D.A. Western Regional lab (*33, 39, 40*) except for 4-allyl anisole (*41*). Hexanal, not present in the 'Gala' sample analyzed by using Osme, was ranked at the 12th position, similar to the ranking based on calculated odor units. Comparison of odor intensity and odor unit data for 'Gala' apple headspace indicated that the first five compounds with an odor unit above one were also those with the highest odor intensity (Table II). Except for compounds 13 to 16 that had the exact same ranking order and 4-allyl anisole that was ranked last by the odor unit value, there were inversions in the ranking of some of the compounds, but the inversions did not exceed two positions. For example, methyl-2-methyl butyrate was ranked in the fourth position by Osme intensity and in the second position according to the odor unit value. The odor threshold value of 4-allyl anisole was obtained from a different group of researchers (*41*), and this may explain the discrepancy with values obtained from the U.S.D.A. Western Regional laboratory. Headspace samples used to obtain Osme data and calculate odor units were taken from different groups of apples (same orchard, same storage type but stored for one versus five months) which might account for the slight discrepancy between the two ranking methods. The ranking obtained from the perceived odor peak areas, which combine odor intensity and time during the perception of the odorous stimulus (*15*), also resulted in a few inversions from odor intensities (data not shown).

Overall, the ranking of aroma-active esters present in 'Gala' apple according to the information obtained from the GCO technique Osme resulted in an order comparable with ranking based on odor units. However, we found limitations to both approaches in determining which compounds contributed most significantly to 'Gala' headspace aroma. The use of odor units requires the knowledge of an odor threshold value. Odor threshold determination is very time consuming, and threshold values were found to vary considerably between laboratories and methods used (*42-45*). For example, the method of presenting compounds (100 mL glass jars with lids or in

Teflon squeeze bottles) was found to significantly affect thresholds and reproducibilities (*44*). Compounds presented in squeeze bottles had 100 fold lower threshold values than glass jars with lid. Odor thresholds used in this study were generated with the method using squeeze bottles (*8, 33, 39, 40*); this may explain the discrepancy between the concentrations analyzed from apple headspace (in µg/L) and the concentration (in mg/L) necessary to attain a similar level of odor in the experimental solutions presented in glass jars with lids (Table I). Additionally, odor units or OAV, like Charm values, ignore the power function of the response to stimulus concentrations and slope differences between different odorants (*46*). The limitation in using GCO data to prepare mixtures of the odor-active compounds stands in the fact that unidentified odor active compounds are not accounted for (*24*). In 'Gala' apple, 19 of the 44 odor active compounds were unknown (*31*). Among those, compounds that had mushroom, earthy, or skunk descriptors had high odor intensities and probably contributed to the apple aroma. One comment from a panelist confirmed that hypothesis: this panelist rated some solution mixtures similar to the apples, but commented the mixtures were missing a sulfury component perceived in the fruit. The lack of duplication of apple aroma by combining apple-like odor active compounds in a decreasing order of odor activities led to the use of an experimental screening design to identify those odor active compounds contributing most to 'Gala' apple aroma. The volume of headspace and concentration of compounds in the jars were adjusted after some panelists mentioned that some solutions were "too weak" in the first experiment (Table I).

Screening Design. The first replicate test indicated that hexyl acetate, butyl acetate, and hexanal were necessary to impart the least difference between the solutions and apples. Pentyl acetate and hexyl-2-methyl butyrate contributed the most to differences between solutions and apples. In the second replicate test performed one week later, hexyl acetate and hexanal were found again to contribute to the least difference from apples, as did 2-methyl butyl acetate and methyl-2-methyl butyrate. Similarly, pentyl acetate and hexyl-2-methyl butyrate contributed to the largest difference, along with butyl hexanoate and 4-allyl anisole. The difference between the two replicate tests again may be due to variation between apples on the same day of the experiment as mentioned above, and differences in ripening between apples from one week to the other. Nevertheless, common results from both replicate tests indicated that hexyl acetate and hexanal contributed to 'Gala' aroma. The combination of results of both tests showed that the four esters having the highest Osme value and an odor unit greater than one (Table II) contributed the most to 'Gala' aroma. Unfortunately, odor intensity of hexanal was not available from the samples sniffed by Osme. Published threshold values for hexanal gave an odor unit of less than one for our apples; however, the screening design experiment showed that hexanal in mixtures contributed significantly to 'Gala' aroma. This confirms that no definite conclusion can be drawn from the odor activity of compounds alone. Results regarding pentyl acetate and hexyl-2-methyl butyrate led to the same conclusion. Both compounds individually have a definite apple odor, but it seems that when present in the combinations of mixtures, they enhanced the difference from the apple control. 4-Allyl anisole imparted a similar effect to the mixtures. This could be the effect of the chemical aromaticity of that compound or the result of a miscalculation of the concentration used, because the air/water partition coefficient K could not be theoretically calculated (see materials and methods).

About Odor Mixtures. It is generally admitted and has been experimentally demonstrated that odor-active compounds with a certain odor characteristic do not create a novel odor when mixed, and the perceived intensity of the mixture is less than the sum of intensities of individual compounds (*47*). All compounds in the mixtures in this study were esters with fruity, apple-like odors, and one aldehyde with a green apple odor (hexanal), and an allyl phenol compound with the odor of anise (4-allyl anisole). Comments that were generated from some mixtures were either fruity, pear, banana, apple-like, or tended towards descriptors like "artificial apple", "bubble gum", "solventy, nail polish". At the concentrations tested, butyl acetate and 2-methyl butyl acetate alone had those qualities of descriptors (Table II). It was expected that adding other compounds would attenuate the solventy note to give a descriptor closer to apple, but there was no agreement between panelists as to which combination was closer to the apples. Part of the variation between panelists might be an effect of the carryover from one solution to the next, because the instructions did not specify resting time between jars within a subset of five. Olfactory adaptation occurs between odorants having a similar aroma (*48,49*) and similar chemical structure (*50*). 4-Allyl anisole, a compound structurally different from the esters, was occasionally perceived in some but not all solutions containing it. It was believed to contribute to the unique aromatic character of 'Cox's Orange Pippin' apple (*41*), and we also hypothesized that it might contribute to 'Gala' aroma. However, 4-allylanisole enhanced the difference from apples at the concentration used in this study. Butyl acetate was present in the largest amount in 'Gala' apple headspace, followed by hexyl acetate and 2-methyl butyl acetate. Those same compounds were chosen by Young et al. (*25*) as having the highest Charm value for 'Gala' apple. Those authors also included butanol, present in the largest proportion (*25*). However, they used vacuum steam distillation to isolate flavor volatiles. Butanol was not included in our study. We did not believe it would contribute significantly to 'Gala' aroma because it has a high odor threshold: 500 ppb (*33*), and the concentration found in whole 'Gala' headspace was 0.698 μg/L. Young and co-workers measured the effect of compound interactions on a few sensory descriptors that were used to describe 'Gala' apple flavor and aroma. They found negative interactions between hexyl acetate and butanol, and between 2-methyl butyl acetate and butanol; the former affected "Red apple aroma" and the latter "characteristic apple flavour". However, their method did not compare the mixtures with whole apples. Dacremont and Vickers (*24*) used a concept matching technique with partial factorial designs to screen for the compounds contributing to cheddar cheese odor. Similar to the design we used, they questioned the reliability of the information obtained for the main effects when the main effects (compounds) were included in interactions with other compounds. Nevertheless, their technique optimized mixtures of compounds whose odor matched the concept of Cheddar and other cheeses (*24*). In the end, whichever method and design is used, the making of mixtures relies on the previous step of GC analysis. Different recoveries observed in methods used for flavor isolation are well-documented (*51,52*). We used a headspace technique with purge and trap on Tenax because this technique captured the volatile profile of the sample with good recovery and without artifacts. However, the method used for Osme previously revealed that low odor threshold sulfur compounds were present in the samples but were not identified, therefore these compounds could not be included in the mixture experiments.

Conclusion

Mixing 'Gala' odor active compounds in proportions found in apple headspace and in combinations selected by a screening design has confirmed results obtained by the Osme GC-olfactometry technique. Hexyl acetate, hexanal, butyl acetate, 2-methyl butyl acetate and methyl-2-methyl butyrate were found to contribute to overall 'Gala' aroma. The use of a D-optimal linear screening design gave interesting information. The advantage of this design was that it was easily implemented since the number of compound combinations were limited, and there was no need to train panelists. Further experiments using response surface methodology will be necessary to determine 1) the level of interactions between compounds, and 2) how the odor mixtures change when compounds vary in different proportions. The latter determination would be very useful to post harvest physiologists because volatiles produced by apples vary in different proportions when stored in CA as opposed to air. Reduced oxygen and high CO_2 in CA affect straight chain acetate esters more than branched esters and aldehydes (*53*).

Acknowledgements

Dave Buchanan is acknowledged for technical assistance with GC-MS analyses. Financial support for this research was provided by the Washington State Tree Fruit Research Commission.

Literature Cited

1. Acree, T.E. *Anal. Chem. News & Features.* **1997,** 170A - 175A.
2. Mistry, B.S.; Reineccius, T.; Olson. L.K. In *Techniques for Analyzing Food Aroma*; Marsili, R., Ed.; Marcel Dekker: New York, **1997,** pp 265-292.
3. Guadagni, D.G.; Okano, S.; Buttery, R.G.; Burr, H.K. *Food Technology.* **1966,** *20,* 166-169.
4. Cunningham, D.G.; Acree, T.E.; Barnard, J.; Butts, R.M.; Braell, P.A. *Food Chemistry.* **1986,** *19,* 137-147.
5. Patton, S.; Josephson, D.V. *Food Research.* **1957,** *22,* 316-318.
6. Rothe, M.; Thomas, B. *Z. Lebensm. Unters. Forsch.* **1963,** *119,* 302-310.
7. Keith, S.E.; Powers J.J. *J. Food Sci.* **1968,** 33, 213-218.
8. Teranishi, R.; Buttery, R.G.; Stern, D.J.; Takeoka, G. *Lebensm. Wiss. Technol.* **1991,** *24,* 1-5.
9. Grosch, W. *Flavour and Fragrance J.* **1994,** *9,* 147-158.
10. Acree, T. E.; Barnard, J.; Cunningham, D.G. *Food Chem.* **1984,** *14,* 273-286.
11. Grosch, W. *Trends in Food Sci. & Technol.* **1993,** *4,* 68-73.

12. Steven's S.S. *Psychol. Rev.* **1957,** *64,* 153-181.
13. Cain, W.S. *Perception & Psychophysics.* **1969,** *6(6A),* 349-354.
14. McDaniel, M. R.; Miranda-Lopez, R.; Watson, B.T.; Micheals, N.J.; Libbey, L.M. In *Flavors and Off-Flavors*; Charalambous, G., Ed.; Elsevier:Amsterdam, **1990,** pp 23-36.
15. Da Silva, M.A.A.P.; Lundahl, D.S.; McDaniel, M.R. In *Trends in Flavour Research;* Maarse, H.; Van Der Heij, D.G., Eds.; Elsevier: Amsterdam, **1994,** pp 191-209.
16. Young, S.L. Master's thesis, Oregon State University, Corvallis, OR. Sept. **1997**
17. Sanchez, N.B.; Lederer, C.L.; Nickerson, G.B.; Libbey, L.M.; McDaniel, M.R. In *Food Science and Human Nutrition;* Charalambous, G., Ed.; Elsevier Science: Amsterdam, **1992,** pp 371-402.
18. Sanchez, N.B.; Lederer, C.L.; Nickerson, G.B.; Libbey, L.M.; McDaniel, M.R. In *Food Science and Human Nutrition*; Charalambous, G., Ed.; Elsevier Science: Amsterdam, **1992,** pp 403-426.
19. Da Silva, M.A.A.P.; Elder, V.; Lederer, C.L.; Lundahl, D.S.; McDaniel, M.R. In *Shelf Life Studies of Foods and Beverages;* Charalambous, G., Ed.; Elsevier Amsterdam, **1993,** pp 707-738.
20. Abbott, N.; Etievant, P.; Issanchou, S.; Langlois, D. *J. Agric. Food Chem.* **1993,** *41*, 1698-1703.
21. Schieberle, P.; Gassenmeier, K.; Guth, H.; Sen, A.; Grosch, W. *Lebensm. Wiss. Technol.* **1993**, *26*, 347.
22. Schieberle, P. In *Trends in Flavour Research*; Maarse, H.; Van der Heij, D.J., Eds.; Elsevier: Amsterdam, **1994,** pp 345-351.
23. Schieberle, P.; Hofmann T. *J. Agric. Food Chem.* **1997,** *45,* 227-232.
24. Dacremont, C.; Vickers, Z. *J. Food Sci.* **1994,** *59,* 981-985.
25. Young, H.; Gilbert, J.M.; Murray, S.H.; Ball, R.D. *J. Sci. Food Agric.* **1996,** *71,* 329-336.
26. Buttery, R.G.; Teranishi, R.; Ling L.C. *J. Agric. Food Chem.* **1987,** *35,* 540-544.
27. Buttery, R.G.; Teranishi, R.; Ling, L.C.; Turnbaugh J.G. *J. Agric. Food Chem.* **1990,** *38,* 336-340.
28. White, A.G. *Fruit Var. J.* **1991,** *45,* 2-3.
29. Green, D.W.; Autio, W.R. *Fruit Var. J.* **1990,** *22,* 1-4.
30. Patterson, B.D.; Hatfield, S.G.S.; Knee, M. *J. Sci. Food Agric.* **1974,** *25,*843-849.
31. Plotto, A.; McDaniel, M.R.; Mattheis, J.P. Presented at the 92nd Annual meeting of the American Society for Horticultural Science, Montréal, QUE, August 1995; *HortScience.* **1995**, *30*, 889.
32. Mattheis, J.P.; Buchanan, D.A.; Fellman, J.K. J. *Agric. Food Chem.* **1991**, *39*,1602-1605.
33. Flath, R.A.; Black, D.R.; Guadagni, D.G.; McFadden, W.H.; Schultz T.H. *J. Agric. Food Chem.* **1967,** *15,* 29-35.
34. Buttery, R.G.; Bomben, J.L.; Guadagni, D.G.; Ling, L.C. *J. Agri. Food Chem.* **1971,** *19,* 1045-1048.
35. Jordan, T.E. *Vapor Pressure of Organic Compounds.* Interscience: New York, **1954.**

36. Lyman, W.J.; Reehl, W.F.; Rosenblatt, D.H. *Handbook of chemical property estimation methods. Environmental Behavior of Organic Compounds;* Mc Graw-Hill Book Company: New York, **1982**.
37. Pierotti, G.J.; Deal, C.H.; Derr E.L. *Ind. Eng. Chem.* **1959**, *51(1)*,95-102.
38. Poll, L.; Hansen K. *Lebensm.-Wiss.u.-Technol.* **1990,** *23,* 481-483.
39. Takeoka, G.R.; Flath, R.A.; Mon, T.R.; Teranishi, R.; Guentert, M. *J. Agric. Food Chem.* **1990,** *38,* 471-477.
40. Takeoka, G.R.; Buttery, R.G.; Flath, R.A.; Teranishi, R.; Wheeler, E.L.; Wieczorek, R.L.; Guentert, M. **1989,** In *Flavor Chemistry: Trends and Developments*; Teranishi, R.; Buttery, R.G.; Shahidi, F., Eds.; ACS Symposium Series 388; American Chemical Society: Washington, DC. **1989,** pp 223-237.
41. Williams, A.A.; Lea, A.G.H.; Timberlake, C.F. In *Flavor Quality: Objective Measurement*; Scanlan, R.A., Ed.; ACS Symposium Series 51; American Chemical Society: Washington, DC. **1977,** pp 71-88.
42. Pangborn, R.M.; Berg, H.W.; Roessler, E.B.; Webb, A.D. *Perceptual and motor skills.* **1964,** *18,* 91-103.
43. Larsen, M.; Poll L. *Z. Lebensm. Unters. Forsch.* **1990,** *191,*129-131.
44. Guadagni, D.G.; Buttery, R.G.; Okano, S. *J, Sci. Food Agric.* **1963,** *14,* 761-765.
45. Takeoka, G.; Buttery, R.G.; Ling, L.C. *Lebensm. Wiss. Technol.* **1996,** *29,* 677-680.
46. Dravnieks, A. In *Flavor Quality: Objective Measurement*; Scanlan, R.A., Ed.; ACS Symposium Series 51; American Chemical Society: Washington, DC, **1977,** pp 11-28.
47. Laing, D.G.; Panhuber H. In *Progress in Flavour Research;* Land, D.G.; Nursten, H.E., Eds.; Applied Science Publishers: London, **1979,** pp 27-46.
48. Moncrieff, R.W. *J. Physiol.* **1956,** *133,* 301-316.
49. Cain, W.S.; Polack, E.H. *Chem. Senses.* **1992,** *17,*481-491.
50. Pierce, J.D., Jr.; Zeng, X.-N.; Aronov, E.V.; Preti, G.; Wysocki, C.J. *Chem. Senses.* **1995,** *20,* 401-411.
51. Reineccius, G. In *Flavor Measurements;* Ho, C.-T.; and Manley, C.H., Eds.; Marcel Dekker: New York, **1993**, pp 61-76.
52. Weurman, C. *J. Agric. Food Chem.* **1969,** *17,* 370-384.
53. Brackmann, A.; Streif, J.; Bangerth. F. *J. Amer. Soc. Hort. Sci.* **1993,** *118,* 243-247.

Chapter 25

Chemical and Sensory Analysis of Off-Flavors in Citrus Products

M. Naim[1], R. L. Rouseff[2], U. Zehavi[1], O. Schutz[1], and E. Halvera-Toledo[1]

[1]Institute of Biochemistry, Food Science and Nutrition, Faculty of Agricultural, Food and Environmental Quality Sciences, The Hebrew University of Jerusalem, Rehovot 76-100, Israel

[2]Citrus Research and Education Center, University of Florida, Lake Alfred, FL 33850

p-Vinylguaiacol (PVG), 4-hydroxy-2,5-dimethyl-3-(2*H*)-furanone (DMHF, Furaneol) and α-terpineol are known off-flavors in stored citrus products that exceed the taste threshold level under practical processing and storage conditions. PVG is formed from ferulic acid; fortification by thiols such as L-cysteine, N-acetyl-L-cysteine, or glutathione shifts ferulic acid degradation from PVG to vanillin formation. Furaneol is formed under acidic conditions of citrus juices from rhamnose interacting with arginine or γ-aminobutyric acid (GABA); thiol fortification reduces Furaneol content. α-Terpineol is formed from *d*-limonene or linalool but thiol fortification does not reduce its content. Aroma similarity results were preferred over acceptance data for determination of aroma quality of citrus juice in the laboratory. Acceptance of pasteurized juice and GC-Olfactometry of stored juice suggest that important off-flavors other than the above may be formed in processed citrus products.

Color deterioration and appearance of undesirable volatile compounds in citrus products during storage is well documented and the sensory impact of these compounds on quality is negative (*1,2*). Flavor usually becomes unacceptable before color becomes objectional (*3*). Some flavor degradation, along with a significant loss of volatile compounds, occurs when citrus juice is concentrated. Aqueous essence and cold pressed peel oil are sometimes added to restore the original flavor and to mask off-flavors. Three major off-flavors, p-vinylguaiacol (PVG), 4-hydroxy-2,5-dimethyl-3-(2*H*)-furanone (DMHF, Furaneol) and α-terpineol, which can exceed their taste thresholds under practical processing and storage conditions were suggested to be main off-flavors formed during processing and storage of canned orange juice (*4*). Hence, studies were initiated to reveal their precursors and appropriate methodology has been developed to determine these compounds and their precursors in citrus fruit and products (*5-11*). Furthermore, pathways by which PVG and Furaneol are formed in citrus products were suggested (*7,12*); the possible use of natural thiols such as L-cysteine and N-acetyl-L-cysteine to inhibit the formation of these off-flavors was proposed (*13*). In this paper we discuss the results of recent studies related mainly to PVG and Furaneol in citrus juice and evaluate sensory tools to determine their significance to product acceptance. The effect of natural thiols on off-flavor formation in citrus products under simulated industrial processing conditions was evaluated.

Precursors for p-Vinylguaiacol (PVG), Formation and Determination

PVG and other vinyl phenols are produced from free phenolic acids and contribute to either desirable or objectional aroma of important food products (*14*). It has been proposed that in stored orange juice, PVG is a major detrimental off-flavor with a taste threshold of 50 ppb (*2,4*). PVG formation increased during storage of orange juice (*6,9*) and this formation is accelerated if the juice is fortified with ferulic acid. The resultant juice had greatly inferior aroma quality when compared to controls. A mechanism by which cinnamic acids are decarboxylated to vinyl phenols has been proposed (*15*), and the pathway in which PVG is formed from ferulic acid has been delineated (*7*). In all citrus fruit parts, only minor amounts of cinnamic acids occur in the free state while most are present in bound forms (*16*). A concentration gradient of phenolic acids appears to exist from the flavedo towards the juice vesicles; suggesting that, processing technologies that minimize juice contact with peel components during juice extraction will reduce the amount of phenolic acids transferred into the juice. Potential sources of free ferulic acid in oranges and grapefruits are the bound forms of the acid, five of them are currently known: feruloylputrescine, feruloylglucose, feruloylglucaric, diferuloylglucaric, and feruloylgalactaric acids (*17-20*). Bound or conjugated forms of ferulic acid, depending on the type of linkage with ferulic acid in each case, may differ significantly in their ability to serve as precursors for ferulic acid (*21*) and the resulting objectional aroma of PVG. Nevertheless, the amount of free ferulic acid present in citrus fruit after harvesting (*16*) exceeds the amount needed to produce PVG at levels above its taste threshold during juice processing and storage.

Current procedures for the extraction and determination of PVG in citrus juice are based on the procedure described by Lee and Nagy (*9*). With some modifications, this procedure was found to be simple and accurate (*22*). Briefly, following centrifugation of juice samples (10 min, 14,000 g, 4 °C), 1 mL aliquots of the supernatants were applied to C-18 Sep-Pak cartridges (Waters Associates, Medford, MA) that had been conditioned with 3 mL methanol and 5 mL water. Cartridges were then washed with 1 mL water followed by 2 mL hexane. The hexane fraction was filtered before being applied to the HPLC equipped with a Lichrosorb RP-18 column (7 μm, 250 mm x 4 mm, Merck) with a Lichrosphere 100 RP-18 pre-column (25 mm x 4 mm, Merck). 100-μL Aliquot of each extract was injected and the column was eluted isocratically, at room temperature, with 1.5% acetic acid in water/methanol (60:40 v/v) at a flow rate of 1 mL/min. The separated chromatographic peaks were identified and quantified using a F-1050 Merck-Hitachi fluorescent detector (excitation 300 nm, emission 340 nm). PVG could be clearly detected at concentrations of 1 ppb and above (RT=14.45±0.25 min). The recovered yield of an authentic PVG sample (0.25 μg) via this procedure was 89%. Rouseff et al. (10), modified this procedure. They used a water-miscible solvent such as tetrahydrofuran (THF), that has sufficient solvent strength to elute both PVG and its precursor, ferulic acid. This allows determination of both from a single sample preparation and chromatographic analysis.

Formation of PVG During Pasteurization and Storage under Simulated Commercial Conditions

The use of natural thiols such as *L*-cysteine to reduce PVG formation under accelerated storage conditions of commercial orange juice was proposed (*23*). The objective of the following study was to evaluate the effect that thiol fortification of orange juice has on PVG formed under pasteurization and storage conditions designed to approximate those in the marketplace. Fresh orange juice was fortified with either *L*-cysteine, *N*-acetyl-*L*-cysteine or glutathione, pasteurized in a pilot plant (90-92 °C, 30 sec), hot filled into 250-mL glass bottles, and immediately cooled to room temperature in a water bath (200-fold volume) (*22*). For the pasteurization experiment, samples were kept at -18 °C and then subjected to chemical analyses with no prior storage. For the storage experiment, bottles were then stored for 12 weeks at 25 or 35 °C. Control samples were stored at 4 °C. An hedonic taste test (1-9 scale, 1 for very aversive, 9 for very appealing, and other ratings in between) (*24*) was conducted to determine juice acceptance.

Although pasteurization stimulated PVG formation, its level did not reach the 50 μg/L taste threshold (Table I). It should be noted, however, that pasteurization of orange juice has been found to release ferulic acid from its bound forms (*6*), thereby increasing the precursor content for subsequent PVG formation during storage. On the other hand, thiol fortification significantly improved the acceptance of pasteurized juice, with *N*-acetyl-*L*-cysteine and glutathione appearing to be the most effective (Table I). Thus, acceptance of pasteurized juice was improved by thiol fortification under conditions where the PVG content is below its taste threshold. This suggests that during pasteurization thiols inhibit the formation or induce the masking of off-flavors other than PVG.

Juice stored at 4 °C for 12 weeks contained 3.6 μg/L PVG. Juice stored at 25 °C contained 75 μg/L PVG (Figure 1), a value above the 50 μg/L taste threshold (*4*). Thus, PVG content increased by 20-fold during storage at 25 °C and reached a level that could reduce juice quality. The use of PVG concentration as an indicator for quality deterioration under normal storage conditions is therefore suggested. Similar storage at 35 °C elevated PVG level about 10-fold above the taste threshold (Figure 1), undoubtedly contributing significantly to reduced quality.

Thiol concentrations in these experiments were selected to be below those found in previous experiments to form detectable, objectionable odors of their own. The low cysteine levels did not significantly inhibit PVG formation in juice stored at 25 °C, but reduced PVG content by 20% in juice stored at 35 °C. *N*-Acetyl-*L*-cysteine and glutathione, which can be used at higher concentrations without forming sulfur

Table I. The effect of L-cysteine (Cys), N-acetyl-L-cysteine (AcCys) and glutathione (GSH) fortification on PVG formation and acceptance of orange juice during pasteurization

Additives (mM)	PVG (μg/L)	Acceptance (1-9 scale)
AcCys 4.0	8.4±1.9	5.42±1.9 [a]
AcCys 3.0	14.6±3.4	5.38±1.9 [ab]
GSH 2.0	16.8±2.5	5.19±2.1 [ab]
GSH 3.0	19.3±0.2	5.02±2.0 [ab]
AcCys 2.0	20.2±4.1	4.96±2.2 [ab]
Cys 1.0	10.5±1.5	4.75±2.2 [ab]
GSH 4.0	15.6±0.8	4.65±2.0 [ab]
Cys 0.5	14.9±2.5	4.58±2.2 [b]
No additive	8.1±0.9	3.70±2.5 [c]

PVG values are the means ± SEM of three replicates, each tested twice by HPLC. Values for the hedonic test are the means ± SEM of 24 subjects. Values not sharing the same superscript letter are different at $p < 0.05$ (Adapted from ref. 22).

Containing off-flavors, were more effective; PVG formation was reduced by up to 50% as compared to the unadulterated juice. Glutathione (1.0-1.5 mM) also inhibited PVG formation.

Pathways for Furaneol Formation in Citrus Juice

As indicated above, Furaneol, along with PVG and α-terpineol, has been proposed to be responsible for the objectionable flavor in aged canned orange juice, and it was found to mask the original aroma of orange juice with a detection threshold of 0.1 ppm (*4,25*). Furaneol appears to be formed at relatively high storage temperatures (*8*), and its accumulation under accelerated storage can be reduced by thiol fortification (*26*).

Furanones, produced mostly via Maillard reactions, generally impart caramel-like, sweet, fruity flavors. It is reported that Furaneol is formed in heat-processed foods containing hexoses (*27*). In processed fruit juices, its presence is proportional to the content of methyl pentoses, suggesting that it is a product of a carbonyl and amine interaction during the final stage of ripening (*28*) or during thermal processing. Furaneol has been reported to be produced during base-catalyzed fructose degradation (*29*) and at pH 3.5 from rhamnose and alanine (*30*). A unified sample preparation and a modified chromatographic procedure for the determination of Furaneol and PVG was recently developed (*11*). It was possible to isolate both PVG and Furaneol by C_{18} solid phase extraction in a single extract. A 30 min reversed-phase HPLC

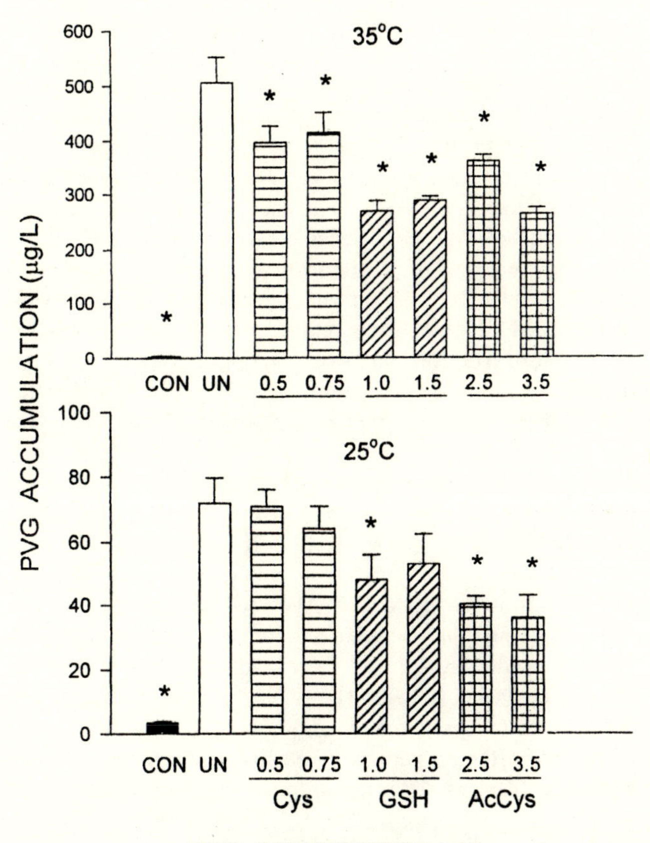

Figure 1. Effect of L-cysteine (Cys), glutathione (GSH), and N-acetyl-L-cysteine (AcCys) on PVG formation in orange juice stored at 25 or 35 °C for 12 weeks. Control samples (CON) were kept at 4 °C. UN indicates unfortified juice samples. Values are the mean and SEM of three replicates, each tested twice by HPLC (Adapted from ref. 22).

gradient method employing UV and fluorescence detectors in series allowed the determination of both off-flavors from a single injection.

Glucosidically bound Furaneol has been reported in strawberry (*31*), pineapple (*32*), raspberry (*33*), and tomato (*34*). One may assume that the presence of Furaneol in stored citrus products might be due to a release of Furaneol from putative Furaneol glycosides, as found in other fruits. However, glycosidically bound Furaneol, which might be formed from reducing sugars if they are present in significant concentrations, has not been reported in citrus fruit and is unlikely to be present in citrus juice. We observed no release of free Furaneol after applying β-glucosidase to grapefruit and orange juice extracts from which free sugars and free volatiles had been removed using Amberlite XAD-2. Therefore, sugar-amine (Maillard) reactions occurring during processing and storage are likely to be the source for Furaneol in citrus juice.

The presence of rhamnose in orange and grapefruit juices, as the likely sugar precursor for Furaneol formation via a sugar-amine interaction, was verified in a recent study (*12*). Rhamnose was identified and determined in citrus juice by GC of the trimethylsilyl (TMS) derivatives. Presumably the galacturonic acid and rhamnose found in citrus juice results from the enzymic degradation of pectin which occurs during processing and storage. Apparently, the enzymic release (e.g., via naringinase) of rhamnose from the flavonoids naringin (grapefruit) or hesperidin (orange) may contribute additional free rhamnose (*35*). Although overall rhamnose content is low, it is higher in commercial juices than in fresh juice (*12*), probably due to processing and pasteurization of the former.

Furaneol was not formed during storage of acidic model solutions of orange juice (MOJ) which contained the main sugars and amino acids except rhamnose (*12*). The addition of rhamnose to MOJ resulted in Furaneol after storage. Experiments employing modified MOJ solutions, in which amino acids were added separately, showed arginine and γ-aminobutyric acid (GABA) to be the most reactive in terms of Furaneol formation. The fact that the basic amino acids arginine and GABA, which are abundant in citrus products (*36*), were very reactive is thus relevant to Furaneol formation in citrus juice during storage. The small rhamnose concentrations in citrus juices reported elsewhere (*37*), and the range of 2-3 ppm found in commercial juice (*12*), may be sufficient for the formation of the above-taste-threshold level (0.1 ppm) of Furaneol occurring in citrus juices stored for long periods at temperatures above 25 °C (*8,26*).

Additional experiments using buffered solutions indicated that under acidic conditions Furaneol was formed only in the presence of rhamnose and arginine, and its level increased with increasing pH (Figure 2a). Neither rhamnose alone nor other sugars added to acidic solutions (pH 3.5) resulted in Furaneol formation. At pHs above 6, rhamnose, glucose and fructose with or without amino acids, were active substrates for Furaneol formation (*12*). In fact, the combination of rhamnose and arginine resulted in Furaneol formation which was 40- to 50-fold higher than any other sugar-amine combination. During the Maillard reaction, the rhamnose-amine interaction, which is apparently an important stage for Furaneol formation via the 2,3-enolization (Figure 3), is favored at higher pH values (Figure 2a). On the other hand,

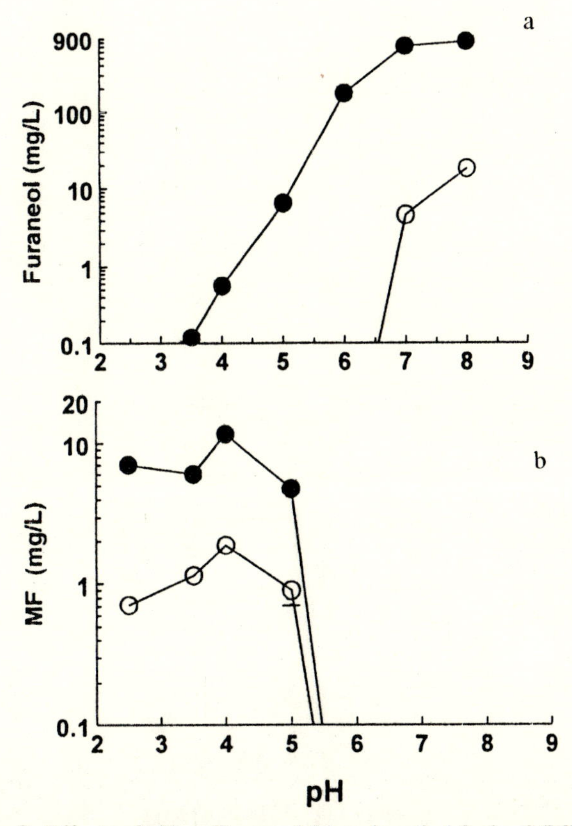

Figure 2. Effects of pH on Furaneol (a) and methyl furfural (MF, b) formation in buffered solutions containing rhamnose alone (open circles) or rhamnose and arginine (closed circles). Values are the means of two samples determined by HPLC. SEM values were too small to be presented (Adapted from ref. 12).

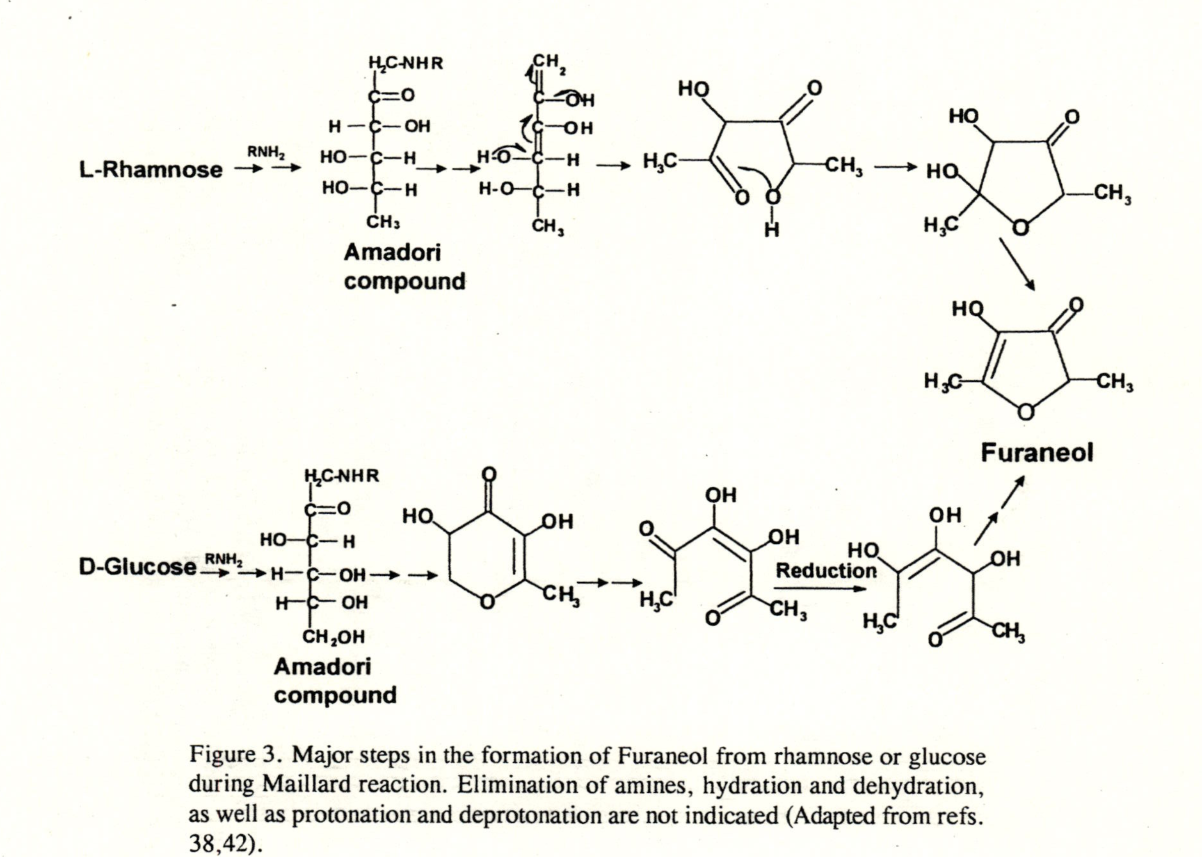

Figure 3. Major steps in the formation of Furaneol from rhamnose or glucose during Maillard reaction. Elimination of amines, hydration and dehydration, as well as protonation and deprotonation are not indicated (Adapted from refs. 38,42).

1,2- enolization with further dehydration of glucose and fructose to hydroxymethyl furfural (HMF) (*12*) and rhamnose to methyl furfural (MF) is favored at lower pH (Figure 2b). The formation of Furaneol from hexoses must involve reduction (*38*), a step not needed with 6-deoxyhexsoses such as rhamnose (Figure 3). Reducing agents required for Furaneol production can be formed from glucose or fructose-amine interactions in the presence of ascorbic acid, or other unsaturated carbonyl compounds (reductones) formed via Maillard reactions through 2,3 enolization (*39-41*).

Indeed, the addition of the natural reductone, ascorbic acid, which structurally is closely related to the Maillard reductones, significantly enhanced Furaneol formation from glucose and arginine at pH 6, 7, and 8 in incubated buffered solutions (*12*). In contrast, the addition of the same amounts of ascorbic acid to rhamnose-arginine solutions adjusted to pH 3.5 did not enhance Furaneol formation above that observed in absence of ascorbic acid (data not shown). No Furaneol was detected in stored glucose-arginine solutions under acidic conditions after the addition of ascorbic acid.

Thus, in Furaneol formation, one needs to differentiate between hexoses and 6-deoxyhexoses (*38,42*). Steps of the Maillard reaction (in the presence of amino acids), dehydration, and cyclization are suggested in both cases (Figure 3); hexoses, however, also require a reduction step. The small amount of rhamnose present in citrus juices appears to be the substrate for Furaneol formation during Maillard reactions in citrus products. Ascorbic acid and other reductones, may further protect Furaneol as it is generally unstable in air and in aqueous solutions (*43,44*).

Similarity, Hedonic Tests and Gas Chromatography-Olfactometry (GC-O) in Evaluation of Aroma Quality of Citrus Products

A variety of parametric and nonparametric statistical taste tests are available to determine food sensory quality. Evidently, organoleptic tests are the best tool to evaluate the overall product acceptance by consumers. Normally, parametric tests (due to their strength), such as hedonic-scale sensory tests (*24*), are recommended to determine product acceptance. In some cases, hedonic scaling procedures may even be better in detection of sensory differences in citrus juice samples than discrimination tests (e.g., pair comparisons and triangle tests) (*6*). When a reference for a desirable acceptance level of a product is available (e.g., the aroma of fresh orange juice compared to stored juice), the choice of similarity tests (*45,46*) may be considered. These tests, which basically indicate correlation coefficient rather than regular statistical analysis, may sometimes be more sensitive and provide more relevant information on quality than the hedonic scaling tests. In the following study, the effect of thiol fortification on aroma similarity and aroma acceptance of stored orange juice was evaluated. Pasteurized control and thiol-fortified orange juice samples were stored for 12 weeks at 35 °C (*22*).

Aroma Similarity. Aroma similarity was determined in fortified and unadulterated juice samples after storage using aroma similarity cluster and multidimensional scaling (MDS) analyses (*26,45,46*). Panelists opened the vials of two samples, smelled them, and were requested to rate the similarity level of their aromas on a scale of 1 to 20 (1 for no similarity, 20 for identical). During a 2-h aroma session, four 15-min aroma similarity tests were conducted. Eight sample pairs were presented in coded, randomized order per test, resulting in 32 pairs tested during each aroma session. Two sessions were conducted with at least 4-h interval, so that each panelist rated the similarity levels of 64 treatment combinations. A data matrix representing the aroma similarity results was obtained which was analyzed by the cluster program ADDTREE (*45*), yielding a tree structure of branches and subdivisions with aromas located at the ends of the branches (Figure 4a). Sets of aromas connected to the same node at relatively short (horizontal) distances from each other were highly similar. MDS which is another way for analysis and presentation of sensory similarity data in a space diagram (*46*) was applied to the same similarity data (Figure 4b).

Cluster analysis of the aroma similarity data showed that the aroma of juice fortified with 1.0 mM glutathione (branch 4) was connected to the same node as the control juice (branch 8) kept at 4 °C (r^2 = 0.98), indicating that both exhibited a similar aroma (Figure 4a). This suggests that fortification with 1.0 mM glutathione is most effective, compared with other treatments, at retaining the aroma of the original juice. MDS of the same aroma similarity data (Figure 4b) clearly shows that, in the space diagram, the aroma of juice fortified by 1 mM glutathione (column 4) and that by 3.5 mM *N*-acetyl-*L*-cysteine (column 7) were located close to the aroma of the control juice (column 8) kept at 4 °C. Coefficient alienation of the MDS analysis was 0.0075, much below the 0.15 reliability border indicating a very high correlation of the space diagram with original matrix data. On the other hand, the aroma of the unfortified stored juice (column 1) was located far away in the space diagram, and in all cases, the aroma of thiol-fortified samples was located closer to the aroma of the control (4 °C) juice than that of the unadulterated stored juice. It thus appears that juice stored in the presence of small concentrations of these thiols had an aroma that was much more similar to fresh juice aroma than samples stored without thiol fortification.

Hedonic Test. To compare the aroma similarity results with acceptance data, juice aroma acceptance was evaluated by an hedonic test (1-9 scale described above) conducted with the same 8 taste panelists who participated in the aroma similarity experiment. The aroma of the unadulterated juice received the lowest acceptance score, the control juice kept at 4 °C received the highest score whereas the aromas of thiol-fortified juices received significantly higher scores than that of the unadulterated juice (Table II). It may, therefore, be concluded that the sensory similarity results are in agreement with those of the hedonic ratings. Furthermore, the sensory results are compatible with the chemical analyses suggesting the use of low doses of natural thiols to retain the quality of citrus juice.

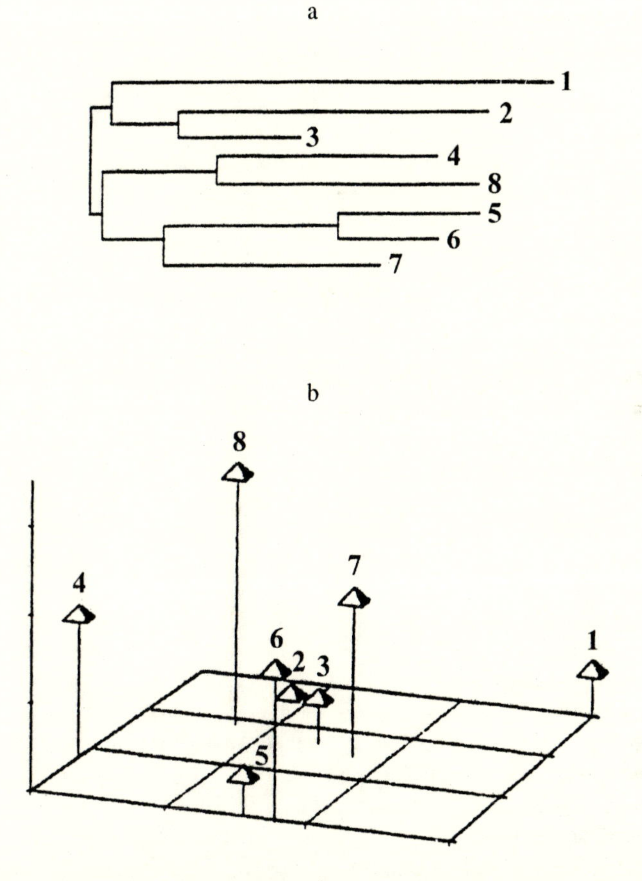

Figure 4. Cluster analysis by ADDTREE (a) and MDS analysis (b) of aroma similarity values of orange juice stored at 35 °C for 12 weeks. Branches or columns: 1 - Juice stored at 35 °C for 12 weeks; 2 - fortified with 0.5 mM *L*-cysteine; 3 - with 0.75 mM *L*-cysteine; 4 - with 1.0 mM glutathione; 5 - with 1.5 mM glutathione; 6 - with 2.5 mM *N*-acetyl-*L*-cysteine; 7 - with 3.5 mM *N*-acetyl-*L*-cysteine; 8 - control juice stored after pasteurization at 4 °C. Each value in the original matrix represents the mean of eight 8 panelists (each tested twice) (Figure 4a: Reproduced with permission from ref. 22. Copyright 1997 ACS).

Table II. Effect of thiol fortification on aroma acceptance of juice stored at 35 °C for 12 weeks

Additives (mM)	Acceptance (1-9 hedonic scale)
1. No Additive	3.56±0.47 ‡
2. Cys 0.5	4.63±0.45 ‡
3. Cys 0.75	5.09±0.42 *‡
4. GSH 1.0	4.91±0.64 *‡
5. GSH 1.5	4.94±0.48 *‡
6. AcCys 2.5	4.94±0.41 *‡
7. AcCys 3.5	5.25±0.50 *‡
8. Control (4°C)	7.38±0.35 *

Values are the means ± SEM of 8 subjects. * indicates a significantly ($p<0.02$, t-test) higher hedonic score than that given to "no additive" samples. ‡ indicates a significantly ($p<0.01$, t-test) lower hedonic score than that given to control juice kept at 4 °C (Adapted from ref. 22).

GC-Olfactometry. The possibility that, in addition to PVG, Furaneol, and α-terpineol, other detrimental volatiles may develop has not been investigated systematically. The GC-Olfactometry (GC-O technique) combines the improved separation powers of modern capillary chromatography with the sensitivity and selectivity of the human nose. This technique detects only those compounds that produce aroma sensation. The Charm or dilution value is a measure of the relative intensity of each aroma detected (*47*). Charm and Aroma Extraction Dilution Analysis (AEDA) are two of the best known GC-O techniques that employ dilution to determine intensity. Osme is a slightly different GC-O procedure which is basically a time-intensity technique that requires a human assessor to estimate the aroma intensity of each aroma peak (*48*).

The GC Flame Ionization Detection (FID) and human nose response, (GC-O, Charm procedure) chromatograms (aromagrams) of orange juice stored at 40 °C for four months are shown in Figure 5 (a and b, respectively). These results indicate that the most intense peaks in the GC-O aromagrams (see b) are bread-, sulfur-, rotten-, and two floral-like aromas. These aroma compounds were not observed in similar aromagrams of juice stored at 4 °C for two weeks. The aromagram represents the results from at least six separate GC runs. Each injection is 1/3 as dilute as the previous injection and is replicated at least once. Over 73 different aromas were detected in the most concentrated extract. Twenty five aromas were detected in the first dilution. Only five aroma compounds were detected in the most dilute sample and their aroma qualities are shown next to each peak. The bread-, sulfur-, rotten-, and two floral-like aromas at Kovat's retention index values of 865, 1283, 1327,

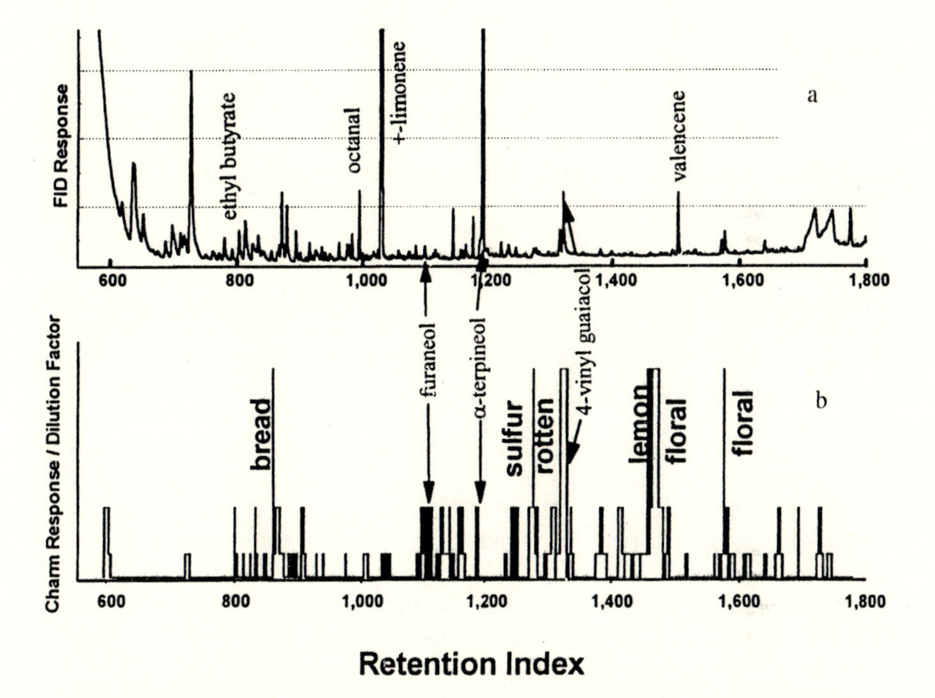

Figure 5. Comparison of GC-Olfactometry (b) with the GC-FID response (a) from an identical sample of orange juice stored at 40 °C for 4 months.

1473, and 1582, in the aromagram were the most intense peaks. Similar aromas appeared in GC-O aromagrams of juice stored at 20 °C for four months but were not as intense. As shown in the chromatogram and aromagram labeled a and b, respectively, only the rotten aroma (later identified as PVG) was detected by FID. The other four are new and so potent that their FID response was extremely small or in the case of bread aroma, lost in a grouping of other small peaks. Some components with high aroma detection threshold such as *d*-limonene, have a large FID peak but little if any odor response (Charm value). Other components, as for example the sulfur peak at retention (Kovat's) index 1283, gave a very intense aroma response but produced little if any FID response. This suggests that some bioactive compounds had extremely low aroma thresholds and that they would need specialized techniques to detect and quantify them. Furaneol and α-terpineol exhibited only mid level aroma intensity. α-Terpineol was easily detected with FID. Its peak height increased with increasing storage time and temperature and is a good quality control index for storage abuse. Valencene, a positive aroma component, decreases with increasing storage temperature but only produces a mid level odor response from the aromagram from the juice stored at 40 °C for four months.

Concluding Remarks and Research Needs

PVG, a major detrimental flavor in citrus products, is formed at above taste-threshold levels under simulated commercial storage conditions. This compound can be determined at ppb levels using a HPLC equipped with a fluorescent detector. Fortification of citrus products with low doses (0.5-4 mM) of L-cysteine, N-acetyl-L-cysteine or glutathione significantly inhibit PVG formation (in some cases to below taste threshold) and Furaneol accumulation in juice stored under simulated industrial conditions. Fortification with low doses of thiols also reduced browning and ascorbic-acid degradation during pasteurization and storage of orange juice. Thus, fortification of citrus juices to maintain quality during pasteurization and storage is strongly recommended. Such fortification should be considered as an enrichment since some endogenic L-cysteine and glutathione are already present in citrus juice. Furaneol is accumulated under the acidic conditions of citrus juices to above taste-threshold levels only during storage at temperatures above 25 °C, and it can be determined concomitantly with PVG by a HPLC method employing UV and fluorescent detectors in series. Furaneol accumulation in citrus juice is related to reportedly small amounts of rhamnose interacting with arginine and GABA via a Maillard reaction.

Evidently, although chemical analyses are useful for the determination of off-flavors, only sensory evaluation will determine acceptance. When a desirable reference for acceptance is available is available (e.g., fresh citrus juice), the use of sensory similarity procedures is recommended for laboratory purposes prior to consumer testing. The results of GC-Olfactometry and the fact that the presence of thiols during pasteurization improved orange juice acceptance regardless of PVG content suggests that unidentified off-flavors are formed in citrus juice during processing. The identification of these newly discovered off-flavors and determination of their precursors is a significant research need.

Acknowledgments

This research was supported in part by Grant No. I-1967-91 from BARD, The United States-Israel Binational Agricultural Research & Development Fund. Comments by I. Zuker, S. Nagy and the statistical advise by Dr. H. Voet are highly appreciated. We also thank the excellent technical assitance by Mrs. M. Levinson.

Literature Cited

1. Dinsmore, H. L.; Nagy, S. *J. Agric. Food Chem.* **1971,** *19*, 517-519.
2. Handwerk, R. L.; Coleman, R. L. *J. Agric. Food Chem.* **1988,** *36*, 231-236.
3. Shaw, P. E.; Tatum, J. H.; Berry, R. E. *Dev. Food Carbohydr.* **1977,** *1*, 91-111.
4. Tatum, J. H.; Nagy, S.; Berry, R. E. *J. Food Sci.* **1975,** *40*, 707-709.
5. Moshonas, M. G.; Shaw, P. E. *J. Agric. Food Chem.* **1989,** *37*, 157-161.
6. Naim, M.; Striem, B. J.; Kanner, J.; Peleg, H. *J. Food Sci.* **1988,** *53*, 500-503.
7. Peleg, H.; Naim, M.; Zehavi, U.; Rouseff, R. L.; Nagy, S. *J. Agric. Food Chem.* **1992,** *40*, 764-767.
8. Lee, H. S.; Nagy, S. *J. Food Sci.* **1987,** *52*, 163-165.
9. Lee, H. S.; Nagy, S. *J. Food Sci.* **1990,** *55*, 162-163 & 166.
10. Rouseff, R. L.; Dettweiler, G. R.; Swaine, R. M.; Naim, M.; Zehavi, U. *J. Chromatog. Sci.* **1992,** *30*, 383-387.
11. Walsh, M.; Rouseff, R.; Naim, M. *J. Agric. Food Chem.* **1997,** *45*, 1320-1324.
12. Haleva-Toledo, E.; Naim, M.; Zehavi, U.; Rouseff, R. L. *J. Agric. Food Chem.* **1997,** *45*, 1314-1319.
13. Naim, M.; Zehavi , U.; Zuker , I.; Rouseff , R. L.; Nagy, S. In *Sulfur Compounds in Foods; ACS Symposium Series 564*; Mussinan , C. J., Keelan, M. E., Eds.; American Chemical Society: Washington DC, 1994, pp 80-89.
14. Maga, J. A. *CRC Critical reviews in Food Science Nutrition* **1978,** *10*, 323-372.
15. Pyysalo, T.; Torkkeli, H.; Honkanen, E. *Lebensem-Wiss. u.Technol.* **1977,** *10*, 145-147.
16. Peleg, H.; Naim, M.; Rouseff, R. L.; Zehavi, U. *J. Sci. Food Agric.* **1991,** *57*, 417-426.
17. Wheaton, T. A.; Stewart, I. *Nature* **1965,** *206*, 620-621.
18. Risch, B.; Herrmann, K.; Wray, V.; Grotjahn, L. *Phytochemistry* **1987,** *26*, 509-510.
19. Risch, B.; Herrmann, K.; Wary, V. *Phytochemistry* **1988,** *27*, 3327-3329.
20. Reschke, A.; Herrmann, K. *Z. Lebensm Unters-Forsch* **1981,** *173*, 458-463

21. Peleg, H.; Striem, B. J.; Naim, M.; Zehavi, U. In *Proceeding of the Sixth International Citrus Congress Middle-East*; Goren , R., Mendel , K., Eds.; Weikersheim: Balaban, Rehovot and Margraf, 1988; Vol. 4, pp 1743-1748.
22. Naim, M.; Schutz, O.; Zehavi, U.; Rouseff, R. L.; Haleva-Toledo, E. *J. Agric. Food Chem.* **1997,** *45*, 1861-1867.
23. Naim, M.; Zuker, I.; Zehavi, U.; Rouseff, R. L. *J. Agric. Food Chem.* **1993,** *41*, 1359-1361.
24. Moskowitz, H. R. In *The Chemical Senses and Nutrition*; Kare, M. R., Maller, O., Eds.; Academic Press: New York, 1977, pp 71-96.
25. Nagy, S.; Rouseff, R. L.; Lee, H. S. In *Thermal Generation of Aromas; ACS Symposium Series 409*; Parliment, T. H., McGorrin, R. C., Ho, C. T., Eds.; American Chemical Society: Washington DC, 1989, pp 331-345.
26. Naim, M.; Wainish, S.; Zehavi, U.; Peleg, H.; Rouseff, R. L.; Nagy, S. *J. Agric. Food Chem.* **1993,** *41*, 1355-1358.
27. Schieberle, P. In *Flavor Precursors-Thermal and Enzymatic Conversions; ACS Symposium Series 490*; Teranishi, R., Takeoka, G. R., Guntert, M., Eds.; American Chemical Society: Wshungton, DC, 1992, pp 164-175.
28. Pisarnitskii, A. F.; Egorov, I. A.; Egofarova, R.; Erygin, G. D. *Applied Biochemistry and Microbiology* **1979,** *15*, 673-678.
29. Shaw, E. P.; Tatum , H. J.; Berry, E. R. *J. Agric. Food Chem.* **1968,** *16*, 979-982.
30. Shaw, P. E.; Berry, R. E. *J. Agric. Food Chem.* **1977,** *25*, 641-644.
31. Mayerl, F.; Naef, R.; Thomas, A. F. *Phytochemistry* **1989,** *28*, 631-633.
32. Wu, P.; Kuo, M. C.; Hartman, T. G.; Rosen, R. T.; Ho, C. T. *J. Agric. Food Chem.* **1991,** *39*, 170-172.
33. Pabst, A.; Barron, D.; Etievant, P.; Schreier, P. *J. Agric. Food chem.* **1991,** *39*, 173-175.
34. Krammer, E. G.; Takeoka, G. A.; Buttery , G. R. *J. Agri. Food Chem.* **1994,** *42*, 1595-1597.
35. Romero, C.; Manjon, A.; Bastida, J.; Iborra, J. L. *Anal. Biochem.* **1985,** *149*, 566-571.
36. Rouseff, R. L.; Ting, S. V. In *Citrus fruits and their products analysis and technology.*; Tannenbaum, S. R., Walsta., P., Eds.; Marcel Dekker, Inc.: New York, NY, 1986.
37. Stepak, Y.; Lifshitz, A. *Journal of the Association of Official Analytical Chemists* **1971,** *54*, 5.
38. Kim, M. O.; Baltes, W. *J. Agric. Food Chem.* **1996,** *44*, 282-289.
39. Mills, F. D.; Hodge, J. E. *Carbohydr. Res.* **1976,** *51*, 9-21.
40. Rhee, C.; Kim, D. *J. Food Chem.* **1975,** *40*, 460-462.
41. Ninomiya, M.; Matsuzaki, T.; Shigematsu, H. *Biosci. Biotech. Biochem.* **1992,** *56*, 806-807.
42. Belitz, H.-D.; Grosch, W. ; Hadziyev, D., translator, Ed.; Springer-Verlag: Berlin, 1987, pp 774.

43. Shu, C. K.; Mookherjee, B. D.; Ho, C. T. *J. Agric. Food Chem.* **1985,** *33*, 446-448.
44. Hirvi, T.; Honkanen, E.; Pyysalo, T. *Lebensm. Wiss. Technol.* **1980,** *13*, 324-325.
45. Sattath, S.; Tversky, A. *Psychometrika* **1977,** *42*, 319-345.
46. Schiffman, S. S.; Reynold, M. L.; Young, F. W. *Introduction to Multidimensional Scaling*; Academic Press Inc.: London, 1981.
47. Acree, T. A.; Barnard, J.; Cunningham, D. *Food Chem.* **1984,** *14*, 273-286.
48. McDaniel, M. R.; Miranda-Lopez, R.; Watson-Micheals, B. T.; Libbey, L. M. In *Flavors and Off-flavors*; Charlambous, G., Ed.; Elsevier Science Publishers: Amsterdam, 1990, pp 23-36.

Chapter 26

Characterization of Key Odorants in Dry-Heated Cysteine–Carbohydrate Mixtures: Comparison with Aqueous Reaction Systems

P. Schieberle and T. Hofmann

Deutsche Forschungsanstalt für Lebensmittelchemie, Lichtenbergstrasse 4, D-85748 Garching, Germany

The most odor-active compounds in dry-heated mixtures of cysteine/ribose (I) and cysteine/rhamnose (II) were identified by application of aroma extract dilution analyses and the odor activity value concept. A comparison with the key odorants generated in I with those present in the same mixture thermally treated under aqueous conditions revealed that dry-heating yielded higher amounts of the key odorants 2-furfurylthiol (roasty, coffee-like), 2-acetyl-2-thiazoline (roasty, popcorn-like), 2-ethyl- and 2-ethenyl-3,5-dimethylpyrazine (roasty, potato-like), whereas the yields of 5-acetyl-2,3-dihydro-1,4-thiazine (roasty, popcorn-like) and 3-mercapto-2-pentanone (sulfury) were significantly decreased. In II, the amount of 5-methyl-2-furfurylthiol (roasty, coffee-like) was significantly increased, whereas 4-hydroxy-2,5-dimethyl-3(2H)-furanone (caramel-like) and 3-hydroxy-6-methyl-2(2H)-pyranone (seasoning-like) were drastically decreased under dry-heating conditions. The odorants 2-ethenyl- and (Z)-2-propenyl-3,5-dimethylpyrazine were shown to be formed exclusively by dry-heating in I or II, respectively. A formation mechanism indicating the respective 1-desoxyosones as important intermediates in pyrazine formation is presented.

The amino acid cysteine is used as an essential building-block in the production of processed flavors eliciting meat-like, roasty odors (*1*). Until now, studies performed on the volatile fractions of thermally processed cysteine/carbohydrate mixtures have succeeded in identifying more than 200 volatile constituents, and, it has become obvious that the sugar moiety significantly influences the pattern of volatiles formed. By application of Aroma Extract Dilution Analysis (AEDA), a method based on HRGC/Olfactometry of stepwise diluted aroma extracts (*2*), we identified the most odor-active compounds formed during thermal treatment of aqueous solutions of cysteine in the presence of ribose (*3*) and glucose or rhamnose, respectively (*4*). Although the intense odorants 2-furfurylthiol or 5-acetyl-2,3-dihydro-1,4-thiazine were found among the most odor-active compounds in each of the three reaction systems (Table I), several other key odorants were shown to be formed predominantly from one sugar, e.g. the 5-methyl-2-furfurylthiol from rhamnose (Table I).

Table I. Most intense odorants (FD ≥ 256 in at least one mixture) generated by thermal treatment of cysteine (C) and carbohydrates in aqueous solution (ribose: Rib, rhamnose: Rha, glucose: Glc). Data from (3, 4)

Odorant	*Odor quality*	*Flavor Dilution (FD) Factor in*		
		C/Rib	*C/Rha*	*C/Glc*
2-Furfurylthiol	roasty, coffee-like	1024	512	1024
3-Mercapto-2-pentanone	catty, sulfury	512	128	512
2-Methyl-3-furanthiol	meat-like	256	<4	>4
5-Acetyl-2,3-dihydro-1,4-thiazine	roasty, popcorn	256	512	1024
3-Mercaptobutanone	sulfury	128	32	512
2-(1-Mercaptoethyl)furan	burnt	<1	<1	256
4-Hydroxy-2,5-dimethyl-3(2H)-furanone	caramel-like	32	65536	512
5-Methyl-2-furfurylthiol	roasty, coffee-like	<4	2048	<4
3-Hydroxy-6-methylpyran-2-one	seasoning-like	<4	16384	<4

It has previously been reported (*5*) that the overall odor profiles of such reaction flavors are not only influenced by the sugar moiety, but also by the reaction parameters applied, such as temperature or the presence or absence of water. The following investigation was performed to gain insight into the changes in the key odorants of processed cysteine/carbohydrate mixtures induced by application of dry-heating conditions.

Cysteine/Ribose

Aroma Extract Dilution Analysis. A mixture of cysteine (3.3 mmol) and ribose (10 mmol) was mixed with silica gel (3 g) containing 10 % of an aqueous sodium phosphate solution (0,5 mol/L; pH 5.0) and heated for 6 min at 180°C in a closed vessel. Phosphate was used as a catalyst but not to stabilize the pH. The overall odor of the reaction mixture was described by a sensory panel as roasty and meat-like. The volatiles formed were isolated by extraction with diethyl ether followed by sublimation in vacuo (*3*). The most odor-active compounds were then evaluated by application of AEDA and identified by means of reference flavor compounds. The results of the identification experiments are summarized in Table II. Based on their high Flavor Dilution (FD) factors, 2-furfurylthiol followed by 2-acetyl-2-thiazoline were characterized as the most odor-active compounds in the dry-heated mixture. Other important odorants were 2-methyl-3-furanthiol, furan-2-aldehyde, 2-ethyl-3,5-dimethylpyrazine and 2-propionyl-2-thiazoline. In previous studies we had reacted cysteine (3.3 mmol) with the three sugars (10 mmol each) under aqueous conditions (100 mL of phosphate buffer; 0.1 mol/L; pH 5.0) (*3, 4*). A comparison of the results obtained (cf. Table 1) with the data discussed above revealed that especially the FD factors of 2-furfurylthiol and 2-acetyl-2-thiazoline as well as those of the three pyrazines were increased, whereas the FD-factors of the popcorn-like 5-acetyl-2,3-

Table II. Most odor-active volatiles identified in the dry-heated cysteine/ribose mixture[a]

Odorant	*FD-factor*	*Odorant*	*FD-factor*
2-Furfurylthiol	16384	2,3-Diethyl-5-methylpyrazine	128
2-Acetyl-2-thiazoline	1024	2-Ethenyl-3,5-dimethylpyrazine	128
2-Methyl-3-furanthiol	256	4-Hydroxy-2,5-dimethyl-3(2H)-furanone	128
Furan-2-aldehyde	256	3-Mercapto-2-pentanone	64
2-Ethyl-3,5-dimethylpyrazine	256	3-Hydroxy-4,5-dimethyl-2(5H)-furanone	64
2-Propionyl-2-thiazoline	256		

[a] Eleven of the 24 most odor-active compounds identified in the Flavor Dilution (FD) factor range of 2 to 16384 are displayed (Hofmann and Schieberle, J. Agric. Food Chem., 1998, submitted).

dihydro-1,4-thiazine and the sulfury 3-mercapto-2-pentanone were decreased in the dry-heated system.

Quantitative measurements. Stable isotope dilution assays (SIDA) based on either deuterium or carbon-13 labelled internal standards were then used to quantify selected key odorants showing significant differences in their FD-factors in the two reaction systems. The structures of the labelled odorants used as the internal standards are displayed in Figure 1 (syntheses are reported by Hofmann and Schieberle, J. Agric. Food Chem., 1998, submitted). Although the application of SIDA ideally compensates for losses during the work-up procedures, this is not the case if the flavor compound and the internal standard are completely degraded. This behavior was observed for 2-methyl-3-furanthiol (MFT) when an extraction/distillation/concentration process was applied to isolate the MFT and its internal standard (*6*). To overcome this problem, we used the "purge and trap"-technique (TCT-system, Chrompack, The Netherlands) to enrich the volatiles for mass spectral measurements. The equipment is shown in Figure 2. In combination with SIDA, this enrichment technique was proven to be very effective for the quantification of flavor-active thiols. As shown in Table III, the values determined for 7 thiols in a model mixture differed not more than 5 % from the actual amounts. It should be noted that using this technique, exact quantitative data are obtained very quickly, e.g., within one hour.

SIDA was then used to determine the concentrations of several key odorants in the dry-heated mixture, and the data were compared with the amounts generated in the aqueous system. The concentrations displayed in Table IV are based on 10 mmol of ribose; the odor activity values are based on dissolving each of the processed mixtures in 100 mL of water. The quantitative data corroborated the results of the AEDA and indicated an increase of 2-furfurylthiol, 2-acetyl-2-thiazoline and three pyrazines in the dry-heated (I; Table IV) compared with the aqueous system (II; Table IV). On the

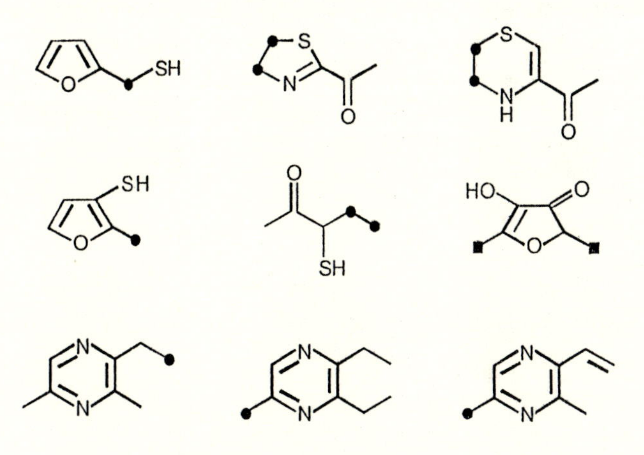

Figure 1. Structures of labelled internal standards used in the stable isotope dilution assays (•: deuterium label; ■: carbon-13 label)

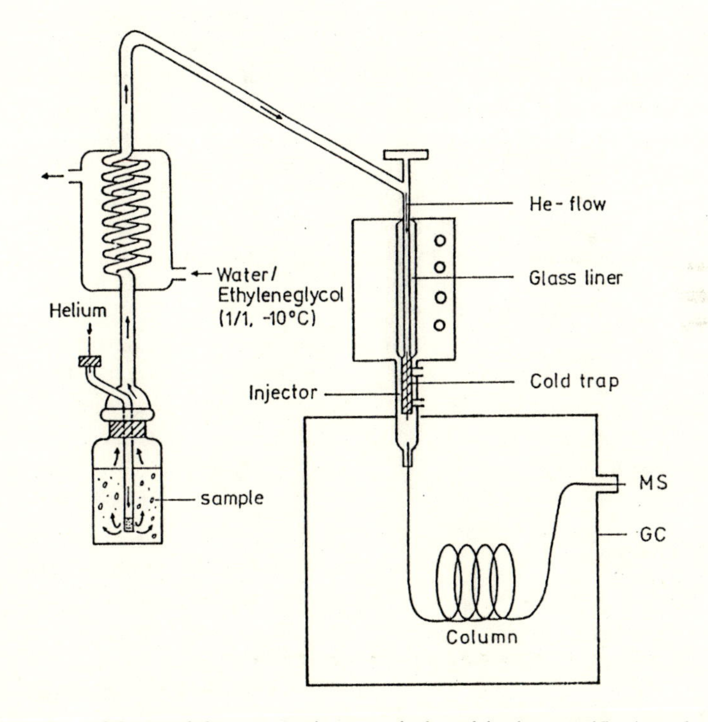

Figure 2. Scheme of the purge and trap method used in the quantification of odor-active thiols

Table III. Recoveries of 7 thiols determined by stable isotope dilution assays and using the „purge and trap“ technique as the enrichment procedure[a]

Thiol	*Recovery (%)*[b]
3-Mercapto-2-butanone	99
3-Mercapto-2-pentanone	95
2-Furfurylthiol	96
2-(1-Mercaptoethyl)-furan	97
2-Methyl-3-furanthiol	104
2-Thiophenmethanethiol	96
2-(1-Mercaptoethyl)-thiophene	98

[a] Known amounts of the thiols (2-5 μg of each odorant) and the corresponding labeled internal standards (2-5 μg of each labeled standard) were dissolved in phosphate buffer (15 mL; 0.5 mol/L; pH 5.0) and subsequently swept onto the trap (see Figure 2) by flushing with helium for 20 min.

[b] The measurements were done by stable isotope dilution assays using mass chromatography of selected ions (cf. (2)).

other hand, the amounts of 5-acetyl-2,3-dihydro-1,4-thiazine and 3-mercapto-2-pentanone were significantly decreased.

Odor activity values showed that in both reaction systems 2-furfurylthiol (FFT) and 2-methyl-3-furanthiol (MFT) are by far the most important odorants (Table IV). However, in the dry-heated mixture FFT is higher than MFT, whereas the reverse was true for the aqueous system. 2-Ethyl-3,5-dimethyl-, 2-ethenyl-3,5-dimethyl- and 2,3-diethyl-5-methylpyrazine contributed with comparatively low OAVs to the overall odor of the dry-heated mixture and did not participate in the flavor of the aqueous system. Although present in high concentrations, due to its high odor threshold, furan-2-aldehyde showed a very low odor contribution to the dry-heated system.

Based on the OAV data it might be speculated that the overall odors of both mixtures do not differ significantly. This assumption was corroborated for both types of processed flavors using a sensory panel of 10 members. The results, shown as a spider web diagram in Figure 3, indicated that in both systems the roasty and meat-like notes predominated. An earthy note was only detectable in the dry-heated mixture. This result is in good agreement with the higher OAVs of the three earthy, potato-like pyrazines in the dry-heated mixture (Table IV).

Rhamnose/Cysteine

The same concept was then applied to processed rhamnose/cysteine mixtures. First, the most-odor active compounds generated by dry-heating rhamnose and cysteine were identified based on the results of AEDA. The data (Table V) indicated that 5-methyl-2-furfurylthiol (roasty, coffee-like) and 4-hydroxy-2,5-dimethyl-3(2H)-furanone (caramel-like) are the most important odorants. (Z)-2-Propenyl-3,5-dimethylpyrazine

Table IV. Concentrations and odor activity values (OAV) of selected key odorants generated from ribose and cysteine - Influence of the reactions conditions

Odorant (*Odor threshold: µg/L of water*)[b]	*Conc. (µg/100 mmol ribose)* *I*[c]	*II*[d]	*OAV*[a] *in* *I*	*II*
2-Furfurylthiol (0.01)	972	121	97200	12100
2-Acetyl-2-thiazoline (1.0)	49	7	49	7
2-Propionyl-2-thiazoline (1.0)	18	<1	18	<1
2-Ethyl-3,5-dimethylpyrazine (0.1)	11	<0.1	110	<1
2,3-Diethyl-5-methylpyrazine (0.1)	3	<0.1	30	<1
2-Ethenyl-3,5-dimethylpyrazine (0.1)	5	<0.1	50	<1
Furan-2-aldehyde (12000)	79000	52	7	<1
5-Acetyl-2,3-dihydro-1,4-thiazine (1.25)	10	424	8	340
3-Mercapto-2-pentanone (0.7)	101	599	144	856
2-Methyl-3-furanthiol (0.007)	251	198	35857	28286

[a] Odor activity values (OAVs) were calculated by dividing the concentrations (based on 1 L of water) by the odor thresholds.

[b] Odor thresholds were determined by the triangle test using a seven-membered sensory panel.

[c] Cysteine (3.33 mmol) and ribose (10 mmol) were intimately mixed with silica gel (2.7 g containing 0.3 g of the phosphate buffer) and were reacted for 6 min at 180°C.

[d] Cysteine (3.33 mmol) and ribose (10 mmol) were dissolved in phosphate buffer (100 mL; 0.5 mol/L; pH 5.0) and reacted for 20 min at 145°C.

(roasted potato, earthy) and 2-acetyl- as well as 2-propionyl-2-thiazoline (each smells popcorn-like) followed with slightly lower FD-factors. A comparison of the key odorants in the dry-heated and the aqueous rhamnose/cysteine system (cf. Table V with Table I and (*4*)) indicated an increase in 5-methyl-2-furfurylthiol, 2-acetyl-2-thiazoline, 2-propionyl-2-thiazoline and three pyrazines amongst which the (Z)-2-propenyl-3,5-dimethylpyrazine showed the highest difference in the FD-factor. On the other hand, a drastic decrease in the FD-factors of 4-hydroxy-2,5-dimethyl-3(2H)-furanone (caramel-like) and 3-hydroxy-6-methyl-2(2H)-pyranone (seasoning-like) was detected. By quantitative measurements these results were confirmed (Table VI).

A calculation of the odor activity values of selected odorants in both reaction systems indicated that in the dry-heated system, the roasty, coffee-like 5-methyl-2-furfurylthiol was the key odorant followed by 2-methyl-3-furanthiol and 4-hydroxy-2,5-dimethyl-3(2H)-furanone (HDMF). In the aqueous reaction system HDMF and 3-hydroxy-6-methyl-2(2H)-pyranone (HMP) showed by far the highest odor activities.

These differences in the odor activities are very well reflected by the odor profiles of the reaction systems (Figure 4). While in the aqueous mixture, the caramel-

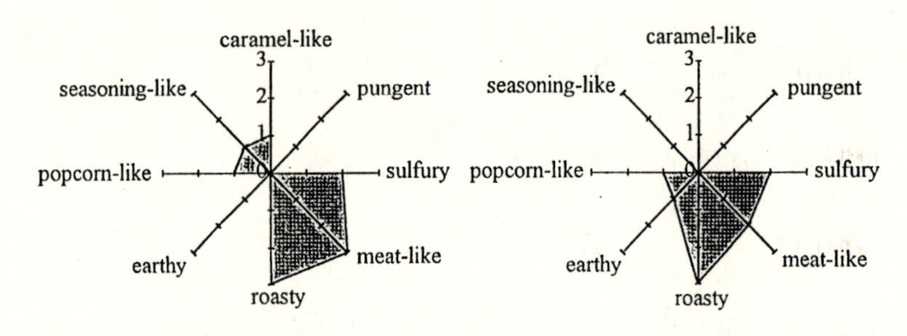

Figure 3 Odor profiles of the thermally treated ribose/cysteine mixtures; left: aqueous conditions (20 min; 145°C); right: dry-heating conditions (6 min; 180°C)

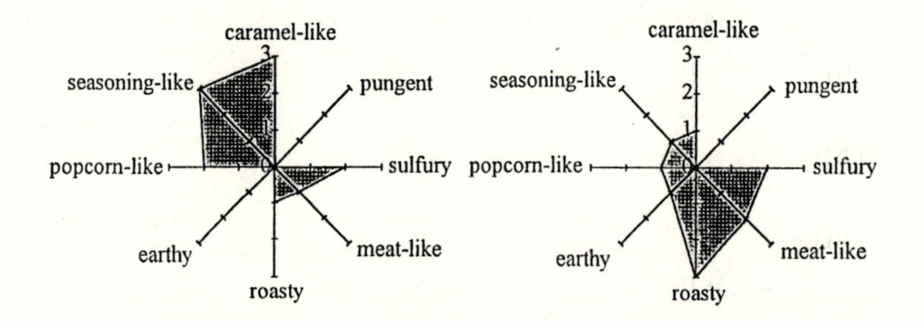

Figure 4. Odor profiles of the thermally treated rhamnose/cysteine mixtures; left: aqueous conditions (20 min; 145°C); right: dry-heating conditions (6 min; 180°C)

Table V. Most odor-active volatiles identified in the dry-heated cysteine/rhamnose mixture[a]

Odorant	*FD-factor*	*Odorant*	*FD-factor*
5-Methyl-2-furfurylthiol	4096	5-Methylfuran-2-aldehyde	512
4-Hydroxy-2,5-dimethyl-3(2H)- furanone (Furaneol)	2048	2-Ethyl-3,5-dimethylpyrazine	256
		2,3-Diethyl-5-methylpyrazine	128
(Z)-2-Propenyl-3,5-dimethylpyrazine	1024	Unknown (sweet, roasty)	128
		2-Furfurylthiol	32
2-Acetyl-2-thiazoline	1024	3-Hydroxy-6-methyl-2(2H)-pyranone	32
2-Propionyl-2-thiazoline	1024	3-Hydroxy-4,5-dimethyl-2(5H)-furanone	32

[a] Twelve of the 22 odorants detected in the Flavour Dilution (FD) factor range of 2 to 4096 are displayed (Hofmann and Schieberle, J. Agric. Food Chem., 1998, submitted).

Table VI. Concentrations and odor activity values (OAVs) of selected key odorants generated from rhamnose and cysteine - Influence of the reaction conditions

Odorant *(Odor threshold: µg/L of water)*[b]	*Conc. (µg/100 mmol rhamnose)* *I*[c]	*II*[d]	*OAV*[a] *in* *I*	*II*
5-Methyl-2-furfurylthiol (0.05)	1156	65	24083	1354
2-Acetyl-2-thiazoline (1.0)	48	6	48	6
2-Propionyl-2-thiazoline (1.0)	18	<1	18	<1
2-Ethyl-3,5-dimethylpyrazine (0.1)	12	<0.1	120	<1
(Z)-2-Propenyl-3,5-dimethylpyrazine (0.1)	3	<0.1	30	<1
2,3-Diethyl-5-methylpyrazine (0.1)	1	<0.1	10	<1
5-Methylfuran-2-aldehyde (4500)	147000	<10	331	<1
4-Hydroxy-2,5-dimethyl-3(2H)-furanone (10)	35600	198000	3560	19800
3-Hydroxy-6-methyl-2(2H)-pyranone (15)	1355	245300	90	16353
2-Methyl-3-furanthiol (0.007)	31	8	4429	1143
2-Furfurylthiol (0.01)	4	8	400	800

For footnotes see Table IV.

Table VII. Amounts of odor-active pyrazines formed in cysteine/carbohydrate mixtures[a]

Pyrazine	*Amount (μg) formed in the presence of*		
	Ribose	*Rhamnose*	*Glucose*
2-Ethyl-3,5-dimethyl-	11	12	15
2-Ethenyl-3,5-dimethyl-	5	<0.1	<0.1
(Z)-2-Propenyl-3,5-dimethyl-	<0.1	3	<0.1
2,3-Diethyl-5-methyl-	3	1	2

[a] Cysteine (33 mmol) and the carbohydrate (100 mmol) were mixed with silica gel (27 g containing 3 g of phosphate buffer [0.5 mol/L; pH 5.0]) and were reacted for 6 min at 180°C.

and seasoning-like odor notes of HDMF and HMP prevailed, in the overall odor of the dry-heated mixture the roasty note of 5-methyl-2-furfurylthiol predominated. The less intense earthy note detected in the latter mixture is undoubtedly due to the higher odor activities of the three earthy, potato-like pyrazines in the dry-heated mixture (cf. Table VI).

Formation of odor-active pyrazines

The results displayed in Tables IV and VI indicate that the four odor-active pyrazines 2-ethyl-3,5-dimethyl- (EDMP), 2-ethenyl-3,5-dimethyl- (EnDMP), (Z)-2-propenyl-3,5-dimethyl- (PDMP) and 2,3-diethyl-5-methylpyrazine (DEMP) are predominantly formed in the dry-heated systems. This result is well in line with literature data showing that higher temperatures favor pyrazine formation (*7*).

In Table VII the amounts of the four pyrazines formed by dry-heating cysteine in the presence of ribose, rhamnose and glucose are contrasted. The data indicate that the EDMP is formed from the three sugars in nearly equal amounts. Recently (*8*), it has been shown that this pyrazine is formed in relatively high amounts when 2-oxopropanal and the amino acid alanine are reacted for 7 min at 180°C. The authors suggested a reaction mechanism indicating 2-aminopropanal, aminoacetone and acetaldehyde as the key intermediates in EDMP formation. This assumption was recently verified by labelling experiments (*9*). 2-Oxopropanal and acetaldehyde may be formed by cleavage of the carbon skeleton of each of the three carbohydrates under investigation, explaining the formation of the EDMP in the three reaction systems (Table VII).

In contrast to EDMP, the ethenyl- (EnDMP) and the propenylpyrazine (PDMP) were formed exclusively from either ribose or rhamnose, respectively (Table VII).

EnDMP has recently been identified in roasted coffee (*10*), and it has been speculated that 2-aminopropanal, aminoacetone and acetaldehyde are also intermediates in its formation. As shown in Figure 5, the formation of EnDMP can be explained by an alternative route from the respective 1-desoxyosone: the *Strecker* reaction of the α-diketones 1-oxopropanal and 1-desoxyribosone may lead to 2-aminopropanal and 3-amino-4,5-dihydroxyhexane-2-one, respectively. Condensation of both, followed by

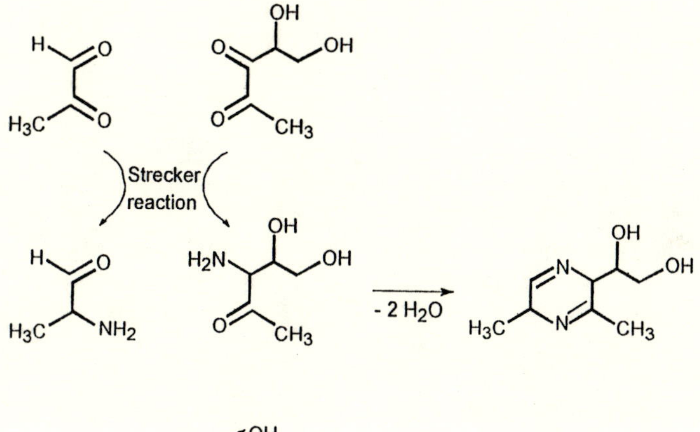

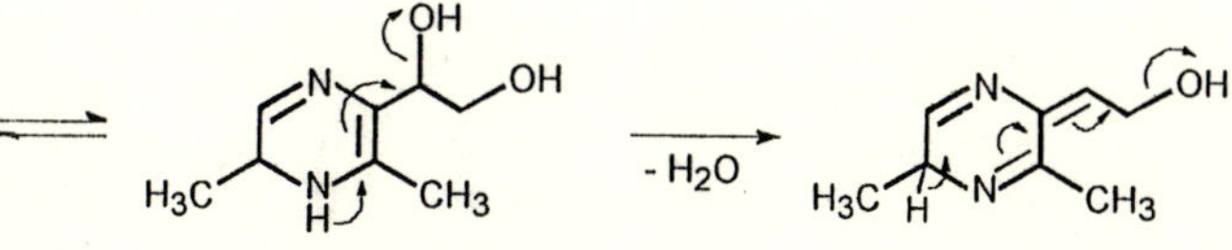

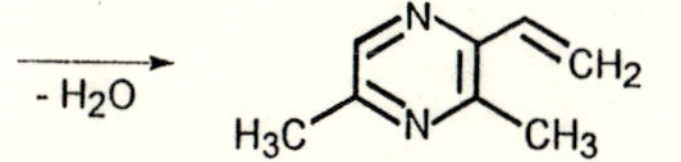

Figure 5. Reaction pathway leading from the 1-desoxyosone of ribose and 2-oxopropanal via a *Strecker* reaction to 2-ethenyl-3,5-dimethylpyrazine

elimination of two molecules of water from the dihydropyrazine moiety affords the 2-ethenyl-3,5-dimethylpyrazine. A similar reaction of the 1-desoxyosone of rhamnose might be responsible for PDMP formation.

Conclusions

The data show that processing conditions used in the production of reaction flavors significantly influence the yields of key odorants. Based on the odor value concept, was shown that changes in the overall odors of the mixtures are well reflected by changes in the odor activity values of the key odorants present. Extended knowledge on how their yields can be influenced by the reaction conditions will enable manufacturers to more selectively generate desired odor notes in processed flavors.

Literature cited

1. MacLeod, G.; Seyyedain-Ardebili, M. *Crit. Rev. Food Sci. Nutr.* **1981**, 14, 309-437.
2. Schieberle, P. In *Characterization of Food: Emerging Methods*, Goankar, A.G., Ed.; Elsevier, **1995**, pp. 403-431.
3. Hofmann, T.; Schieberle, P. *J. Agric. Food Chem.* **1995**, 43, 2187-2194.
4. Hofmann, T.; Schieberle, P. *J. Agric. Food Chem.* **1997**, 45, 898-906.
5. Lane, M.J.; Nursten, H. In *The Maillard Reaction in Foods and Nutrition*, ACS Symposium Series 215; American Chemical Society: Washington, DC, **1983**, pp. 141-158.
6. Schieberle, P.; Hofmann, T. In *Flavour Science - Recent Developments*, Taylor, A.J.; Mottram, D.S., Eds.; The Royal Society of Chemistry, Cambridge, UK, **1996**, pp. 175-181.
7. Grosch, W.; Zeiler-Hilgart, G.; Cerny, C.; Guth, H. In *Progress in Flavour Precursor Studies*, Schreier, P.; Winterhalter, P., Eds.; Allured Publ. Corp., Carol Stream, IL, **1993**, 329-342.
8. Cerny, C.; Grosch, W. *Z. Lebensm. Unters. Forsch.* **1994**, 198, 210-214.
9. Amrani-Hemaimi, M.; Cerny, C.; Fay, L.B. *J. Agric. Food Chem.* **1995**, 43, 2818-2822.
10. Czerny, M.; Wagner, R.; Grosch, W. *J. Agric. Food Chem.* **1997**, 44, 3268-3272.

Chapter 27

Aroma Profile of the Australian Truffle-Like Fungus *Mesophellia glauca*

S. Millington[1], D. N. Leach[1,3], S. G. Wyllie[1], and A. W. Claridge[2]

[1]Centre for Biostructural and Biomolecular Research, University of Western Sydney, Hawkesbury, Richmond, New South Wales 2753, Australia
[2]Centre for Resource and Environmental Studies, Australian National University, Canberra, ACT, Australia

The Australian truffle-like fungus *Mesophellia glauca* is a common mycorrhizal associate of eucalypt trees and other woody shrubs. Its fruit-bodies also provide an important food resource for a variety of ground dwelling marsupials, including the long-nosed potoroo (*Potorous tridactylus*) and the Tasmanian bettong (*Bettongia gaimardi).* At maturity the fruit-bodies of *M. glauca* produce pungent aromas which are superficially similar to those found in the European black truffle (*Tuber melanosporum*). These aromas serve to attract the fungus-feeding animals. A combination of simultaneous distillation extraction with GC-MS, static and dynamic headspace GC and GC-olfactometry was used to determine the aroma profile of *M. glauca.* The major components of the aroma of this fungus included, 1-octen-3-ol, 1-hexen-3-ol, 1,3-octadiene, 1-octene, hexanol, and hexan-3-ol. In the field, wildfires have been found to trigger increased foraging of marsupials for *M. glauca.* Comparison of fruit-bodies collected from recently burned and unburned forest indicated differences in aroma profile: fruit-bodies from burned sites had less of the more volatile constituents. A similar effect was reproduced when fruit bodies from an unburnt site were subjected to 60°C heat for 30min. However there was a major difference between the fire and heat treated fruit-bodies , with high levels of monoterpenes found in the former samples. These differences suggest that in addition to a heating effect caused by fire, other mechanisms contributing to changes in the aroma profile of fruit-bodies in burned habitats. While currently unknown, perhaps these mechanisms relate to physiological changes in the associated host plant.

[3]Current address: ATTORI, Southern Cross University, Lismore, New South Wales 2480, Australia

Fungi are a diverse and interesting food source across the animal kingdom. Their unique morphology, textures, flavors and aromas have fascinated gourmets for centuries and over the past few decades some of their mystique has been unraveled by chemical analysis.

Taxonomically, fungi are currently divided into four main classes; Phycomycetes, Ascomycetes, Basidiomycetes and Fungi Imperfecti (*1*). The Black Perigord Truffle, *Tuber melanosporum*, an ascomycete, is probably the most well known and interesting of all edible fungi. Unable to produce their own carbohydrates, Black Truffles form symbiotic mycorrhizal associations on the roots of a variety of plant hosts, including Oak, Hazel, Chestnut and Cedar (*2*). Their unique odor quality has led to many studies on the volatile composition of the fruit-body (*3-6*). Sulfur compounds, especially thiobismethane, were found to be a major component by headspace analysis and are assumed to be the main source of the characteristic odor of the truffle. However, Talou *et al.*, (*3*) stated that none of the individual compounds identified were characteristic of truffle aroma. Freytag *et al.*, (*5*) detected a range of C_8 alcohols, characteristic of mushroom aroma, in truffles but these were not reported by Talou, *et al.*, (*3*) nor by Pacioni *et al.*, (*4*). The more recent work of Pelusio, *et al.*, (*6*) using SPME confirmed the significance of the sulfur compounds in their aroma profile. Removal of these compounds by evaporation followed by reanalysis revealed a suite of C_8 "mushroom" alcohols.

In Australia a large variety of truffle-like fungi commonly fruit in late autumn and winter in native bushlands. These fungi are of considerable nutritional importance to a variety of plants which typically grow in extremely old, nutrient deficient soils (*7*). Without these nutrient accumulators, many mycorrhiza-forming plants would either perform very poorly or would not survive at all. In addition, many native fauna would be seriously affected without fungi as they form a very important part of their diet (*8*).

Two classes of fungi that are most commonly encountered in native bushland are the ascomycetes and basidiomycetes, the latter being by far the most common and highly evolved of the two. The basdiomycete genus *Mesophellia* is comprised of fourteen species one of which, *M. glauca,* is the focus of this study. *M. glauca,* first described as *Diploderma glaucum*, is a hypogeous mycorrhizal truffle-like basidiomycete and is the most widely distributed, highly variable and misidentified species in the genus *Mesophellia* (*9*). This fungus may be mistaken for other similar species which posses slight morphological differences, although their characteristic odor assists in differentiation.

The odor emitted by this fungus is the primary vector for detection by a range of ground dwelling mammals including, rat-kangaroos such as the long-nosed potoroo (*Potorous tridactylus)* and Tasmanian bettong *(Bettongia gaimardi),* as well as a variety of bandicoots, rodents and possums. Once detected, the fruit-bodies of these fungi represent a nutritionally valuable food source for these animals (*8*). In contrast there does not appear to be any reported ethnobotanical use of this fungus as a human food source, although other Australian fungi have been utilized for food and medicinal purposes (*10*).

While recent studies have shed light on the ecology and nutritional properties of *M. glauca* (*8,11*), there is virtually no information on the odor chemistry of this widespread and important fungus. In the only work so far done, Donaldson *et al.*, (*12*) observed that the odor profile of mature fruit-bodies of *M. glauca* comprised a complex of alcohols, aldehydes, ketones, alkenes, and esters. However, no data on the amount or identity of the compounds was reported. Given this deficiency, the aim of our study was to better define the chemical profile of the volatiles produced by mature fruit-bodies of *M. glauca*. To do this, we used a variety of analytical techniques, dictated by the need to carefully control the temperature exposure of the samples. Thus, methods used included static and dynamic headspace-GC and simultaneous distillation-extraction (SDE). The important odor compounds were explored using GC-Olfactometry and Aroma Extraction Dilution Analysis (AEDA). As an adjunct to the main study, we also investigated the role of fire and heat in altering the aroma profile of fruit-bodies of the same fungus: this was necessary because *M. glauca* is often found in recently burned habitats, where it is a preferred food resource for ground-dwelling mammals.

Experimental Procedures

A. Sampling

Sampling for fruit-bodies of *M. glauca* was undertaken at several forested sites in south-eastern mainland Australia during autumn and winter in 1995 and 1996. Briefly, fruit-bodies were collected by raking over the soil from around the bases of mature eucalypt trees, in both burnt and unburnt sites, using a four-tined garden cultivator. Soil types at both sites were coarse loams developing on tertiary sediment. Fruit-bodies of *M. glauca*, weighing between 1g and 3g with a mean weight of 1.4g, were excavated from the soil to a depth of approximately 10cm below the soil-leaf litter interface. Multiple fruit-bodies (1-2) were collected from locations within an unburned forest site and within a site burned 1 week previously. All samples were stored at harvest in plastic vials placed on dry ice: these were later transferred to a freezer (-20°C) prior to further analysis.

B. Analysis of the Headspace Volatile Components from *M. glauca* by Dynamic Headspace Gas Chromatography

i. Sample Preparation

Charcoal filtered laboratory air was passed over a single, whole sample held in a glass container (50mL) at ambient temperatures and the effluent trapped on the injector liner filled with Tenax TA 35/60 mesh (80mg). A total of 300 mL of headspace sample was collected (20 mL/min for 15 min).

ii. Instrumentation

Gas Chromatography was performed on a Hewlett-Packard 5890 Series II chromatograph fitted with a BP1 capillary column (SGE, 25m x 1.0μm x 0.22mm i.d.) and an Ai Cambridge Programmable Temperature Vaporizer Injector. The trap containing the sample was inserted into the PTV and desorbed under the following conditions: initial temperature, 40°C; ramp rate, 16°C/sec, final temperature 170°C. GC conditions for analysis were as follows: initial column temperature, 45°C; initial time, 5.0 min; program rate, 5°C/min; final temperature, 170°C; detector temperature, 240°C; carrier gas, helium, column head pressure, 10 psi; injector split ratio 10:1. Data was stored and processed using a Hewlett-Packard computer installed with HP Chemstation software (HP3365).

C. Analysis by Static Headspace Gas Chromatography

i. Sample Preparation

A single sample was cut to fit into a 20 mL headspace vial which was then tightly sealed with a Teflon-lined rubber septum. The sample was incubated at 40°C for the duration of one GC run (35 min).

ii. Instrumentation

Static Headspace Gas Chromatography (SHGC) analyses were carried out on a Hewlett-Packard 5890 Series II chromatograph equipped with a Hewlett-Packard (HP 19395A) headspace sampler fitted with a 3 mL sample loop. Data was acquired under the following conditions: column, SGE 25mm x 0.22mm i.d. 1.0 μ BP1 capillary column; initial temperature, 45°C; initial time, 5.0 minutes; program rate, 5°C/min. final temperature, 170°C, injector temperature, 200°C, detector temperature, 240°C, carrier gas nitrogen at 10 psi. The headspace sampler was operated under the following conditions; bath temperature, 40°C; valve\loop temperature, 90°C, sample probe delay, 3 secs; pressurization, 10 secs; venting, 20 secs and injection, 1 minute. Data was stored and processed using a Hewlett-Packard computer installed with HP Chemstation software (HP3365).

D. Analysis Simultaneous Distillation Extraction

i. Sample Preparation

A pre-weighed sample of whole fungus in water (approx. 100mL) was subjected to simultaneous distillation extraction (SDE) for 2 hours using pentane as the extracting solvent. A known quantity of tridecane as the internal standard was added to the mixture prior to distillation. The extract was concentrated in a Kuderna-Danish flask apparatus attached to a Snyder column using a bath temperature of 45°C.

ii. Gas Chromatography/Mass Spectroscopy (GCMS)

The concentrated extracts were chromatographed using a Hewlett-Packard 5890 Series II gas chromatograph fitted with an SGE BP1, 25m x 0.22mm i.d. capillary column of 2.0 μm film thickness. Data was acquired under the following conditions: initial temperature, 0°C, 5 mins, program rate 45°C/min and 40°C hold for 5 minutes, program rate 5°C/min, final temperature, 200°C; final time,1 min; injector temperature 220°C, carrier gas, helium at 10 psi, injection volume of 1.0 μL. The column terminated at a Hewlett-Packard Mass Selective Detector (MSD)(HP5971A).

The ion source was run in the EI mode at 170°C using an ionization energy of 70eV and scan rate of 0.9 scans/sec.

Data was stored and processed using a Hewlett-Packard computer installed with HP Chemstation software and the Wiley Mass Spectral library. Compound identifications were based on comparison with reference compounds, matched with compounds in the Wiley spectral library, published spectra and comparison of calculated Kovat's indices with published Kovat's indices (*13*). Quantitative data is uncorrected for response factor variations.

E. Analysis by GC-Olfactometry and Aroma Extraction Dilution Analysis

The concentrated extracts from Lickens-Nickerson extraction were analyzed using a Pye Unicam GCV Gas Chromatograph. The outlet from the column (J&W DB-1, 30m x 0.32mm i.d. x 0.32μ film thickness) was divided 1;1 using an outlet splitter (SGE, Australia) with one arm connected to an FID detector and the other to a sniffing port (SGE, Australia) flushed with humidified air at 500mL/min. Data was acquired under the following conditions: initial temperature, 45°C; initial time, 5.0min; program rate, 5°C/min; final temperature, 200°C; injector temperature, 220°C; detector temperature, 240°C; carrier gas, nitrogen at 10psi. The sensory response to the column effluent was recorded as previously reported (*14*). The responses were determined by a single subject and given an odor description which was recorded then synchronized with the aromagram. The sensory significance of each odorant was evaluated by aroma extraction dilution analysis (AEDA) by injecting decreasing amounts of the extract into the chromatograph.

Results and Discussion

SDE Technique

Volatile composition of the SDE of bulked *M. glauca* samples from an unburnt site, a burnt site and from natural stand fungi subjected to controlled heat treatment are shown in Table I. The extract from the natural site fruit-bodies from the unburnt site exhibits an interesting mix of alkenes, alcohols, esters and furans. The major components include 1-octen-3-ol, 1-octanol, 1-hexen-3-ol, 1,3-octadiene, linalool, α-terpineol, ethyl 3- methylbutanoate and 3-(4-methyl-3-pentenyl)furan as well as a number of unidentified compounds. The large proportion of alcohols in the volatile profile is the major source of the "mushroom-like" component of *M. glauca* aroma, in particular the eight carbon compounds, 1-octen-3-ol, and 1-octanol. They are formed as a result of enzymatic lipid degradation reactions (*15*), especially 1-octen-3-ol "mushroom alcohol", the characteristic impact flavor and odor compound of mushrooms (*16*). The *M. glauca* extract also contains other eight carbon compounds found in mushrooms including, 1-octenal, 2-octen-1-ol, 1,3-octadiene and 1-octene.

The presence of 3-(4-methyl-3-pentenyl)furan may contribute to the musty odor of *M. glauca* since certain molds also contain this compound (*17*). The earthy, mushroom odor in this fungus is balanced by the presence of typical fruit esters, aldehydes and floral terpene alcohols. Other components also found in both edible mushrooms and *M. glauca* include, limonene, 1-hexanol and 1-pentanol. These compounds, however, were not reported to impart any characteristic mushroom aromas or flavors (*18*). Compounds known to occur in many eucalyptus oil, including, limonene, terpinen-4-ol, α-terpineol, *p*-cymen-8-ol, and β-eudesmol were also identified. Considered individually, however, none of the compounds identified could be termed character impact compounds of this truffle-like fungus. Indeed this fungus bears little chemical similarity to the European truffle because of its distinct lack of sulfur volatiles, although the presence of the C_8 alcohols together with 2- and 3-methylbutanal in both fungi contributes in some way to the overall odor perception.

Field observations suggest that mycophageous (fungus-feeding) ground-dwelling mammals will increase their foraging activity for *M. glauca* dramatically after fire has swept through a forested area (*19*). It is possible that the heat from the fire causes direct or indirect changes in the aroma of this truffle-like fungus and it is this factor which triggers the feeding activity. In our study, the aroma of fruit-bodies from a site which experienced fire was compared with that from fruit-bodies from a site that had been unburnt (Table I). In general, concentrations of the more volatile components have decreased where samples of *M. glauca* experienced fire. This is most simply explained by the direct effect of heat and presumably the emitted volatiles are used as an aid to truffle location by the marsupials. However, there is an apparent increase in concentration of less volatile compounds, most notably the terpenes. This may be a result of the close synergy with the eucalypt through the root system reflecting the impact on the foliage canopy and oil glands by fire. This is supported by the data in Table I obtained from natural fungi exposed to heat in a laboratory oven. The *M. glauca* samples were treated at 60°C for 30 minutes to mimic the conditions experienced after the passage of fire. These conditions were based on the work by Auld *et al.*, (*20*), who found post-fire temperatures of up to 60°C 10-15 cm or more below the soil surface. From the data in Table I for heat-treated fungi, it is again apparent that the more volatile compounds have been lost whilst concentrations of the less volatile compounds have increased.

The marked difference between the heat-treated samples and those recovered from a burnt site is the non-appearance of terpenes in the former. This suggests that the response to the *in situ* heating process undergone by fruit-bodies in a burned site may be mediated by the host tree (*12*). However, other factors may have slight influences on the aroma composition of *M. glauca*, including, nutrient facilitation (chemical changes resulting from altering conditions in the host tree, i.e. alterations as a result of fire), smoke-induced changes rather than heat-induced, or factors resulting from soil or leaf litter alterations as a result of fire. These factors are more difficult to investigate and were not investigated in the present study.

Static and Dynamic Headspace Analysis

The demonstrated influence of heat on the aroma profile of *M. glauca* implies that extraction methods such as SDE which expose the sample to relatively high temperatures may not be appropriate for making comparisons between heated and nonheated samples. Accordingly static and dynamic headspace extractions which

Table I. Volatiles Profile in *M. glauca* from Natural and Burnt Sites and after Heat Treatment

Peak No	I_k[a]	Compound	Unburnt ppm	Burnt ppm	Heated ppm	ID
1	622	3-methyl-1,3-pentadiene	35.22	5.5	-	MS
2	635	3-methylbutanal	27.7	12.0	-	MS,RT
3	645	2-methylbutanal	9.7	2.6	-	MS,RT
4	661	unidentified	-	4.1	-	-
5	674	pentanal	8.6	7.8	-	MS,RT
6	689	unidentified (C_8 hydrocarbon)	54.1	26.0	24.0	MS
7	700	heptane	59.3	30.1	29.1	MS,RT
8	719	unidentifiedC_7H_{12}	12.4	-	-	-
9	723	methylcyclohexane (impurity)	190.6	90.3	-	MS
10	749	ethyl 2-methylpropanoate	81.8	-	32.1	MS,RT
11	770	1-hexen-3-ol	532.3	87.8	369.7	MS,RT
12	779	hexanal	98.6	54.0	115.3	MS,RT
13	786	unidentified	-	10.9	-	-
14	789	3-hexanol	109.4	25.8	64.4	MS,RT
15	795	unidentified (C_7 alcohol)	-	30.6	-	-
16	799	1-octene	34.6	tr	-	MS,RT
17	815	furfural	tr	7.7	-	MS,RT
18	820	1,3-octadiene	514.0	116.3	283.0	MS,RT
19	829	unidentified	-	24.9	26.0	-
20	840	ethyl 2-methylbutanoate	30.7	3.4	-	MS,RT
21	842	ethyl 3-methylbutanoate	145.2	-	20.9	MS,RT
22	857	1-hexanol	135.5	28.6	194.8	MS,RT
23	864	3-methylbutyl acetate	-	-	21.8	MS,RT
24	868	1-hepten-3-ol	47.5	-	-	MS,RT
25	874	unidentified	tr	-	51.8	-
26	882	*o*-xylene (+ 2-butylfuran)	tr	9.4	21.2	MS,RT
27	900	unidentified	tr	-	13.5	-
28	905	2-hexenyl acetate *	32.6	-	-	MS
29	917	unidentified	-	5.4	-	-
30	923	ethyl tiglate	15.7	7.5	14.5	MS,RT
31	925	unidentified	22.4	11.4	-	-
32	934	α-pinene	22.7	17.0	28.6	MS,RT
33	937	propyl 3-methylbutanaote *	17.7	-	-	MS
34	948	camphene	14.0	24.4	10.3	MS,RT
35	960	(5Z)-octa-1,5-dien-3-ol *	20.7	15.5	25.0	MS
36	967	1-octen-3-ol	700.5	198.5	925.9	MS,RT
37	981	2-pentylfuran	48.5	99.3	5.9	MS,RT
38	983	ethyl hexanoate	36.5	40.0	48.0	MS,RT
39	1007	phenylacetaldehyde	49.8	27.3	34.1	MS,RT
40	1024	limonene	30.4	51.2	11.5	MS,RT
41	1029	ocimene*	tr	-	-	MS

Continued on next page.

Table I. Volatiles Profile in *M. glauca* from Natural and Burnt Sites and after Heat Treatment - Continued

Peak No	I_k[a]	Compound	Unburnt ppm	Burnt ppm	Heated ppm	ID
42	1037	3-octen-1-ol	tr	23.4	25.5	MS,RT
43	1055	2-octen-1-ol *	30.7	16.2	81.2	MS
44	1057	1-octanol	165.8	77.2	225.4	MS,RT
45	1063	*cis*-linalool oxide	tr	22.7	-	MS,RT
46	1070	pentyl butanoate	tr	-	-	MS,RT
47	1077	*trans*-linalool oxide	tr	20.1	-	MS,RT
48	1081	α-terpinolene	-	23.9	-	MS,RT
49	1088	linalool	162.9	112.0	219.3	MS,RT
50	1090	3-(4-methyl-3-pentenyl) furan	124.9	128.0	380.6	MS,RT
51	1093	octenyl acetate *	12.3	-	18.4	MS
52	1105	methyl octanoate	48.5	18.8	-	MS,RT
53	1113	unidentified	tr	-	-	MS
54	1124	camphor	-	14.2	-	MS,RT
55	1139	2-nonenal *	tr	55.1	-	MS
56	1150	*trans-p*-menth-2-ene-1,8-diol*	-	25.4	-	MS
57	1155	1-borneol	-	23.4	-	MS,RT
58	1160	unidentified	-	20.7	-	-
59	1162	*p*-cymen-8-ol *	tr	13.4	-	MS
60	1167	terpinen-4-ol	tr	31.4	-	MS,RT
61	1173	*cis-p*-menth-2-ene-1,8-diol *	-	37.9	-	MS
62	1176	α-terpineol	175.5	220.4	78.1	MS,RT
63	1183	ethyl octanoate	tr	-	-	MS,RT
64	1203	unidentified	-	31.8	-	-
65	1236	unidentified	-	4.5	-	-
66	1240	2-decenal	30.0	32.4	30.0	MS
67	1269	unidentified	-	8.5	-	-
68	1271	*cis*-2,4-decadienal *	-	12.8	-	MS
69	1292	*trans*-2,4-decadienal *	31.1	21.1	44.4	MS
71	1311	unidentified	tr	-	-	-
72	1343	2-undecenal *	25.3	26.0	26.7	MS
73	1402	unidentified	14.4	-	-	-
74	1410	unidentified	82.7	124.9	33.1	-
75	1423	unidentified	-	21.8	-	-
76	1496	2,6-bis-(1,1-dimethylethyl)phenol*	109.5	-	-	MS
77	1696	tetradecanal	tr	-	-	MS
78	1720	unidentified	357.3	-	tr	-

[a] Calculated Kovat's Indices

* Tentative Identification

utilize much lower sample temperatures were employed. The identity and relative composition of the odor profiles generated by these two techniques is summarized in Table II.

Table II. Volatiles in *M. glauca* from Unburnt and Burnt Sites after Heat Treatment

I_k[a]	Compound	SHS[b] Unburnt (%)	DHS[c] Unburnt (%)	SHS Burnt (%)	DHS Burnt (%)	DHS Heated (%)
-	ethanol	4.88	12.59	14.59	19.23	11.26
-	unidentified	3.60	9.89	14.60	13.25	-
531	3-methyl-1,2-butadiene	-	-	-	2.92	0.30
600	1-hexene	6.29	0.38	2.57	1.61	0.70
622	2-methyl-1,3-pentadiene	14.45	1.39	1.35	19.63	1.22
637	3-methylbutanal	0.44	4.62	tr	1.88	0.98
774	1-hexen-3-ol	15.31	15.33	4.47	1.66	5.02
795	1-octene	9.38	10.39	18.74	8.12	20.09
828	1,3-octadiene	35.95	23.54	26.39	10.04	33.23
848	ethyl 3-methylbutanoate	0.72	1.73	tr	tr	-
948	camphene	tr	-	tr	-	tr
962	1-octen-3-ol	2.53	1.54	4.46	1.53	0.61
1038	limonene	0.43	1.96	1.04	0.46	0.50
1099	3-(4-methyl-3-pentenyl)furan	0.93	0.45	-	-	0.35

a. Kovat's Indices on BP-1
b. Static Headspace System - 40°C for 35 minutes
c. Dynamic Headspace System-ambient temperatures, 15 minutes @ 20 mL/min)

Data presented here was from the analysis of a single fruiting body only and must therefore be regarded as indicative only. Components detected using both SHS and DHS analysis and also found in the SDE of *M. glauca* included, 2-methyl-1,3-pentadiene, 1-hexen-3-ol, 1-octene, 1,3-octadiene and 1-octen-3-ol, 3-methylbutanal, ethyl 3-methylbutanoate, limonene and 3-(4-methyl-3-pentenyl)furan. The key characteristic eight carbon compounds identified in the SDE extracts were also predominant in the headspace profiles and confirm their contribution to the mushroom-like odor. Whilst direct comparisons between the data are tenuous because of inherent biological variability associated with small sample size, some trends have emerged. First, there is a good correlation between the dynamic and static headspace composition data for fungi from the same site. There are some quantitative discrepancies between the techniques especially with the more volatile components that could be attributed to a combination of technique selectivity and sample inhomogeneity. Burnt site truffles showed a consistent decrease in concentrations of 3-methylbutanal, 1-hexen-3-ol, 1,3-octadiene, ethyl 3-

methylbutanoate and 3-(4-methyl-3-pentenyl)furan using both headspace techniques. This is also consistent with the SDE data.

Heat treated truffles showed a dramatic increase in the amounts of 1-octene and 1,3-octadiene and losses of alkene alcohols. Opposing trends observed for 1-octene, 1,3-octadiene and 2-methyl-1,3-pentadiene between burnt site and heat treatment would suggest that significant biochemical changes are induced by the fire that cannot be induced by simple heating. Losses of hexen-3-ol, 3-methylbutanal, ethyl 3-methylbutanoate and 1-octen-3-ol observed in both fire and heat affected truffles may provide a clue for the principal detection volatiles used by marsupials. Those compounds found to decrease in concentration are assumed to be released into the surrounding environment and as such may influence the behavior of the marsupials.

GC-Olfactometry and Aroma Extraction Dilution Analysis

Analysis of the natural site *M. glauca* SD extract by AEDA is summarized below in Table III. The undiluted extract presents a broad range of odor descriptors from floral through earthy and mushroom to rotting vegetation. Some of these descriptors are at odds with the known odor profile of the assigned compound suggesting coelution of trace components that are not readily discernible from mass spectral analysis. For example the hydrocarbons heptane, methyl cyclohexane and o-xylene are not usually associated with earthy, sweet and rotting vegetable odors and this is indicative of some coelution. At a ten fold dilution the aroma profile is dominated by the mushroom notes of 1-octen-3-ol, methyl octanoate and an unidentified constituent. Contributing to the overall perception are the floral grassy components 1-hexen-3-ol, 3-hexanol and 1-hexanol. Only two compounds could be detected at a twenty fold dilution, namely 1-hexen-3-ol and o-xylene, the latter considered to be coeluting with a very low odor threshold component.

Conclusion

There are discernible quantitative and qualitative differences in the odor profiles of fruit-bodies of *M. glauca* collected from unburned and burned forest sites. Volatiles analyzed from fungi heat treated under controlled laboratory conditions confirmed a direct physical effect of heat. However, the profile differed in some important respects from fungi from the burned site, suggesting that other (unknown) mechanisms may also help regulate changes in the aroma profile of fruit-bodies of this fungus. AEDA confirmed a strong contribution to the overall odor profile of *M. glauca* from C_6 and C_8 alcohols, however the sensory perception of the marsupials to such compounds may be quite different.

Table III. Aroma Profile of *M. glauca* by AEDA

Compound	SDE /20	SDE /10	SDE /5	SDE /1	Aroma Description
unidentified				+	Sweet
3-methylbutanal			++	+++++	Chocolate
unidentified				++	Floral
heptane + ?				++++	Mushroom/Earthy
methyl cyclohexane + ?			++	+++	Floral
ethyl isobutyrate			++	++++	Sweet
1-hexten-3-ol	+	+	++	++	Floral
3-hexanol		++	+++	+++++	Grassy
ethyl 3-methylbutanoate				++	Floral
1-hexanol		++	+++	++	Herbal
unidentified			+++	+++	Mushroom
o-xylene + ?	+	+	++	+++	Off vegetable
unidentified				++	Mushroom
propyl-2-methyl propanoate				++	Floral
ethyl tiglate				++++	Earthy
unidentified				+++	Rotten
camphene			++	+++++	Sweet
1-octen-3-ol		+	+++	++++	Raw mushroom
limonene				++	Earthy
β-ocimene				++	Stale
1-octanol				+	Pine
linalool				++++	Floral/sweet
octenyl acetate *				++	Mushroom
methyl octanoate		++	++	+++	Mushroom
2-nonenal *				+++	Chemical
p-cymen-8-ol				+++	Green/earthy
α-terpineol				+++	Mushroom
ethyl octanoate				++++	Chemical
unidentified				+++++	Earthy
tetradecanal *				+++	Woody/musty

* Tentative identification

+, ++, +++, ++++, +++++ intensity of response

Acknowledgments

One of us (SM) thanks the University of Western Sydney, Hawkesbury for the 1996 Honors Scholarship (Research) and another (DNL) gratefully acknowledges the financial support of the Australian Tea Tree Oil Research Institute to attend the ACS Symposium.

Literature Cited

1. *The Fungi*, Carlile, M. J.; Watkinson, S. C. Eds.; Academic Press, London, **1994**.
2. Mushrooms and Truffles, Botany, Cultivation and Utilization., Singer, R., Interscience Publishers Inc., Leonard Hill, London, **1961**, 272.
3. Talou, G.; Delmas, M.; Gaset, A. Aroma. *J. Agric. Food Chem.*, **1987**, 35, 774-777.
4. Pacioni, G.; Bellina-Agostinone, C.; D'Antonio, M. *Mycol. Res.*, **1990,** 94, 201-204.
5. Freytag, W.; Ney, K. H., Gordian, **1980**, 80, 214.
6. Pelusio, F.; Nilsson, T.; Montanarella, L.; Tilio, R.; Larsen, B.; Faccetti, S.; Madsen, J.O.; *J. Ag. Food Chem.*, **1995**, 43, 2138-2143.
7. Young, T. *Common Australian Fungi, A Naturalist's Guide*, University of NSW Press, Sydney, **1994**.
8. Claridge, A. W.; Cork, S. J. *Aust. J. Zool.*, **1994,** 42, 701-710.
9. Trappe, J. M.; Castellano, M. A.; Malajczuk, N. *Aust. Syst. Bot.* **1995**, 9, 773-802.
10. Kalotas, A. C. In *Fungi of Australia, Introduction-Fungi in the Environment*; Orchard, A. E., Ed.; ABRS/CSIRO, Canberra, Australia, **1996**, Vol. 1B, 269-295.
11. Claridge, A. W. *Aust. J. Ecol.*, **1992,** 17, 223-225.
12. Donaldson, R.; Stoddart, M. *J. Chem. Ecol.*, **1994**, 20, 1201-1207.
13. Jennings, W.; Shibamoto, T. In *Qualitative Analysis of Flavor and Fragrance Volatiles by Glass Capillary Gas Chromatography.* Academic Press, New York, 1980.
14. Wyllie, S. G.; Leach, D. N.; Wang, Y.; Shewfelt, R.L. In *Sulfur Compounds in Foods*, Mussinan, Keelan., Ed; ACS Symposium Series 564, ACS, Washigton DC, 1994, 36-48.
15. De Lumen, B. O.; Stone, E. J.; Kazeniac, S. J.; Forsyth, R. H. *J. Food Sci.*, **1978**, 43, 698.
16. Karahadian, C.; Josephs, D. B.; Lindsay, R. C. *J. Agric. Chem.*, **1985**, 33, 339-343.
17. Boerjesson, T. S.; Stoellman, U. M.; Schnuerer, J. L. *J. Agric. Food. Chem.*, **1993,** 41 2104-2111.
18. Chen, C. C.; Wu, C. M. *J. Food. Sci.*, **1984,** 49, 1208-1209.
19. Taylor, R. J. *Aust. J. Ecol.*, **1991**, 6, 409-411.
20. Auld, T. D.; Bradstock, R. A. *Aust. J. Ecol.*, **1996,** 21, 106-109.

Chapter 28

Analysis of Aroma-Active Components of Light-Activated Milk

Keith R. Cadwallader and Cameron L. Howard

Department of Food Science and Technology, Mississippi Agricultural and Forestry Experiment Station, Mississippi State University, Box 9805, Mississippi State, MS 39762

Milks of varying fat levels (skim, 2% and whole) were exposed to fluorescent light at 200 FC. Control (not exposed) and light-activated flavored (LAF) milks were subsequently evaluated by sensory and instrumental methods. Results of sensory evaluation demonstrated that flavor of LAF milk is impacted by the fat level of the milk. The combined results of gas chromatography-olfactometry (GC-O) of volatiles isolated by static (SHS) and dynamic (DHS) headspace sampling and by vacuum distillation-solvent extraction (VDSE) from control and LAF milks of varying fat contents revealed that odorants of low, intermediate, and high volatility are involved in LAF. Furthermore, these compounds are derived from both lipid and non-lipid precursors. Specific components involved the aroma of LAF milk can be quantified by DHS.

Freshly pasteurized and homogenized milk of good quality has a subtle but distinctive clean and fresh flavor. Badings (*1*) stated that there are three basic elements responsible for the flavor of milk: (1) pleasant mouthfeel due to presence of macromolecules such as colloidal proteins and fat globules, (2) sweet and salty taste due to lactose and milk salts, respectively, and (3) a weak and delicate aroma due to numerous volatile compounds present at near or below their odor threshold levels. Patton and coworkers (*2*) were first to study the aroma constituents of fresh milk and found dimethylsulfide to be an important constituent. Since then, carbonyl compounds, alcohols, free fatty acids, and various sulfur compounds also have been found to play important roles in fresh milk flavor. These compounds are derived mostly through normal metabolism of the cow or from the feed by either direct transfer or release during digestion (*1,3*). During and after processing, the mild flavor of milk can be negatively impacted by many processes and chemical reactions (*3*). These include both oxidative and hydrolytic rancidity, thermal degradation, packaging interactions, microbial contamination, and exposure to light. Probably the most common and severe flavor problem encountered in milk is caused during its exposure to light (*4*). This flavor defect has been studied for over 40 years and is commonly referred to as "sunlight", "oxidized", "light-induced", and "light-activated" (5).

Exposure of milk to light results in the development of off-flavor and causes the destruction of several key nutrients such as riboflavin, ascorbic acid and the essential amino acid methionine (5). Factors affecting the extent of off-flavor formation or nutrient loss include wavelength and intensity of light, duration of exposure, type of packaging material (e.g., light transmission properties), and product temperature (*5*). Extensive work has been carried out on the effects of light on the flavor and nutritional quality of milk and numerous reviews have been published *(5-8*). Patton (*9*) was first to propose the involvement of two mechanisms in the development of light-activated flavor (LAF) in milk, but it wasn't until some time later that these mechanisms were confirmed (*10-12*). The key player in both mechanisms is the photocatalyst riboflavin. Photoreduction of riboflavin in milk results in the Strecker degradation of methionine to form the potent odorant 3-(methylthio)propanal (methional), and also leads to the photogeneration of superoxide anion (*1,3,13*). Superoxide anion can subsequently undergo dismutation to form singlet oxygen (and other "activated" oxygen species) which can initiate oxidation of polyunsaturated fatty acids, leading to formation of numerous volatile carbonyl compounds (*13*). It is the combination of the products of the two types of reactions that give typical LAF in milk (*1, 3,5-8*).

Early investigators employed distillation-based methods (i.e. vacuum distillation followed by either solvent extraction or static headspace sampling of the distillate) for the analysis of volatile constituents of LAF milk (*13-15*). Distillation methods have the advantage of allowing for isolation of flavor compounds with a wide range of volatilities and recently have been applied in milk flavor studies (*16-17*). A major disadvantage of these techniques, however, is that they are time consuming and generally required a large amount of product and, therefore, are not well suited for the routine analysis of LAF milk. Several researchers recognized this problem and began using dynamic headspace sampling (DHS)(*18-21*). This method is both rapid and sensitive, requires minimal sample preparation, and can accommodate small sample sizes. Furthermore, with DHS no solvent is required, insuring that only sample volatiles are analyzed. However, the technique is not without limitations, e.g., it will only allow for the determination of sample components of high and intermediate volatility. Therefore, the suitability of DHS for the routine analysis of LAF milk would depend on the volatility of the key odor-active components. However, considerable confusion currently exists in regards to the identities of the predominant odorants in LAF milk.

The objectives of our study were: (1) to identify odor-active components of LAF milk by employing several complimentary isolation techniques and gas chromatography-olfactometry and (2) to evaluate the suitability of DHS for the routine analysis of LAF milk by focusing on the key odorants involved in the development of this off-flavor.

Experimental

Milk. Fresh pasteurized and homogenized milk of varying fat levels (i.e., skim, 2%, and whole) was obtained from a processor in Kosiusko, MS. Milk, in standard 1 gallon plastic containers, was transported on ice to the Department of Food Science and Technology and stored in a walk-in cooler (4°C) until needed. During transport and storage, care was taken to prevent any exposure of the milk to light. Prior to analysis, milk quality (i.e. lack of LAF) was assured by sensory testing by a 3-6 member expert panel employing standard ADSA dairy score card techniques.

For development of LAF, milk in clear flint glass bottles (French square, Qorpak no. 7905) equipped with PTFE lined caps was exposed to fluorescent light at 200 FC. The light source consisted of four 48 in. 34 W fluorescent bulbs (Supersaver Coolwhite, Sylvania, Louisville, KY). Two sets of light banks (Power Products Co., Philadelphia, PA), each comprised of two bulbs horizontally aligned and spaced 3 in. apart on wooden supports, were placed facing each other to create a "light box". A shelf was placed in the center of the light box and the light banks were oriented so that light meter readings taken at the center of the shelf (facing each light bank) registered approximately 200 FC. Sample bottles were vertically placed in random order along the shelf. Light exposure was conducted in a walk-in cooler at 4°C.

Chemicals. Aroma compounds listed in Tables I-IV were obtained from the following commercial sources: nos. **1-7**, **9**, **14**, **17**, **19**, **21**, **25** (Aldrich Chemical Co., St. Louis, MO); no. **11** (Bedoukian Research Inc., Danbury, CT); and nos. **8** and **15** (Lancaster Synthesis, Inc., Windham, NH). Standard no. **13** was obtained from Dr. R. Buttery (USDA, ARS, WRRC, Albany, CA). Compound no. **16** was synthesized according to Ullrich and Grosch (*22*) and compound no. **18** was prepared from 1-nonen-3-ol (Lancaster Synthesis, Inc.) by oxidation with pyridinium chlorochromate (Aldrich Chemical Co.)(*23*). 2-Methyl-3-heptanone (internal standard) was purchased from Aldrich Chemical Co.

Vacuum distillation-solvent extraction (VDSE). The apparatus used for distillation is shown in Figure 1. Prior to use, clean glassware was baked at 160°C for at least 2 h. A 25 mL aliquot of milk was placed in a 250-mL round bottom flask (**a**) and the system connected as shown. The two receiving tubes (**b1** and **b2**) were placed in liquid nitrogen and allowed to cool for 5 min. The sample flask was kept at room temperature and vacuum (**c**) was applied (<0.5 torr) to the system. After 5 min, the sample flask was heated to 40°C and distillation continued for 90 min. The distillate from tube **b1** was adjusted to pH 8 with dilute aqueous NaOH (0.1 N) and then extracted with redistilled dichloromethane (3 x 3 mL). The combined solvent extract was dried over 2 g of sodium sulfate, concentrated to 50 μL under a stream of nitrogen, and stored at -20°C until analysis.

Gas chromatography-olfactometry (GC-O). The GC-O system consisted of a HP 5890 Series II GC (Hewlett-Packard Co., Palo Alto, CA) equipped with a flame ionization detector (FID), a sniffing port, and an on-column injector. Each VDSE extract (2 μL) was injected into a capillary column (DB-WAX or DB-5ms, 30 m length x 0.25 mm i.d. x 0.25 μm film thickness (d_f); J&W Scientific, Folson, CA.).

Table I. Odor-active compounds detected by static headspace sampling-gas chromatography-olfactometry of control and light-activated milk.

No.[a]	Compound	Retention Index[b]		Odor Description[c]	Odor Intensity[d]		
		DB-Wax	DB-5ms		Skim	2%	Whole
8	1-Hexen-3-one[e]	1100	796	plastic, water bottle	**nd**/+	**nd**/+	**nd**/++
11	(Z)-4-Heptenal[e]	1231	- -	rancid, fishy	**nd**/**nd**	**nd**/+	**nd**/+
15	1-Octen-3-one[e]	1304	975	mushroom, earthy	**nd**/+	**nd**/++	+/++
20	Unknown	1439	1098	fresh, creamy	+/+	+/+	**nd**/**nd**

[a]Numbers correspond to those in Tables II-IV. [b]Retention indices from GC-O data. [c]Odor quality as perceived by panelists during GC-O. [d]Odor intensity scored on a 4-point scale for control/light-activated milk, where **nd** = not detected, + = weak, ++ = medium, +++ = strong. Value based on average of duplicate determinations. [e]Compound identified by comparing its RI values on two capillary columns and odor properties with those of the reference compound.

Table II. Odor-active compounds detected by dynamic headspace sampling-gas chromatography-olfactometry of control and light-activated milk.

		Retention Index [b]			Odor Intensity [d]		
No. [a]	Compound	DB-Wax	DB-5ms	Odor Description [c]	Skim	2%	Whole
1	Methanethiol[c]	631	<600	Rotten, cabbage	**nd/+**	**+/nd**	+/+
2	Acetaldehyde[f]	661	<600	Solvent, yogurt	**nd/+**	+/++	**nd/+**
3	Dimethylsulfide[f]	721	<600	Canned corn	+/++	++/++	++/++
4	2-Methylpropanal[f]	816	<600	Dark chocolate	+/++	++/++	+/+
5	2,3-Butanedione[f]	979	620	Buttery	**nd/nd**	**nd/+**	**nd/+**
6	3-Methylbutanal[f]	915	651	Dark chocolate	+/+	+/+	+/+
7	Pentanal[f]	982	701	Sour, cut-grass	**nd/nd**	**nd/++**	**nd/+++**
8	1-Hexen-3-one[f]	1094	806	Plastic, water bottle	**nd/++**	+/+++	+/++
9	Hexanal[f]	1082	821	Green, cut-grass	**nd/++**	**nd/++**	**nd/++**
10	Unknown	1171	832	rotten, potato	**nd/++**	**nd/+**	**nd/+**
14	Dimethyltrisulfide[f]	1383	972	cooked cabbage	**nd/+**	**nd/+**	**nd/+**
15	1-Octen-3-one[f]	1298	980	mushroom, earthy	+/+++	+/+++	+/+++
16	(Z)-1,5-Octadien-3-one [c]	1379	982	metallic, planty	**md/+**	**nd/+**	**nd/+**
17	Octanal[f]	1291	1005	sweet, orange	+/+	+/+	**nd/+**
18	1-Nonen-3-one [c]	1409	1081	Mushroom	**nd/+**	**+/nd**	**nd/nd**
19	Unknown	- -	1100	raw peanut, earthy	**nd/+**	**nd/+**	**nd/+**
20	Unknown	1444	1102	fresh, creamy	+/+	+/+	**nd/+**

[a-c]Footnotes same as in Table I. [f]Compound identified by comparison of its mass spectrum, RI values on two capillary columns and odor properties with those of the reference compound.

Table III. Odor-active compounds detected by gas chromatography-olfactometry of flavor extracts prepared from control and light-activated milk by vacuum distillation-solvent extraction.

No.	Compound	Retention Index		Aroma Property	Odor Intensity		
		DB-Wax	DB-5ms		Skim	2%	Whole
5	2,3-butanedione[f]	- -	624	buttery	**nd/nd**	**nd/+**	**nd/+**
7	Pentanal[f]	- -	700	sour,cut-grass	**nd/nd**	**nd/+**	**nd/+**
8	1-Hexen-3-one[e]	1093	799	plastic, water bottle	**+/+**	**+/++**	**+/++**
9	Hexanal[f]	1080	819	green, cut-grass	**nd/+**	**+/+**	**+/+**
11	Z-4-Heptenal[e]	- -	899	rancid, crabby	**nd/+**	**nd/+**	**nd/+**
12	3-(Methylthio)propanal[e]	1455	907	cooked potato	**+/++**	**+/+++**	**+/+++**
13	2-Acetyl-1-pyrroline[e]	1336	921	popcorn	**+/++**	**++/+++**	**++/+++**
15	1-Octen-3-one[f]	1298	978	mushroom, earthy	**+/+++**	**+/+++**	**+/++**
16	(Z)-1,5-Octadien-3-one[e]	1375	981	metallic, planty	**nd/+**	**Nd/++**	**nd/++**
17	Octanal[f]	1289	1003	sweet, orange	**+/+**	**+/+**	**+/+**
18	1-Nonen-3-one[e]	1406	1078	mushroom	**nd/nd**	**+/nd**	**nd/nd**
20	Unknown	1440	1098	fresh, creamy	**+/+**	**+/+**	**+/+**
21	(E,Z)-2,6-Nonadienal[e]	1585	1154	cucumber	**nd/+**	**nd/+**	**nd/+**
22	Unknown	- -	1161	hay	**+/+**	**+/+**	**+/+**
23	Unknown	- -	1332	fatty, corn tortilla	**nd/+**	**nd/+**	**nd/+**
24	Unknown	- -	1387	apple sauce	**nd/+**	**+/+**	**+/+**
25	3-Methylindole[e]	2501	1420	chlorine, animal	**nd/+**	**+/+**	**+/+**

[a-f]Footnotes same as in Tables I and II.

Table IV. Concentrations of selected volatile compounds in control and light-activated whole milk determined by dynamic headspace sampling-gas chromatography-mass spectrometry.

No.[a]	Compound	Concentration (μg/L)	
		Control	Light-Activated (18 h)
2	Acetaldehyde	1.89 ± 2.14[b] (n/a)[c] 10.2 ± 11.2[d]	21.7 ± 0.9 (38.6) 127.5 ± 4.0
3	Dimethylsulfide	1.27 ± 0.01 (n/a) 70.3 ± 7.9	0.85 ± 0.06 (1.47) 35.4 ± 2.1
4	2-Methylpropanal	0.66 ± 0.54 (n/a) 9.5 ± 8.6	0.87 ± 0.11 (n/a) 9.0 ± 1.0
6	3-Methylbutanal	0.25 ± 0.17 (n/a) 1.4 ± 1.0	0.37 ± 0.09 (0.46) 1.46 ± 0.32
7	Pentanal	1.6 ± 1.1 (n/a) 11.0 ± 8.5	97.0 ± 5.0 (101) 419 ± 26
8	1-Hexen-3-one	Nd[e] (n/a) nd	0.26 ± 0.01 (0.21) 0.98 ± 0.06
9	Hexanal	1.4 ± 0.7 (n/a) 6.0 ± 3.4	27.0 ± 1.6 (28.2) 82.8 ± 5.8
15	1-Octen-3-one	0.06 ± 0.09 (n/a) 0.17 ± 0.24	0.68 ± 0.01 (0.714) 1.209 ± 0.004

[a]Numbers correspond to those in Tables I-III. [b]Value based on internal standardization with calibration data obtained using whole milk as matrix. [c]Value from standard addition plot, n/a = not determined. [d]Value based on internal standardization with calibration data obtained using deodorized water as matrix. [e]nd, not detected.

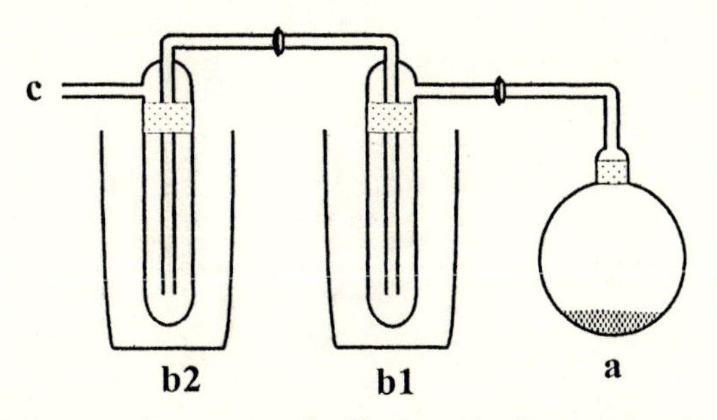

Figure 1. Apparatus for vacuum distillation of milk samples. (Symbols are defined in text).

Column eluent was split 1:1 between FID and sniffing port using deactivated fused silica capillaries (1 m length x 0.25 mm i.d.). GC oven temperature was programmed from 40°C to 200°C at a rate of 8°C/min with initial and final hold times of 5 and 30 min, respectively. FID and sniffing port were maintained at a temperature of 250°C. Sniffing port was supplied with humidified air at 30 mL/min. GC-O was conducted by three experienced panelists. Further details of GC-O have been previously reported (*24*).

Static headspace sampling-gas chromatography-olfactometry of headspace samples (SHS-GCO). A 50 mL aliquot of milk was placed in a 250-mL round bottom flask, and the flask was sealed with a septum. The flask was incubated in a 40°C water bath for 20 min and a headspace sample was drawn from the flask using a preheated (60°C) gastight syringe. The headspace sample was immediately injected at a flow rate of 5 mL/min into an HP5890A GC equipped with a packed column inlet modified for capillary GC injection (Uniliner Sleeve Adaptor; Restek Corp., Bellefonte, PA). Separations were performed on DB-WAX and DB-5ms fused silica capillary columns (30 m length x 0.53 mm i.d. x 1 μm d_f for DB-WAX (or 1.5 μm d_f for DB-5ms); J&W Scientific). Prior to injection a 15 cm section of column was cooled in liquid nitrogen in order to cryofocus the volatiles. The GC oven was rapidly heated and the run started when the oven temperature reached 40°C. Other GC conditions were the same as described for GC-O.

Dynamic headspace sampling-gas chromatography-olfactometry of (DHS-GCO). A Tekmar 3000 Purge and Trap Concentrator/Cryofocusing Module (Tekmar Co., Cincinnati, OH) coupled with an HP5890 series II GC was employed for DHS-GCO. A 20 mL aliquot of milk in a 25-mL purge tube (equipped with foam breaker) was prepurged with helium (40 mL/min) for 2 min. The sample was then preheated to 40°C for 5 min. The volatiles were purged at 40°C for 20 min onto a Tenax TA trap (part no. 12-0083-303, Tekmar Co.) maintained at 0°C. After sampling, the trap was dry purged for 5 min and then the volatiles were desorbed (180°C for 1 min) and subsequently cryofocused (-120°C) onto a 15 cm section of 0.53 mm i.d. deactivated fused silica capillary column. Transfer lines and valves were maintained at a temperature of 150°C. Trap pressure control was set at 4 psi. Helium flow during thermal desorption of the Tenax TA trap (20 mL/min) and cryofocusing trap (1.4 mL/min) was controlled by the split/splitless electronic pressure control pneumatics of the GC (Figure 2). Cryofocused volatiles were thermally desorbed (180°C for 1 min) directly into the analytical GC column. All other GC and GC-O conditions were the same as described for AEDA. Between each analysis, the system was purged, clean glassware was installed, and the Tenax TA trap was baked (225°C for 15 min).

Gas chromatography-mass spectrometry (GC-MS). GC-MS system consisted of an HP5890 Series II GC/5972 mass selective detector (MSD, Hewlett-Packard, Co.). Separations were performed on fused silica capillary columns (DB-WAX or DB-5ms, 60 m length x 0.25 mm i.d. x 0.25 μm d_f; J&W Scientific). Carrier gas was helium at a constant flow of 0.96 mL/min. Oven temperature was programmed from

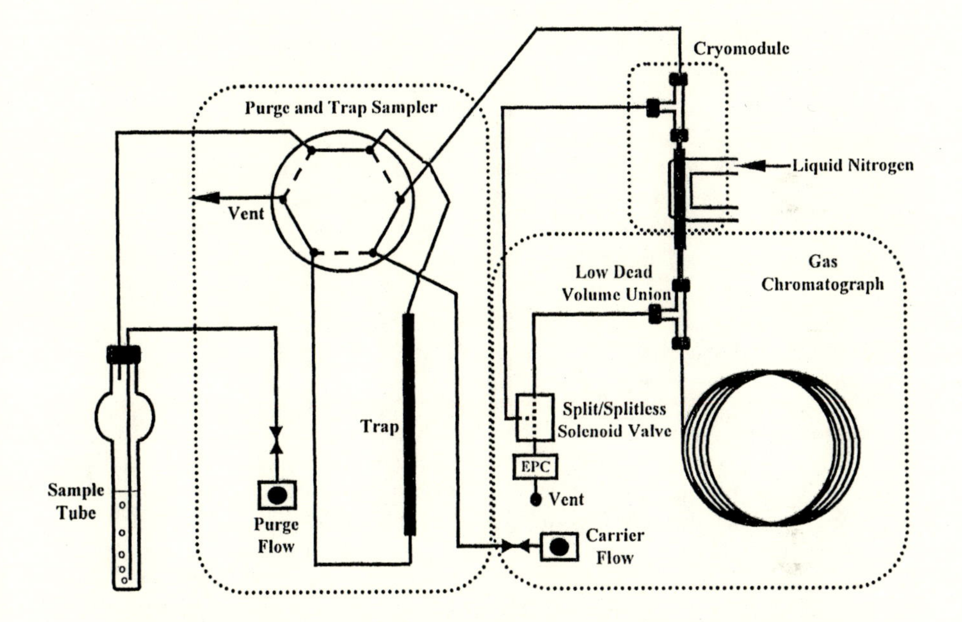

Figure 2. Diagram of apparatus used for dynamic headspace sampling-gas chromatography.

40°C to 200°C at a rate of 6°C/min with initial and final hold times of 5 and 60 min, respectively. MSD conditions were as follows: capillary direct interface temperature, 280°C; ionization energy, 70 eV; mass range, 33-350 a.m.u.; EM voltage (Atune + 200 V); scan rate, 2.2 scans/s.

VDSE extracts (2 µL) were injected in the on-column mode. For DHS, the purge and trap system was connected to the MSD and the above 0.25 mm i.d. capillary columns were used for analysis.

Quantitative analysis by DHS-GC-MS. Control or LAF whole milk was added to a 100-mL volumetric flask and spiked with 10 µL of internal standard solution (512 µg/mL of 2-methyl-3-heptanone in methanol) using a Series 701 syringe (Hamilton Co., Reno, NV). Solutions were vigorously shaken for 5 min and then analyzed in duplicate by DHS-GC-MS as previously described. Calibration by standard addition was accomplished by addition of 0 (blank), 5, 10, or 15 µL of standard stock solution (containing 4.91 mg of no. **2**; 616 µg of no. **3**; 17.1 µg of no. **4**; 14.9 µg of no. **6**; 1.78 mg of no. **7**; 19.7 µg of no. **8**; 907 µg of no. **9**; and 15.3 µg of no. **15** per mL of methanol) to separate 100.0 mL aliquots of LAF whole milk prior to analysis. Solutions for internal standardization were prepared by spiking four 100.0 mL aliquots of control whole milk (or distilled water) with 10 µL of internal standard solution followed by spiking each with either 0, 5, 10, or 15 µL of the standard stock solution.

Descriptive analysis. Triangle difference tests were used to select panelists based on their ability to consistently distinguish LAF milk from control milk. All panelists were routine consumers of milk and had previous experience in descriptive analysis. Roundtable discussions were used to provide group feedback for the panelists. The first task was to establish a list of flavor attributes for control and LAF milks. Panelists tasted milk samples, which were representative of LAF milk of slight, medium, and strong intensities, as well as control milk which had no detectable LAF flavor. Whole, 2%, and skim milk samples were evaluated at each intensity level. Terms were selected to describe the flavor attributes and then standards representative of each term were selected. The final attribute list was determined by the panel and consisted of the following: overall degree of light-activated flavor (LAF), butterscotch (reference=butterscotch candy, Werther's Original Candy, Stork USA, Chicago, IL), buttery (reference=melted butter), mushroom (reference=fresh mushrooms or 1 mg/L aqueous solution of 1-octen-3-one), creamy (reference=cream), and plastic (reference=0.6 mil clear plastic sandwich bags, Kroger, Cincinnati, OH). A 15 cm Spectrum (*25*) line scale was used to record the intensity of each attribute. Panelists received extensive group training using the final score sheet until they felt comfortable with the procedure. Panelists then individually scored milk samples and their scores were compared with those of the other panel members. Panelists whose scores deviated from the group consensus adjusted their evaluation techniques to better match the group. For routine evaluation, samples were presented in styrofoam cups equipped with plastic lids and straws to prevent the appearance of the milk from influencing panelist's decisions. Training lasted for twelve weeks with 10-15 h of total training time.

Results and Discussion

Sensory descriptive analysis of light-activated milks. Exposure of milk to light leads to development of a typical off-flavor that has been previously described by such terms as "burnt protein", "burnt feathers", "cabbage", and "mushroom" (*5, 7, 26*). Many factors affect the intensity of light-activated flavor (LAF), but light intensity and time of exposure are the two most easily manipulated variables. In the present study, milk was exposed to a fluorescent light intensity of 200 FC to produce typical LAF. This light intensity was chosen based on previous reports (*5, 15, 19, 27*) and our observations (local) of the average light intensity within retail milk display cases. The optimum time of light exposure was determined by sensory evaluation of milk (skim, 2%, and whole) exposed to 200 FC for 0, 6, 12, 18, 24, and 48 h. The objective here was to obtain a strong, but typical LAF, thus making it easier to identify odorants involved in LAF. Results from expert ADSA dairy score card judges and a sensory descriptive panel established that an exposure time of 18 h gave the best results for milk at all fat levels tested, while times exceeding 18 h resulted in milk with an off-flavor that was atypical of LAF.

Additional sensory descriptive analyses were conducted to provide some insight into what flavor attributes predominate in LAF milk. This information not only provides a detailed description of LAF, but also is useful to gauge if results of gas chromatography-olfactometry (GC-O) seem reasonable compared to what is actually perceived by human assessors. Results of sensory descriptive analysis of LAF milk of varying fat levels are shown in Figure 3. Terms chosen by the panel to describe controls and LAF milks were mushroom, plastic, butterscotch, creamy, and buttery. Overall degree of LAF also was scored. The creamy and buttery terms did not differentiate control from LAF milks. Instead, they seemed to be more indicative of the fat level of the milk tested, since these terms were negligible for the control skim and LAF skim milk. The remaining terms, mushroom, plastic and butterscotch, comprised the majority of the overall degree of LAF score. LAF milks at all fat levels had strong mushroom and plastic attributes compared with control milks. On the other hand, LAF milks at the whole and 2% fat level had a distinctive butterscotch note, which was weak in the skim LAF milk. The butterscotch characteristic may be due to components arising from oxidation of milk fat, which is why this term predominates in the LAF whole milk and is relatively high in the LAF 2% milk. These results indicate that LAF in milk is impacted by fat level. This could be from participation of fat in both the development of LAF and in the partitioning of the aroma components.

Odorant profiles of light-activated milks. Identification of odorants comprising LAF is an important first step and should be conducted prior to the development of any quantitative analysis procedure. In order to obtain an unbiased sampling of these odorants, volatile components of control and LAF milks were isolated by three complimentary techniques, namely static headspace sampling (SHS), dynamic headspace sampling (DHS) and vacuum distillation-solvent extraction (VDSE). The odor-active components were subsequently established by gas chromatography-olfactometry (GC-O). Furthermore, GC-O was used to perform a side-by-side comparison of the intensities of each odorant in control and LAF milks.

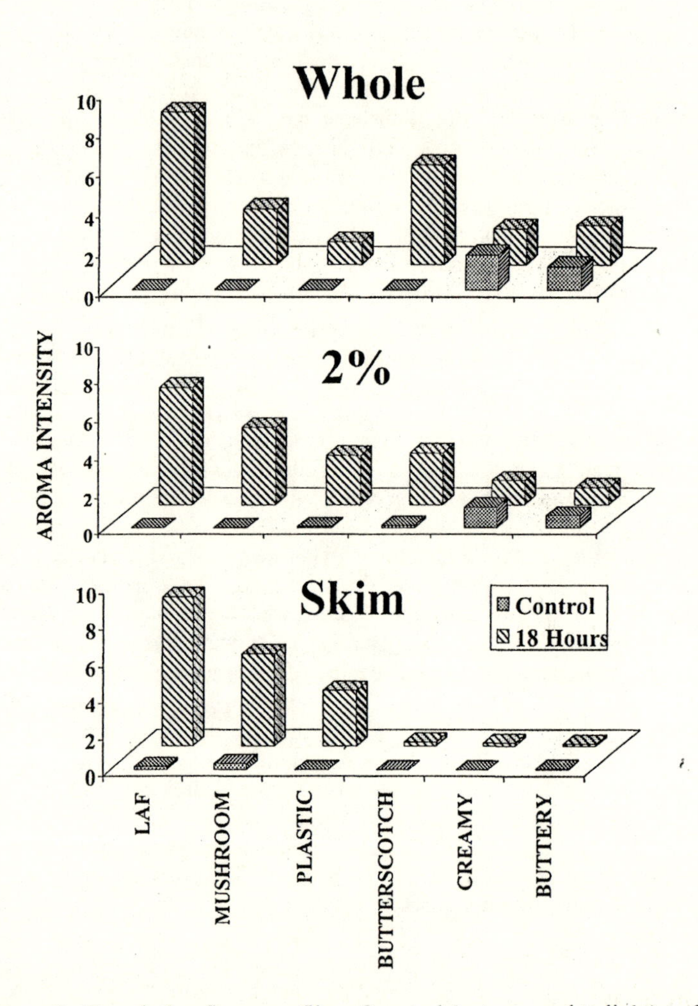

Figure 3. Descriptive flavor profiles of control (not exposed to light) and fluorescent light-activated milks at three fat levels.

Static headspace sampling (SHS). SHS was the least effective among the three methods for isolation of odor-active components (Table I). A total of only four odorants were detected in headspace samples of control and LAF milks. Only one compound, 1-octen-3-one (no. **15**), was consistently detected and was found at higher intensities in LAF milks compared with controls. Poor recovery of odorants by SHS was probably due to the low sample equilibration temperature (40°C) chosen for the analysis, as well as the presence of only low levels of the various odorants. The sensitivity of SHS could be improved by increasing the sample equilibration temperature; however, this was not attempted as it could lead to formation of thermally generated artifacts. Despite its lack of sensitivity, SHS indicated that the mushroom-like odorant, no. **15**, was the predominant contributor to the headspace aroma of LAF milk.

Dynamic headspace sampling (DHS). Like SHS, dynamic headspace sampling (DHS) was conducted using a low sample temperature of 40°C to prevent generation of artifacts. The procedure for DHS was based on numerous preliminary experiments in which variables such as sample volume, purge time, trap desorption time and temperature, etc. were evaluated. Some equipment modifications also were made to achieve high flow rates during thermal desorption and to decrease the dead volume of injector. These modifications are detailed in Figure 1 and were described earlier in the materials and methods section. Final conditions chosen (i.e., 20 mL sample volume and 20 min purge time) were based upon keeping the analysis time short, while at the same time allowing for good recovery of volatiles.

A side-by-side comparison of odorants detected in control and LAF milks of various fat contents is given in Table II. In contrast to SHS, a greater number of odorants were isolated by DHS. A total of 17 odorants were detected in both control and LAF milks. Those found at high intensity were described as sour/cut-grass (no. **7**), plastic/water bottle (no. **8**), green/cut-grass (no. **9**), and mushroom, earthy (no. **15**). Compounds nos. **2**, **7-10**, **14-16**, and **19** were detected at higher odor intensities in LAF milks compared with controls. Effect of fat level was evident, in that several lipid-derived components (e.g., nos. **7**, **9**. **15**, and **16**) increased in odor intensity in LAF milks and their intensities were proportional to the fat content of milk. Notably absent in Table II is 3-(methylthio)propanal (methional) which is generally regarded as a character-impact component of LAF milk (*1,3,9*). Interestingly, all previous studies employing DHS for the analysis of LAF milk failed to detect this important odorant (*19,20*). It is not clear why methional was not detected by DHS-GCO, as this odorant was previously found in cooked shrimp tail meat (*28*) using the same DHS system employed in the present study. It may be that in our previous study a higher sample purge temperature of 60°C was used, which could have allowed for more efficient purging of this compound. In order to determine the presence and/or importance of methional (and possibly other odorants not isolated by DHS) in LAF milk, a distillation-solvent extraction technique was applied for the isolation of volatile constituents prior to GC-O.

Vacuum distillation-solvent extraction (VDSE). The contribution of odorants of intermediate and low volatility to control and LAF milks were established by GC-O of isolates prepared by vacuum distillation followed by direct solvent extraction of the distillates (VDSE). The low temperature used during VDSE allowed for preparation of extracts that were representative of the starting control or LAF milks. A total of 17 odorants were detected in VDSE extracts of both LAF and control milks (Table III). Similar to the results of DHS-GCO, more odorants were detected in LAF (16) than in control milks (11). Predominant odorants isolated by VDSE extracts of LAF milk were described as mushroom/earthy/metallic (nos. **15**, **16**), potato (no. **12**), popcorn (no. **13**), and plastic (no. **8**). Presence of methional (no. **12**) at high odor intensities in LAF milk confirms its importance as a key component of LAF milk. Compounds nos. **5**, **7**, **8**, **11-13**, **15**, **16**, **21**, **23** were detected at higher odor intensities in LAF milks compared with controls. Effect of fat on formation of LAF was evident, since odor intensities of several lipid-derived components (e.g., nos. **15**, **16** and **21**) increased in LAF milks. Furthermore, the intensities of these odorants were proportional to the fat content of the LAF milks.

With the exception of 1-hexen-3-one, 1-nonen-3-one, and 2-acetyl-1-pyrroline, all compounds in Tables I-III have been previously identified in milk. 1-Nonen-3-one has been recently indentified in yogurt (*28*) and 2-acetyl-1-pyrroline was found in cheese (*29*). The plastic attribute detected by the descriptive panel could be due to the presence of 1-hexen-3-one (no. 8, plastic/water bottle). The presence of strong mushroom/earthy (no. 15) and metallic/planty (no. 16) notes in the LAF milks were in agreement with the results of sensory descriptive analysis. The buttery and creamy notes, which were unaffected by light exposure, might be due to the presence of an unidentified odorant with a fresh/creamy note (no. **20**). In agreement with sensory results, the intensity of this odorant also was unchanged in the milks after light exposure. None of the odorants detected by GC-O of volatiles isolated by SHS, DHS or VDSE seem to be directly responsible for the butterscotch note detected by the descriptive panel in the LAF 2% and whole milks. The presence of aldehydes, such as acetaldehye (no. **2**), pentanal (no. **7**) and hexanal (no. **9**), at higher odor intensities in the LAF 2% and whole milks compared with LAF skim milk might explain the occurrence of this attribute. However, the odor properties (solvent/sour/green/cut-grass) of these odorants in no way agree with this suggestion. Another possibility is that the butterscotch attribute was due to many odorants present at near or below detection limits.

Quantitative analysis by dynamic headspace sampling

Despite the fact that 3-(methylthio)propanal could not be effectively isolated by DHS, this method could be useful for monitoring levels of several other key odorants (e.g. nos. **2-4**, **6-9**, and **15**) in control and LAF milks. Table IV contains concentration data for selected odorants determined by DHS-GC-MS of whole milk before and after 18 h of exposure to fluorescent light (200 FC). Three calibration methods were employed; (1) standard addition, (2) internal standardization with standard curves generated in a milk matrix; and (3) internal standardization with standard curves generated in a water matrix. Concentrations determined by methods 1 and 2 were in good agreement. The standard addition method should give the most accurate results because it accounts for sample matrix effects. However, this technique is very time consuming and results can have greater uncertainty due to the error in estimating the x-intercepts from the least squares regressions (*18*). Internal standardization is probably a better alternative for routine analysis because it accounts for matrix effects while also allowing for greater sample throughput. Furthurmore, results for internal standardization were similar to those of standard addition when a milk matrix was used during calibration. Use of milk matrix during preparation of the standard curve is very important as demonstrated by method 3 in which the concentrations were overestimated. If response factors are not used to normalize the data (as in method 3) then the final results will not be representative of the actual levels of volatile components in the sample; however, relative concentration data may be useful for comparing treatment effects in similar samples.

The data in table IV are in good agreement with results of GC-O and indicate increases in concentrations of acetaldehyde (no. **2**), pentanal (no. **7**), 1-hexen-3-one (no. **8**), hexanal (no. **9**), and 1-octen-3-one (no. **15**) after exposure of the milk to fluorescent light. No change was observed for the *Strecker* aldehydes 2-methylpropanal and 3-methylbutanal, while the level of dimethylsulfide decreased.

Conclusions

Results of sensory evaluation and instrumental analyses demonstrated that the aroma of LAF milk is impacted by fat level. Terms such as creamy and buttery did not differentiate control from LAF milks, but were indicative of the fat level of the milk tested. The mushroom, plastic and butterscotch terms were the predominant attributes in the LAF milks. The butterscotch characteristic was intense in LAF whole and 2% milks but not in LAF skim milk, indicating the importance of fat in LAF development and/or in the partitioning of the aroma components.

Compounds of low, intermediate and high volatility are involved in the aroma of control and LAF milks. Furthermore, these compounds are derived from both lipid and nonlipid precursors. Aldehydes and ketones are the major odorants in LAF milk and can be effectively isolated and analyzed by DHS. However, VDSE was able to isolate several additional components not found by DHS that contribute to the aroma of LAF milk, such as 3-(methylthio)propanal and 2-acetyl-1-pyrroline.

Acknowledgements

This is journal article no. PS9249 of the Mississippi Agricultural and Forestry Experiment Station.

LITERATURE CITED

1. Badings, H.T. Milk. In *Volatile Compounds in Foods and Beverages*, Maarse, H. (Ed.). Marcel Dekker, Inc.: New York, Ch. 4, pp. 91-106.
2. Patton, S.; Forss, D.A.; Day, E.A. *J. Dairy Sci.* **1956**, 39, 1469-1470.
3. Forss, D.A. *J. Dairy Sci.* **1979**, *46*, 691-706.
4. White, C.H.; Bulthaus, M. *J. Dairy Sci.* **1982**, *65*, 489-494.
5. Bradley, Jr., R.L. *J. Food Prot.* **1980**, *43*, 314-320.
6. Bekbölet, M. *J. Food Prot.* **1990**, *53*, 430-440.
7. Dimick, P.S. *Can. Inst. Food Sci. Technol.* **1982**, *15*, 247-256.
8. Sattar, A.; deMan, J.M. *CRC Crit. Rev. Food Sci. Nutr.* **1976**, 7, 13-37.

9. Patton, S. *J. Dairy Sci.*, **1954**, *37*, 446-452.
10. Allen, C.; Parks, O.W. *J. Dairy Sci.* **1975**, *58*, 1609-1613.
11. Aurand, L.W.; Singleton, J.A.; Noble, B.W. *J. Dairy Sci.* **1966**, *49*, 138-143.
12. Korycka-Dahl, M.; Richardson, T. *J. Dairy Sci.* **1978**, *61*, 400-407.
13. Jeng, W.; Basette, R.; Crang, R.E. *J. Dairy Sci.*, **1988**, *71*, 2366-2372.
14. Bassette, R.; Ward, G. *J. Dairy Sci.* **1975**, *58*, 428-429.
15. Mehta, R.S.; Bassette, R. *J. Food Prot.*, **1979**, *42*, 256-258.
16. Moio, L.; Dekimpe, J.; Etievant, P.; Addeo, F. *J. Dairy Res.* **1993**, *60*, 199-213.
17. Moio, L.; Langlois, D.; Etievant, P.; Addeo, F. *J. Dairy Res.* **1993**, 215-222.
18. Imhof, R.; Bosset, J.O. *Lebensm.-Wiss. u.-Technol.* **1994**, *27*, 265-269.
19. Kim, Y.D.; Morr, C.V. *Int. Dairy J.*, **1996**, *6*, 185-193.
20. Señorans, F.J.; Tabera, J.; Herraiz, M.; Reglero, G. *J. Dairy Sci.*, **1996**, *79*, 1706-1712.
21. Vallejo-Cordoba, B.; Nakai, S. *J. Agric. Food Chem.* **1993**, *41*, 2378-2384.
22. Ulrich, F.; Grosch, W. *J. Am. Oil Chem. Soc.* **1988**, *65*, 1313-1317.
23. Furniss, B.S.; Hannford, A.J.; Smith, P.W.G.; Tatchell, A.R. *Vogel's Textbook of Practical Organic Chemistry, 5th ed.* Longman Group UK Limited: Essex, England, 1989, pp. 587-588.
24. Cha, Y.-J.; Lee, G.H.; Cadwallader, K.R. In *Flavor and Lipid Chemistry of Seafoods*, Shahidi, F. and Cadwallader, K.R. (Eds.), ACS Symposium No. 674. American Chemical Society: Washington, DC, 1997, Ch. 13, pp. 131-147.
25. Meilgaard, M.; Civille, G.V.; Carr, B.T. *Sensory Evaluation Techniques, 2nd ed.* CRC Press, Inc.: Boca Raton, FL.
26. Dunkley, W.L.; Pangborn, R.M.; Franklin, J.D. *Milk Dealer* **1963**, *52*, 52.
27. Hansen, A.P.; Turner, L.G.; Aurand, L.W. *J. Milk Food Technol.* **1975**, *38*, 388-392.
28. Ott, A.; Fay, L.B.; Chaintreau, A. *J. Agric. Food Chem.* **1997**, *45*, 850-858.
29. Kubickova, J.; Grosch, W. Int. Dairy J. **1997**, *7*, 65-70.

Chapter 29

High Resolution Gas Chromatography–Selective Odorant Measurement by Multisensor Array

A Useful Technique To Develop Tailor-Made Chemosensor Arrays for Food Flavor Analysis

P. Schieberle[1], T. Hofmann[1], D. Kohl[2], C. Krummel[2], L. Heinert[2], J. Bock[2], and M. Traxler[2]

[1]Institut für Lebensmittelchemie, Technische Universität München und Deutsche Forschungsanstalt für Lebensmittelchemie, Lichtenbergstrasse 4, D-85748 Garching, Germany
[2]Institut für Angewandte Physik, Universität Giessen, Germany

A technique combining a chemosensor array with high resolution gas chromatography (HRGC/SOMMSA) was used to systematically study the responses of several experimental and commercial chemosensors (metal oxides, surface acoustic wave devices, phthalocyanine sensors) to a number of selected food volatiles or aroma compounds. It was shown that the temperatures and the dopants used significantly influence the sensitivity and/or selectivity of the chemosensors. For example, a ZnO/Pt sensor, when operated at 300°C, was able to selectively detect the key bread crust odorant 2-acetyl-1-pyrroline. A copper phthalocyanine sensor selectively detected (E)-2-nonenal and (E,E)-2,4-decadienal in mixtures with several pyrazines. Based on such model studies, a portable sensor array was constructed which was able to detect degree of roasting, e.g., of toasted bread slices. A combination of Headspace-HRGC with the SOMMSA technique is proposed as useful approach to develop chemosensor arrays adapted to special targets in flavor control, based on quantitative correlations of key odorants with indicator volatiles.

Aroma, one of the most important attributes of foodstuffs for attracting the consumer, is evoked by volatile chemicals entering the olfactory system of our nose. While inhaling the vapor of, for example bread, we compare the perceived odor impression with concepts or memories of bread experiences in our brain. Finally, we come to the conclusion that we like or dislike what we smell based on our reaction to this comparison.

The uniqueness of consumer's response poses challenges for the food industry:

- Firstly, the quality concepts present in consumer's brain have usually been created from experiences with products manufactured by small scale producers or individual cooks. As a consequence, we expect industrially manufactured products to meet these "standards".
- Secondly, the characteristic flavors of many foods, such as bread or coffee, are often generated during processing by a cascade of chemical reactions from odorless precursors. Significant flavor changes may, however, result if industrial processes are shortened compared with traditional processes for economic reasons.
- Finally, packaging, transport and storage of the products may also significantly influence their flavor.

Especially the conflict between optimum flavor generation and the need to shorten processes induces a demand in the food industry for objective and fast methods to evaluate the food flavor quality.

It is a well-known phenomenon that the reaction of hot metal oxides, such as SnO_2 or ZnO, with organic compounds induces changes in their conductivities (1). Based on this knowledge, chemosensors have been arranged into arrays called "electronic noses". Shown in Figure 1, the sensors are brought in contact with the volatiles present in the vapor above the food. The signals induced at each sensor are then recorded by a computer where the multivariant data can be processed by pattern recognition software to yield a profile of the food. It is implied in Figure 1 that the sensor array, once it is calibrated, should have a similar capability for distinguishing food flavors as the human nose.

When the human nose is used as a biosensor, such as Charm analysis (2) or aroma extract dilution analysis (3), it has been shown for a certain number of foods (4) that of the hundreds of volatiles present, only a small number are needed to generate an overall food aroma. In other words, our biosensor - the nose - selects a certain number of compounds to create an odor impression in the brain. This selection is, however, not based on quantity. Most of the key aroma compounds are present in only trace amounts, but can still be smelled. In contrast to the selectively operating "sensors" in our nose, the relatively unspecific chemosensors more or less respond to all the chemicals present in the vapor.

Newer "electronic noses" containing surface acoustic wave devices or quartz piezo-electric sensors have been developed, but the continuing difference between nasal and electronic response is undoubtedly the reason why it was concluded that there are discrepancies between "... the proposed ideal machine and what is available ...." (5). It was, claimed that improvement in chemosensor arrays would come from the development of more specific sensors (5).

The purpose of the following investigation was to gain insight into chemosensor/structure relationships and, to make chemosensors with greater selectivity for the flavor developed during the toasting of bread.

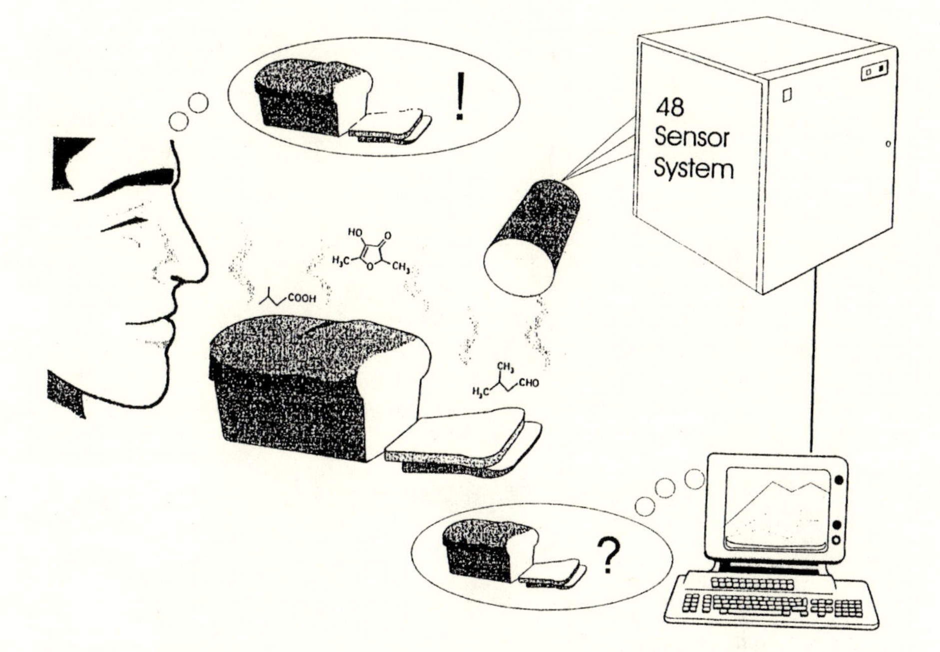

Figure 1. Human olfactory perception vs. electronical perception - How to correlate the different 'sensor' principles for flavor quality evaluation?

EXPERIMENTAL

The ability of single chemosensors to detect food volatiles was evaluated by means of the HRGC/selective odorant measurement multisensor array (SOMMSA) technique recently developed by us (6). The principle of the method is as follows: The compounds under investigation are separated by HRGC and the effluent is split 1:1 into an FID and an array of 7 sensors arranged in a temperature controlled small, self-constructed brass box (1.8 mL internal volume). Using a constant voltage circuit, 0.5 V is applied to the interdigitated electrodes of the metal oxide sensor layer. The d.c. current is measured by an operational amplifier as a voltage and digitized with an analog digital converter. Metal oxide sensors used were either prepared as recently described (6) or supplied by UST (Umweltsensortechnik, Gschwenden, Germany). Commercially available surface acoustic wave devices were also mounted onto the GC outlet, but only heated at 60°C. The types of sensors used are summarized in Table I.

Table I. Types of sensors used

A. Metal oxides (on 3 mm x 3 mm aluminum oxide platelets)	
1. Zinc oxide	4. Tin dioxide
2. Zinc oxide + 5 % palladium	5. Tin dioxide + 5 % palladium
3. Zinc oxide + 5 % platinum	6. Tin dioxide + 5 % platinum
	7. Modified tin dioxide (UST 1000)
	8. Modified tin dioxide (UST 5000)
B. Polymer sensors	
Polyisobutylene applied (2 μm) on a surface acoustic wave device (working frequency: 80 MHz)	
C. Phthalocyanine sensors	
Copper phthalocyanine	
Copper phthalocyanine doped with nickel	

Results

HRGC/selective odorant measurement by multisensor array (SOMMSA). The sensor material, including the addition of dopants, such as palladium or platinum, as well as the temperature of the sensors may influence the responses of the chemosensors (1). However, the principle of such interactions is not yet fully understood nor has a systematic investigation been reported that evaluates the responses of the chemosensors to single food volatiles. To evaluate the capabilities of chemosensors for detection of volatiles, we have recently developed the HRGC/SOMMSA technique (6). The method allows one to check the response of 7 sensors to several volatile compounds simultaneously in only one run.

N-heterocycles formed during processing of foods, such as coffee or bread, via the Maillard reaction are well-known as important odorants, and were, therefore, used to choose sensors via HRGC/SOMMSA. A model mixture of the four pyrazines 2,3,5-trimethyl-, 2-ethyl-3,6-dimethyl-, 2-ethyl-3,5-dimethyl- and 2,3-diethyl-5-methyl-pyrazine, 2-acetylthiazol and 2-acetyl-2-thiazoline was analyzed by the SOMMSA technique. At a temperature of 450°C, about 2-5 ng of all six volatiles could be detected by a ZnO sensor doped with palladium (data not shown). A ZnO sensor doped with platinum showed a slightly lower overall sensitivity, whereas a SnO_2 sensor was not able to detect these volatiles. Decreasing the temperature of the five chemosensors to 300°C, significantly increased the sensitivity of all three sensors, especially for 2-acetylthiazol and 2-acetyl-2-thiazoline (Figure 2). At this temperature, the ZnO sensor doped with platinum (D in Figure 2) showed a distinct selectivity for 2-acetyl-2-thiazoline, whereas the higher oxidized molecule 2-acetylthiazol and, also, the 4 pyrazines did not yield an intense sensor signal.

This sensitivity for 2-acetyl-2-thiazoline could be further increased by lowering the operating temperature of the sensor down to 150°C (data not shown). 2-Acetyl-2-thiazoline belongs to the key odorants of roast meat (7) and, also, processed flavors containing cysteine (8). Therefore, this sensor might be useful in monitoring its formation during thermal processing.

The results obtained for the pair 2-acetylthiazol/2-acetyl-2-thiazoline and the pyrazines suggest that highly oxidized molecules do not react with the ZnO/Pt sensor at lower temperatures. This prompted us to investigate whether this sensor shows the same behavior vs. the homologuous pairs 2-acetylpyrrol/2-acetyl-1-pyrroline and 2-acetylpyridine/2-acetyltetrahydropyridine. The results obtained for the first pair of volatiles are shown in Figure 3. As found for the thiazoline, the ZnO sensor doped with platinum, selectively and very sensitively, detected the 2-acetyl-1-pyrroline (AP) at 300°C (B in Figure 3), whereas at 450°C, both the AP and the 2-acetylpyrrol gave responses (C in Figure 3). It is interesting to note that doping of the ZnO sensor with palladium led to an unselective, but also sensitive detection of both volatiles at 300°C (D in Figure 3). Further data showed that the ZnO/Pt sensor was also able to selectively detect the 2-acetyltetrahydropyridine in mixture with 2-acetylpyridine (data not shown).

These results allow the conclusion that the ZnO sensor doped with platinium discriminates the oxidized moieties of N-heterocycles. However, this assumption has to be confirmed by further systematic studies on homologuous N-compounds.

2-Acetyl-1-pyrroline and 2-acetyltetrahydropyridine are key odorants in wheat bread crust (9) and freshly popped corn (10). Furthermore, both odorants are the key aroma compounds formed from the amino acid proline when heated in the presence of carbohydrates (11). The data above suggest that sensors suitable for monitoring flavor development during manufacturing of processed flavors including baked goods could be made from ZnO/Pt. However, as detailed below, possible other reactions have to be taken into account.

Besides N-heterocycles, compounds formed during lipid peroxidation are often found among the volatiles of food flavor extracts. Among them, (E)-2-nonenal and

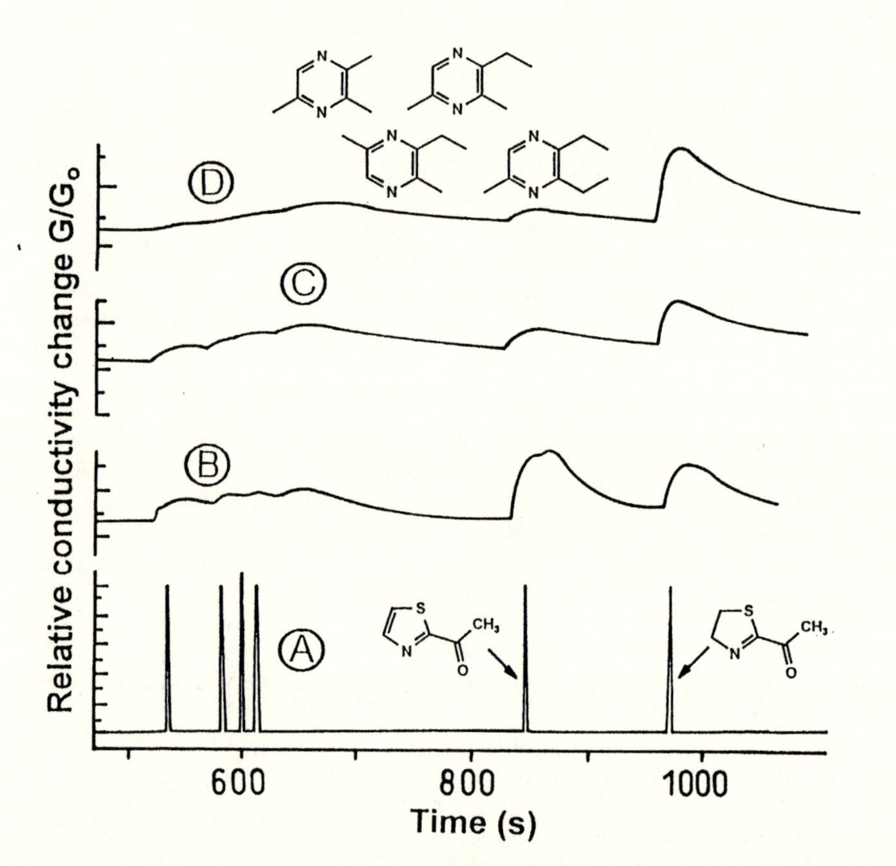

Figure 2. Response of a set of three chemosensors (B-D) operated at 300°C to a mixture of food odorants using HRGC/SOMMSA. A: FID B: SnO_2 C: ZnO/Pd D: ZnO/Pt

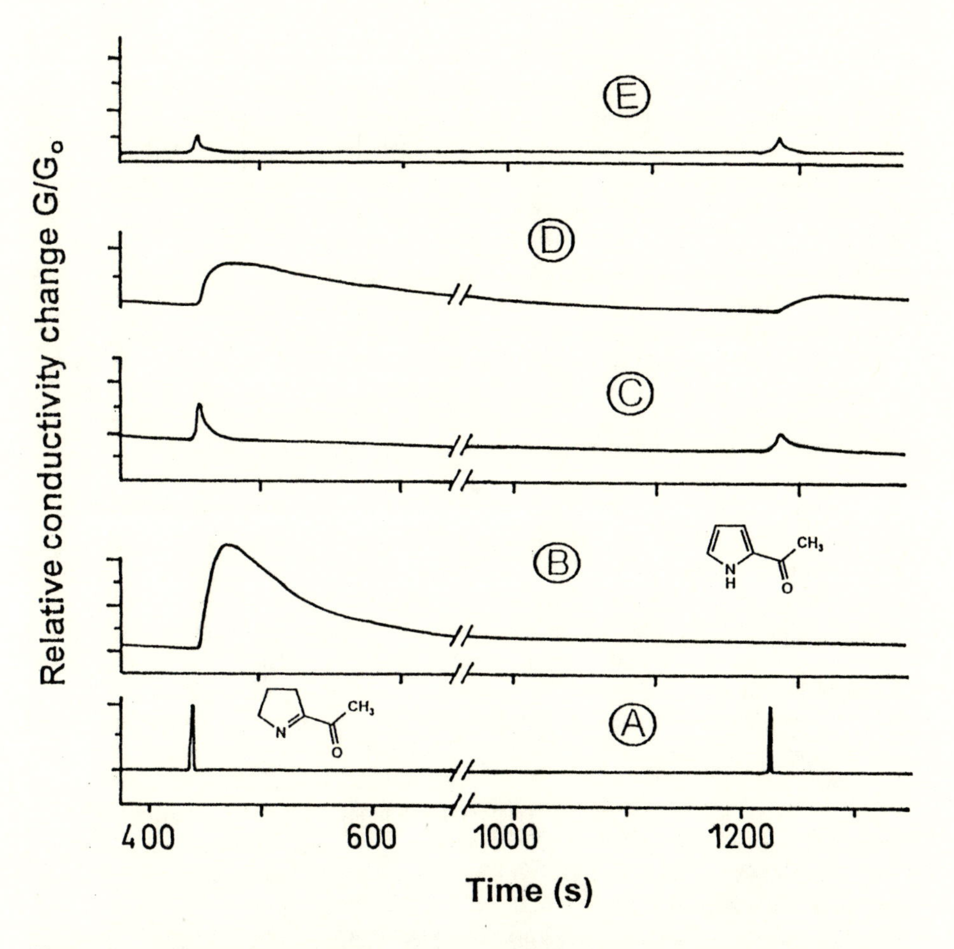

Figure 3. Response of two sensors (B, D) operated at two different temperatures to a mixture of 2-acetyl-1-pyrroline and 2-acetylpyrrol. A: FID B: ZnO/Pt at 300°C C: ZnO/Pt at 450°C D: ZnO/Pd at 300°C E: ZnO/Pd at 450°C

(E,E)-2,4-decadienal have been established as key odorants, e.g., in processed cereal foods (9, 10). In order to find chemosensors to selectively detect these aldehydes in a given mixture of volatiles, a selection of seven food volatiles, containing the two aldehydes, was applied to a surface acoustic wave device (SAW) coated with polyisobutylene. The results indicated that this SAW was a suitable sensor for pyrazines, (E)-2-nonenal and (E,E)-2,4-decadienal, whereas 2-acetyl-1-pyrroline and 4-hydroxy-2,5-dimethyl-3(2H)-furanone (HDF) did not yield a response (data not shown).

Metal-phthalocyanines also respond to gaseous molecules. Application of the same odorant mixture to a copper phthalocyanine sensor doped with nickel gave results similar to those obtained for the SAW device, e.g., 2-acetyl-1-pyroline and HDF were not detected (Figure 4A). However, when the dopant nickel was omitted (Figure 4B), the sensor selectively detects the two aldehydes. A selective and sensitive detection of the (E)-2-nonenal was achieved by using an undoped ZnO sensor operated at 150°C (data not shown). Interestingly, at this temperature, the signal for the (E)-2-nonenal changes from minus to plus. The reason for this behaviour of the sensor is unknown.

The results obtained so far can be summarized as follows: The data have shown that the HRGC/SOMMSA is a very useful method to check the capabilities of a given sensor and to assist studies aimed at developing more selective chemosensors. Furthermore, the method offers the possibility to systematically study reactions occurring at the surface of chemosensors.

Development of a sensor array standardized for baking processes. As reported in previous studies (9, 12), by application of the Aroma Extract Dilution Analysis (AEDA) and the odor activity value concept, we identified the key odorants in wheat bread crust flavor. Among them, 2-acetyl-1-pyrroline, followed by (E)-2-nonenal and 3-(methylthio)propanal showed the highest odor activities (Table II). Further studies revealed (13) that the key odorants in toasted wheat bread crumb were very similar to those of the wheat bread crust. The generation of the overall roasty aroma during toasting correlated well with the increase in the concentration of 2-acetyl-1-pyrroline (13) and, also, with the browning intensity.

As shown above, the sensitivity of a chemosensor and, especially, its selectivity cannot be predicted by theoretical considerations alone. In order to develop a sensor array for measuring the flavor generation during toasting of bread, we used the HRGC/SOMMSA method to select suitable chemosensors and optimum temperatures for the selective detection of the five key crust odorants.

First, a mixture of the crust aroma compounds was separated by HRGC and the effluent was monitored in parallel by the FID and a SnO_2 sensor operated at 400°C. A comparison of both signal patterns revealed (Figure 5A and 5B) that no odorants were detected by this sensor. Lowering the temperature to 200°C, led, however, to a nearly selective detection of the potato-like smelling odorant 3-(methylthio)-propanal (Figure 5C). Using the same mixture, a ZnO sensor doped with platinum was then found to selectively detect 2-acetyl-1-pyrroline and a ZnO sensor doped with palladium and

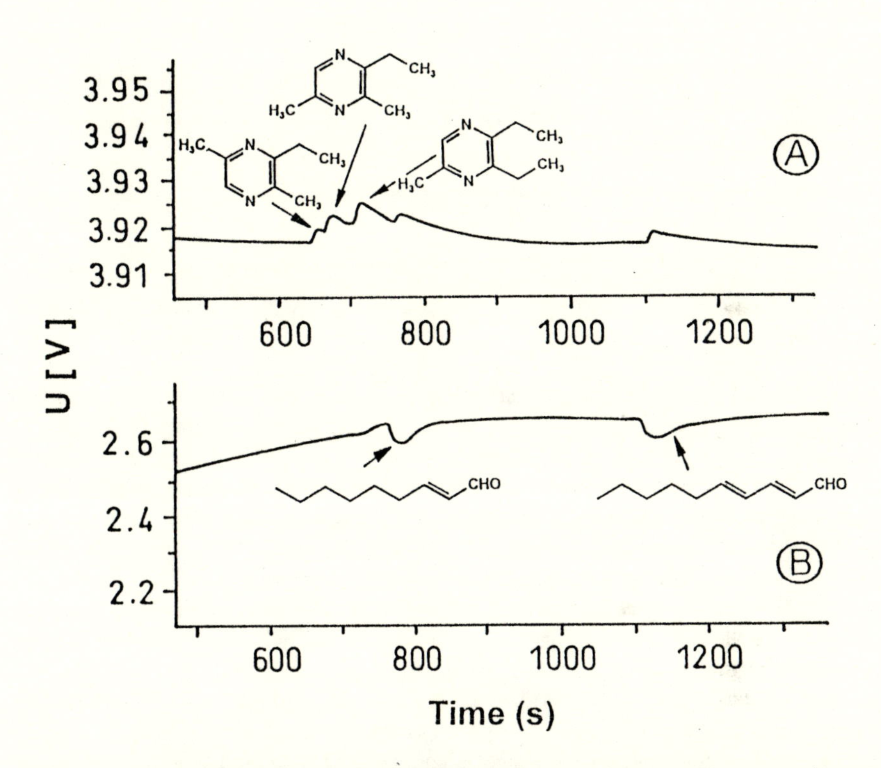

Figure 4. Response of two phthalocyanine sensors to a mixture of three pyrazines, 2-acetyl-1-pyrroline (no response), 4-hydroxy-2,5-dimethyl-3(2H)-furanone (no response), (E)-2-nonenal and (E,E)-2,4-decadienal A: copper phtalocyanine doted with zinc B: copper phtalocyanine

Table II. Concentrations, odor thresholds and odor activity values of the five most-odor active volatiles in wheat bread crust

Odorant	*Odor threshold (µg/kg of wheat starch)*[a]	*Conc. (µg/kg)*	*OAV*[b]
2-Acetyl-1-pyrroline	0.0073	20	2740
(E)-2-Nonenal	0.25	60	240
3-(Methylthio)propanal (methional)	0.27	51	189
4-Hydroxy-2,5-dimethyl-3(2H)-furanone	13.0	1900	146
3-Methylbutanal	32.0	1400	44

a Values are adapted from (13).

b Odor activity values were calculated by dividing the concentrations by the odor thresholds in starch.

operated at 200° C selectively detected (E)-2-nonenal. A SnO_2 sensor doped with platinum and operated at 400°C detected all five odorants with similar sensitivity (data not shown).

Application of a portable, self-constructed 4-sensor array to bread. The final aim in the development of sensor arrays is, of course, to have a small array which can be used to directly measure food flavors without using the HRGC system. Based on the analytical data discussed above, we have, therefore, constructed a sensor system using the four sensors selected, which we call "SPAN" (**s**tandardized **p**rimary **a**roma **n**ose). The principle of the small chemosensor is shown in Figure 6. At the beginning of the measurement, a pump sucks air from the surroundings over the surfaces of the four sensors. Then, the headspace above a toasted or an untoasted slice of bread is sucked for 15 s over the sensors. Finally, the sensors are flushed with air again. The relative conductivity changes of the sensors obtained for the two bread samples are contrasted in Figure 7. The most significant difference was found for sensor 4 (S4) which was shown above to be selective for 2-acetyl-1-pyrroline. This sensor showed a much higher signal for the volatiles in toasted as compared to untoasted bread. This result is well in line with the increase of this odorant during toasting (13).

HRGC/SOMMSA of headspace samples. Although the data reported above recommend, e.g., sensor S2 for a selective detection of 3-(methylthio)propanal (cf. Fig. 5), we have to take into account the relation of the trace amounts of key odorants, such as the methional, to the many non odor-active volatiles which interfere with the sensor signal because they are present in much higher concentrations.

To check which compounds are in fact detected by sensor 2 used in the SPAN, 20 mL of headspace gas were withdrawn from either an untoasted slice or bread slices toasted at two different times. The volatiles were cryofocused on a trap (cf. Figure 8), separated by HRGC and monitored with the SOMMSA device containing the same

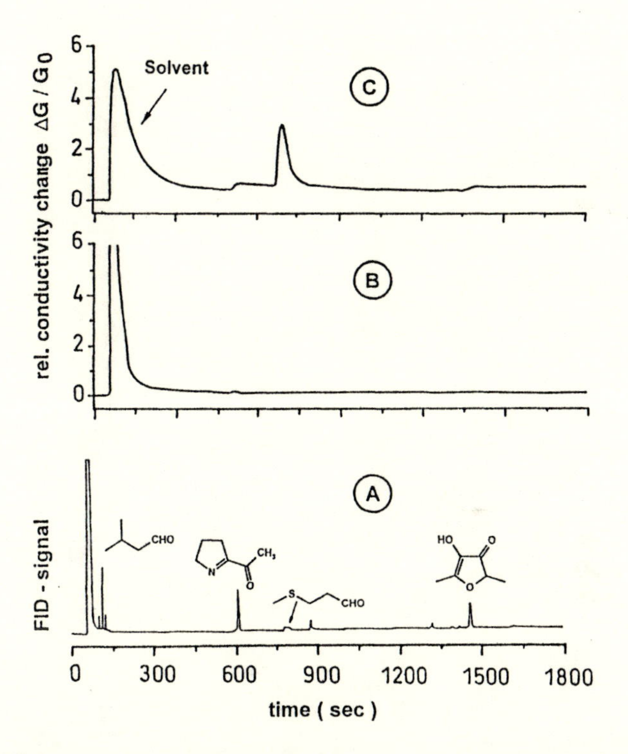

Figure 5. Signal pattern obtained by SOMMSA for a mixture of 4 key bread crust odorants after HRGC separation. A: FID; B: SnO_2 at 400°C; C: SnO_2 at 200°C

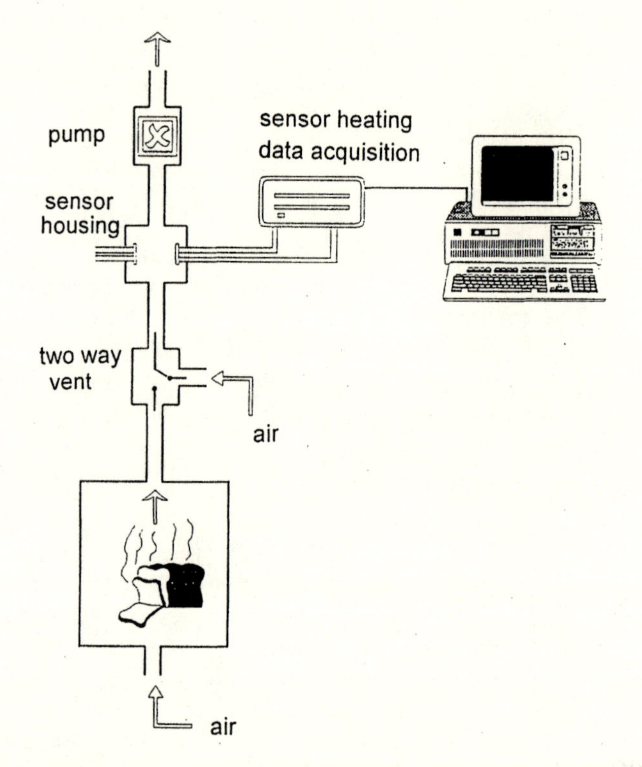

Figure 6. Schematic view of the standardized primary aroma nose (SPAN). For explanation see text

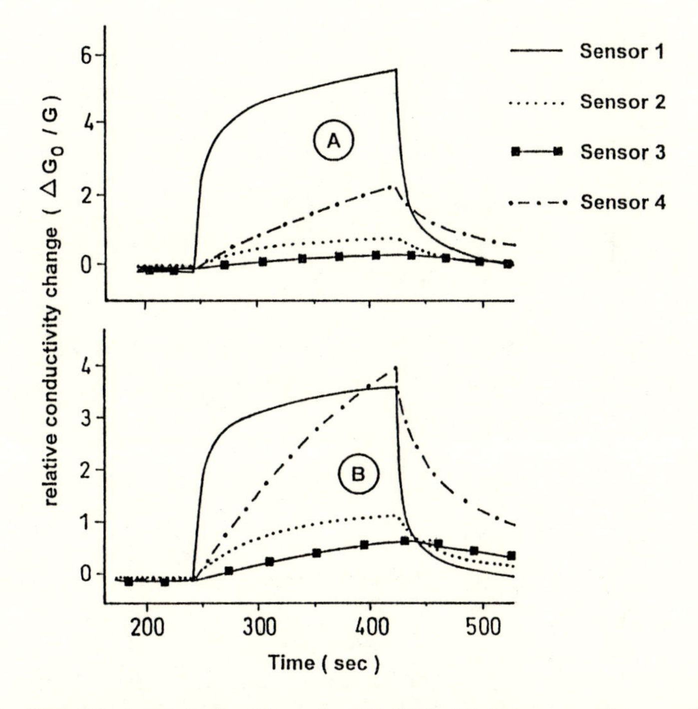

Figure 7. Signal pattern obtained by using SPAN (with four sensors) to monitor the entire set of headspace volatiles of an untoasted (A) and a toasted (B) slice of wheat bread.
S1: SnO_2/Pt at 400°C S2: SnO_2 at 200 °C
S3: ZnO/Pd at 200°C S4: ZnO/Pt at 300°C

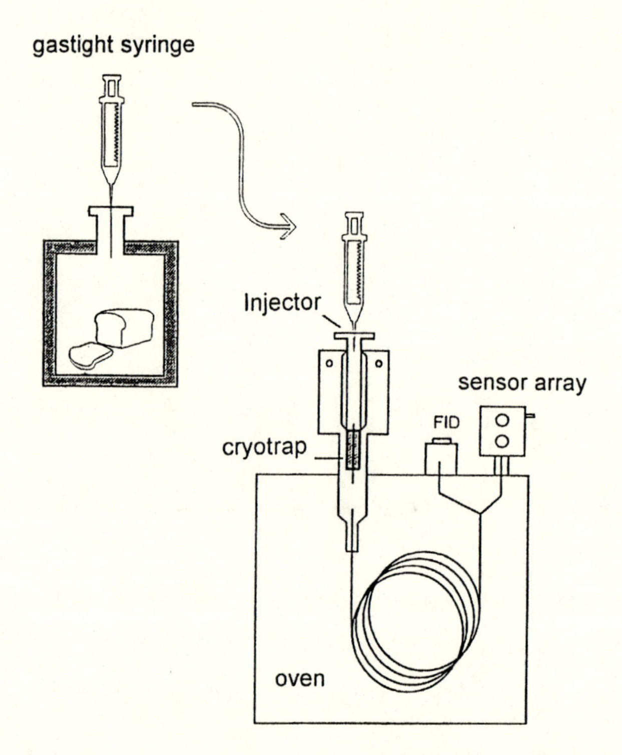

Figure 8. Scheme of the Headspace - HRGC/SOMMSA method. For explanation see text

four sensors mounted in the SPAN. As shown in Figure 9, sensor 2 detected mainly two volatile compounds which both increased with increasing toasting time. However, even in the strongly toasted bread, it was not the methional which was preferentially detected (C in Figure 9).

These results point out the importance of two steps in the correlation of data obtained from the human nose and chemosensors developed for selective sensor arrays:

- First, it is necessary to identify the indicator volatiles present in the headspace of a given food which are sensitively detected by single chemosensors.
- Second, it is important to confirm which indicator volatile (detected by the chemosensor) is quantitatively correlated with the key odorants previously identified, e.g., by ChARM Analysis or Aroma Extract Dilution Analysis.

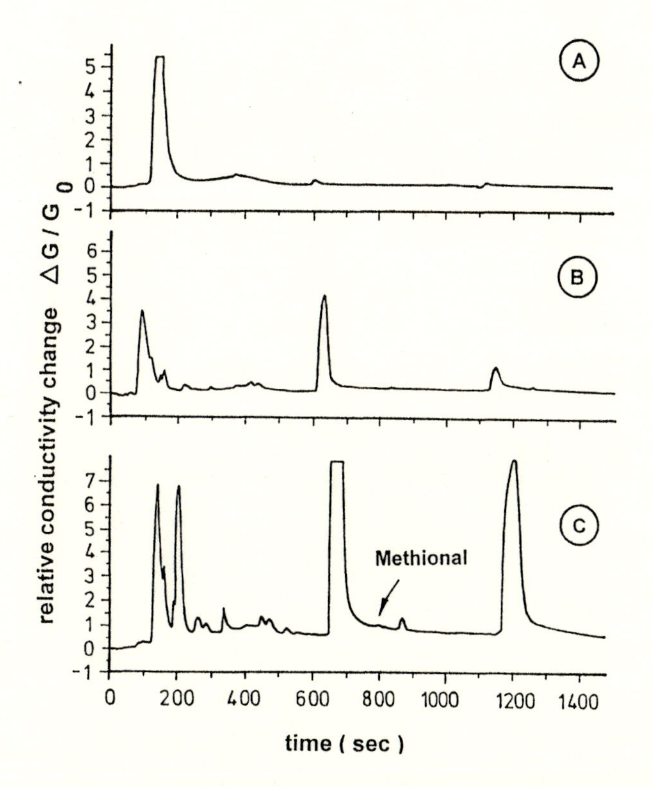

Figure 9 Responses of sensor 2 (cf. Figs. 5 and 7) after HRGC separation of the complete set of volatiles present in the headspace above wheat bread
A: untoasted; B: medium toasted; C: strong toasted

Conclusions

Chemical compounds established as key food odorants based on the human sensory response are the objective standards for flavor analysis. To develop selective chemosensor arrays for fast and objective flavor measurements either the key odorants themselves or indicator volatiles that are quantitatively correlated with the key odorants should be used in sensor selection. Because sensor arrays developed by this procedure correlate the odor response of the human nose with the electronic data obtained from chemosensor arrays using the same chemical standards, a more precise, fast and objective odor evaluation of foods results.

Literature

1. Kohl, D. In *Handbook of Biosensors and Electronic Noses: Medicine, Food and the Environment*, CRC Press, **1997**, pp. 553-561.
2. Acree, T. In *Flavor Science.* Acree, T.E.; Teranishi, R. (eds.) ACS Professional Reference Book, ACS, Washington, DC, **1994**, pp. 1-20.
3. Grosch, W. *Trends Food Sci Technol* **1993**, 4:68-73.
4. Schieberle, P. In *Characterization of Foods: Emerging Methods*, Gaonkar, A.G. (ed.) Elsevier, Amsterdam, **1995**, pp. 403-431.
5. Mlotkiewicz J.; Elmore J.S. In *Flavour Science- Recent Developments*, A.J. Taylor and D.S. Mottram (Eds.) The Royal Society of Chemistry, London, **1996**, pp. 462-463.
6. Hofmann, T.; Schieberle, P.; Krummel, C.; Freiling, A.; Bock, J.; Heinert, L.; Kohl, D. *Sensors and Actuators B Chemical* **1997**, 41: 81-87.
7. Cerny, C.; Grosch, W. *Z. Lebensm. Unters. Forsch.* **1993**, 196:417-422.
8. Hofmann, T.; Schieberle, P. J. *Agric. Food Chem.* **1995**, 43:2187-2194.
9. Schieberle, P.; Grosch, W. *J. Agric. Food Chem.* **1987**, 35:252-257.
10. Schieberle, P. *J. Agric. Food Chem.* **1991**, 39:1141-1144.
11. Roberts, D.; Acree, T.E. In *Thermally Generated Flavors*, Parliment, T.; Morello, M.J.; McGorrin, R.J. (eds.) ACS Symposium Series 543, Washington, DC, **1994**, pp. 71-79.
12. Schieberle, P.; Grosch, W. *Flavour Fragrance J.* **1992**, 213-218.
13. Rychlik, M.; Grosch, W. *Lebensm. Wiss. Technol.* **1996**, 29:515-525.

Author Index

Subject Index

C

D

E

F

G

H

I

J

L

M

P

R

S

T

U

V

W

Y